FIELD GEOLOGIST'S TRAINING GUIDE

The EXLOG Series of Petroleum Geology and Engineering Handbooks

Field Geologist's Training Guide

Written and Compiled by
EXLOG Staff

Edited by Alun Whittaker

International Human Resources Development Corporation • Boston

© 1985 by EXLOG®.* All rights reserved. No part of this book may be used or reproduced in any manner whatsoever without written permission of the publisher except in the case of brief quotations embodied in critical articles and reviews. For information address: IHRDC, Publishers, 137 Newbury Street, Boston, MA 02116.

Library of Congress Cataloging in Publication Data

Main entry under title:

Field geologist's training guide.

Bibliography: p.
Includes index.
1. Petroleum—Geology. 2. Petroleum engineering.
3. Geology—Field work. I. Whittaker, Alun II. EXLOG (Firm).
TN870.5.F48   1985   622'.1828   83-25173
ISBN 0-88746-043-7

Printed in the United States of America

*EXLOG is a registered service mark of Exploration Logging Inc., a Baker Drilling Equipment Company.

CONTENTS

List of Illustrations xi
Preface xvii
Acknowledgments xviii
Acknowledgments for Figures xix

1. INTRODUCTION 1

2. PETROLEUM GEOLOGY & OILFIELD FLUIDS 3

PETROLEUM GEOLOGY 3
  INTRODUCTION 3
    Sedimentary Rocks 3
    Geological Structures 8
    Earth Movements 8
  APPLICATION OF GEOLOGICAL CONCEPTS 10
    Early Exploration 10
    Geophysical Exploration 10
    Surface Exploration 10
    Subsurface Exploration 12
    Exploration Applications 14
    Conceptual Models of Exploration 15
  PETROLEUM ACCUMULATIONS 17
    Origin of Petroleum 17
    Migration 18
      Primary Migration 18
      Secondary Migration 20
    Reservoir Rocks 21
      Sandstones 21
      Carbonates 24
      Estimations of Porosity and Permeability 24
    Traps 26
      Anticlinal Traps 28
      Fault Traps 28
      Stratigraphic Traps 30
      Lenticular Traps 30
  PETROLEUM RESERVOIRS 30
    Physical Characteristics 30
    Reservoir Pressure 31
OILFIELD FLUIDS 32
  RESERVOIR WATER 32
  RESERVOIR OIL AND GAS 35
    Composition of Petroleum 35
      Hydrocarbon Size Distribution 36
      Hydrocarbon Type Distribution 37
    Source of Petroleum 38
    Origin of Gases 41
    Occurrence of Petroleum 41
    Maturation of Petroleum 42

        Factors Which May Affect Crude Oil Gravity 43
        Impurities Associated with Hydrocarbons 44
        Products from Crude Oil 45
    FLUID DISTRIBUTION 46

3. RIG TYPES & THEIR COMPONENTS 47

GENERAL 47
LAND RIGS 47
OFFSHORE RIGS 49
    BARGE 49
    JACK-UP 49
    FIXED PLATFORM 50
        Piled Steel Platforms 50
        Gravity Structures 52
    SEMI-SUBMERSIBLE 52
    DRILLSHIP 54
RIG COMPONENTS 56
    HOISTING SYSTEM 57
        Derrick (or Mast) 57
        Traveling Block, Crown Block, Drilling Line and Drilling Hook 58
        Drawworks (the Hoist) 60
    CIRCULATING SYSTEM 60
        Mud Pumps 61
        Standpipe and Rotary Hose 62
        Drillstem 62
            Kelly and Swivel 63
            Drillstring 63
                1) Drillpipe 63
                2) Drill Collars 64
                3) Subs 66
                4) Bottomhole Assembly (BHA) 67
            Drill Bits 67
                1) Drag Bit 67
                2) Tri-cone Bit 67
                3) Diamond Bits 73
                4) Reamers/Hole Openers 75
    ROTATING SYSTEM 75
      The Rotary Table 77
      Master Bushings 77
      Kelly Drive Bushings 78
      Slips 78
      Tongs 78
      Spinning Wrench (Power Tongs, Pipe Spinners) 78
    MOTION COMPENSATION SYSTEM 78
      Drillstring Compensator 78
      Riser and Guideline Tensioners 80
    BLOWOUT PREVENTION (B.O.P.) SYSTEM 81

4. DRILLING & COMPLETING A WELL   87
ROUTINE DRILLING   87
   CONNECTIONS   87
   TRIPS   88
RELATED SERVICES   89
   DRILLING FLUIDS ENGINEERING   92
      Drilling Fluid Technology   92
         Controlling Subsurface Pressures   92
         Removing and Suspending the Cuttings   99
         Cooling and Lubricating the Bit and Drillstring   100
         Walling the Hole with an Impermeable Filter Cake   100
      Drilling Fluid Chemistry   100
         Water/Clay Muds   100
         Water/Oil/Clay Muds   102
         Compressed Gases   102
      Drilling Fluid Conditioning Equipment   102
         Shale Shaker   104
         Settling Pit (Sand Trap)   104
         Desander, Desilter and Centrifuge   104
         Degasser   106
         Mixing Hopper   106
         Suction Pit (Active Pit)   107
   CASING AND CEMENTING   108
      Casing   108
         Conductor Pipe   108
         Surface Casing   109
         Intermediate Casing   109
         Liner String   110
         Production Casing (Oil String)   110
         Running the Casing   110
         Casing Accessories   110
         The Wellhead   112
      Cementing   113
         Primary Cementing   113
         Secondary Cementing   116
         Cement Classifications   116
         Cement Additives and Their Effects   119
         Mixing and Other Surface Equipment   121
         Job Considerations   122
   CORING   123
      Conventional Coring   123
         Cutting the Core   123
         Receiving the Core   125
         Sampling the Core for Special Purposes   126
         Examining the Core   126
         Packing and Shipping   127

Sidewall Coring 127
VERTICAL CONTROL AND DIRECTIONAL DRILLING 128
    Vertical Control 128
        Measuring Inclination 128
        Preventing and Correcting Deviation 130
    Directional Drilling 130
        Basic Hole Patterns 131
        Deflection Tools 132
        Orientation of Deflection Tools 137
        Drilling the Deviated Section of Hole 137
        Engineering and Formation Evaluation Considerations 138
FISHING 138
    Situations Requiring a "Fishing Job" 138
        Fatigue Failures 138
        Stuck Pipe 139
        Foreign Objects in the Hole 140
    Fishing Tools 140
        Fishing for Junk 140
        Fishing for Pipe 140

5. FORMATION EVALUATION PROCEDURES 145

WELL LOGGING 145

INTRODUCTION 145
DRILL RETURNS LOGGING 147
    Theory 147
    Application 148
    Mud Logging Techniques and Equipment 149
        Lag Determination 150
        Pump Stroke Counters 156
        Depth and Drill Rate Recorder 156
        Gas Determination from the Drilling Mud 162
        The Gas Trap 164
        The Vacuum System 166
        The Gas Detector 166
        The Microgas Analyzer 170
        Recording Gas Information 170
        Trip and Connection Gases 170
        Chromatography 172
        Ultraviolet-Light Box 175
        Mud Press 175
        Pit Level Indicators 176
        Mud Logging Equipment Flowchart 178
        Logging Procedures 179
    Samples 181
        Collection and Preparation 181
        Examination of the Cuttings 182
        Noncarbonate Clastics 185

>>>>Carbonates   187
>>>>Evaporites   190
>>>>Hydrocarbon Evaluation   190
>>>>Oil Determination from the Drilling Mud   191
>>>>Oil Determination from Formation Samples   192
>>>Secondary Equipment and Services   193
>>>>Mudweight Recording   196
>>>>Mud Temperature Recording   196
>>>>Mud Resistivity Recording   197
>>>>Mudflow Monitor   197
>>>>Drill Monitor (System and Panel)   197
>>>>Computational System   199
>>>>Shale Density Determination   199
>>>>Shale Factor Determination   199
>>>>H2S and CO2 Detection   200
>>>>Stain Kit   200
>>>>Calcimetry   200
>>>>Core Analysis   200
>>>Mud Log Presentation and Standardization   201
>>>>Drafting   201
>>>>Typing   203
>>>>Copying   207
>>>Routine Responsibilities of the Logging Geologist   208
>>>>Well Initiation   208
>>>>Specific Responsibilities   209
>>>>Routine Reports   209
>>>>Well Completion   210
>>>>Wellsite Relationships   210
>>PRESSURE LOG   211
>>CORE LOG   212
>>WIRELINE LOGS   212
>>>Classification   214
>>>>Resistivity   214
>>>>Porosity/Lithology   217
>>>>Production Characteristics   219
>>>>Miscellaneous Parameters   219
>>>Interpretation   220
>>>Abnormal Pressure Evaluation   221
>FORMATION TESTS   221
>>DRILLSTEM TESTING   221
>>WIRELINE FORMATION TESTING   226
>>>Repeat Formation Tester   227
>>>Formation Interval Tester   228
>>>Evaluation of Recovered Sample   228
>>WELL STIMULATION   228

Appendix A: Examples Relating to Mud Logs   231

Appendix B: Test Procedures 243
Appendix C: Mud Logging Techniques with Oil-Based Drilling Fluids 254
Appendix D: IADC Bit Classification 259
Appendix E: EXLOG's Role in Drillstem Testing 268
Glossary 277
Index 285

ILLUSTRATIONS

2-1   Sedimentary Rocks   4
2-2   Fundamental Tetrahedron for Classifying Sedimentary Rocks   4
2-3   Depositional Environments   5
2-4   Recent Depositional Environments of Northwestern Gulf of Mexico   6
2-5   Distribution of Surface Sediments of the Bahama Banks   7
2-6   Simple Kinds of Folds   8
2-7   Simple Kinds of Faults   8
2-8   Cross-Section Showing Rotational and Upthrust Faults   9
2-9   Unconformities   9
2-10 Schematic Cross-Section of Seeps   10
2-11 Reflected Seismic Waves   11
2-12 Diffusion of Hydrocarbons   12
2-13 Geochemical Temperature Facies in a Typical Basin   13
2-14 Generation of Hydrocarbons in the Los Angeles Basin   14
2-15 Cross-Section Illustrating Fault Pattern, Quitman Field, Texas   16
2-16 Sand Pinchout   17
2-17 Changes During Normal Compaction of Shales   18
2-18 Fluid Flow During Compaction   19
2-19 Migration Induced by Fixed Water   20
2-20 Unit Cells of Cubic and Rhombohedral Packing   21
2-21 Some Processes Controlling Pore Size Distribution in Sandstones   22
2-22 General Porosity Decrease with Depth   23
2-23 Evolution of Some Dolomite Textures and Pore Types   25
2-24 Spill Point of a Hydrocarbon Trap   27
2-25 Fault Trap Leakage   27
2-26 Basic Reservoir Traps   28
2-27 Potential Hydrocarbon Traps Associated with Rotational Faults   28
2-28 Thrust Traps   29
2-29 Piercement Traps Associated with a Salt Dome   29
2-30 Cross-Section of Pinchout Trap Between Unconformities, East Texas Field   30
2-31 Representative Permeability and Porosity Determinations   31
2-32 Dissolved Solids (Concentrations)   33
2-33 Oilfield-Water Analyses   34
2-34 Chemical Composition of Oil, Asphalt and Kerogen   36
2-35 Composition of a Typical Crude Oil   36
2-36 Typical Hydrocarbon Structures   37
2-37 Hydrocarbons Commonly Detected at the Wellsite   38
2-38 Average Chemical Composition of Natural Substances   39
2-39 Composition of Living Matter   39
2-40 Origin of Organic Matter Deposited with Sediments   40

2-41 Reservoir Age 42
2-42 Properties of Liquid Hydrocarbons 44
2-43 Impurities Associated with Hydrocarbons 45
2-44 Crude Oil Products 45
2-45 Reservoir Fluid Distribution 46
3-1 Rig Types and Offshore Operating Profiles 47
3-2 Land Rig 48
3-3 Jack-Up Rigs 50
3-4 Fixed Platforms 51
3-5a Semi-submersible Rig—Pontoon Type 53
3-5b Semi-submersible Rig—Twin Hull Type 54
3-6 Drillship 55
3-7 Diagrammatical Display of Rig Components Showing Circulating System 56
3-8 Hoisting System Components 58
3-9 Hoisting and Rotary Components 59
3-10 Schematic Cross-Sections of Duplex and Triplex Pumps 61
3-11 API Drillpipe with Weld-On Tool Joints 63
3-12 Drill Collar Weight 65
3-13 Stabilizers 66
3-14 Drag Bit 68
3-15 Water Courses 68
3-16 Milled Tooth Bit 68
3-17 T-Gauge Bit (Cone) 69
3-18 T-Gauge Bit Cone with Inserts 69
3-19 Insert Bit 69
3-20 Journal Configuration 70
3-21 Offset on Soft-Formation Cones 71
3-22 Diamond Bit 74
3-23 Hole Opener and Under-Reamer 75
3-24 Rig Components Used for Pipe Handling 76
3-25 Master and Kelly Drive Bushings 77
3-26 Drillstring Compensation System 79
3-27 Riser and Guideline Tensioner Systems 80
3-28 Schematic Projection Showing the Relationship Between the Circulation and Blowout Prevention Systems 81
3-29 Blowout Preventer (B.O.P.) Stack, in Various Operational Modes 82
3-30 Blowout Preventers 84
3-31 B.O.P. Stack Positions Governed by Rig Type 85
4-1 Making a Connection 87
4-2 Pulling Out of the Hole (Tripping Out) 88
4-3 Separating a Connection 89

4-4   Pipe Handling System   90
4-5   Interrelationship of Operator, Drilling Contractor and Service Companies   91
4-6   Formation Pressures   95
4-7   Relationship Between Piezometric Surface and Ground Level   97
4-8   Mud Conditioning Equipment   103
4-9   Hydroclone   104
4-10 Centrifuge   105
4-11 Vacuum Degasser   107
4-12 Jet Hopper for Mud Mixing   107
4-13 Casing Strings and Hanger   109
4-14 Casing Accessories   111
4-15 Primary (Single-Stage) Cementing   114
4-16 Record of Circulating Pressures While Cementing   115
4-17 Successive Steps for Multi-Stage Cementing   117
4-18 API Cement Classification   118
4-19 Basis for API Well Stimulation Test Schedules   118
4-20 Typical Cement Mixing and Pumping Operations   121
4-21 Temperature Survey Showing Top of Cement   122
4-22 Diamond Bit Core Barrels   124
4-23 Suggested Method of Boxing Cores   125
4-24 Drift Survey Instrument   129
4-25 The Pendulum Effect   130
4-26 Applications of Directional Drilling   131
4-27 Basic Hole Patterns   132
4-28 Schematic Cross-Section of a Turbine Motor   133
4-29 Downhole Motor Deflected with a Bent Sub   134
4-30 Deflection by Jetting   135
4-31 Deviating with a Whipstock   136
4-32 Key Seating   139
4-33 Differential Sticking   139
4-34 "Junk" Fishing Tools   141
4-35 "Pipe" Fishing Tools   142
5-1   Lag Calculation   152
5-2   General Capacities and Displacements of Drillpipe, Collars, Casing and Hole   154
5-3   Kelly Height, Recorder Chart, Example No. 1   157
5-4   Kelly Height, Recorder Chart, Example No. 2   158
5-5   Drill-Rate Chart for Random Intervals   159
5-6   Example Data Worksheet   160
5-7   Kelly Height, Recorder Chart, Example No. 3   161
5-8   Kelly Height, Recorder Chart, Example No. 4   161
5-9   Depth Correction of Mud and Cuttings   163

5-10 Gas Detection Flowchart   164
5-11 Gas Trap   165
5-12 Vacuum System   166
5-13 Schematic Diagram of a Simple Catalytic Gas Detector   167
5-14 Specimen Ditch Gas Recording Chart   169
5-15 Standard (Catalytic) and F.I.D. Chromatographs   173
5-16 Sample Chromatogram   174
5-17 Basic PVT System   177
5-18 Logging Unit Basic Data Inputs   179
5-19 Comparison Charts for Visual Estimation of Percentage Composition   183
5-20 Grain Size Terminology   186
5-21 Electromagnetic Radiation Spectrum   191
5-22 Procedure Upon Encountering a Show   194
5-23 Basic and Secondary (Including GEMDAS) Data Inputs   195
5-24 Drill Monitor System   198
5-25 Standard Recessed Plot   203
5-26 Ladder Plot   203
5-27 Exlog's Standard Lithology Symbols   206
5-28 Wireline Logging Operations   213
5-29 Symbols Used in Wireline Log Interpretation   215
5-30 The Effect of Salinity on SP Response   217
5-31 Principal Types of Drillstem Tests   223
5-32 DST Procedure   224
5-33 Repeat Formation Tester   227
A-1   Typical Mud Logging Program   232
A-2   Suggested Scale for Rate of Penetration   233
A-3   Standard Abbreviations for Descriptions and Reports   234
A-4   Standard Abbreviations for Wireline Logs   237
A-5a Sample Mud Log Heading   239
A-5b Sample Mud Log   239
A-6   Supplemental Log for Core Description and Test Results   241
A-7   Bit Data Record   242
B-1a Carbonate Staining   244
B-1b Flowchart for Stain Procedure, Using Aliziran Red S   245
B-1c Flowchart for Stain Procedures, Using Distilled Water   246
B-1d Flowchart for Stain Procedures, Using Heat   246
B-2   Chloride Test Procedures   247
B-3a Nitrate Ion Test Procedure   248
B-3b Nitrate Ion Test Procedure Flowchart   249
B-4a Calcimetry Procedure   250
B-4b Bernard Calcimeter   251

B-5 Standard Procedure for Shale Factor 252
B-6 Flowchart for Core Analysis Procedures 253
C-1 Set-Up for Chromatographic Analysis of Blender Samples 258
D-1 Tri-Cone Bit Code: Hughes Tool Company, Vendor Code 1 260
D-2 Tri-Cone Bit Code: Varel, Vendor Code 2 261
D-3 Tri-Cone Bit Code: Reed Tool Company, Vendor Code 3 262
D-4 Tri-Cone Bit Code: Security, Vendor Code 4 263
D-5 Tri-Cone Bit Code: Smith Tool, Vendor Code 5 264
D-6 Tri-Cone Bit Code: S.M.F. (Creusot-Loire), Vendor Code 6 265
D-7 Tri-Cone Bit Code: T.S.K. (Tsukamoto Seiki) 266
D-8 Tri-Cone Bit Code: C.P. (Chicago Pneumatics) 267

PREFACE

Written for oil industry professionals, this handbook presents a basic overview of and introduction to petroleum geology, oilfield terminology, and formation evaluation procedures. Designed primarily for beginning logging geologists, this much needed and comprehensive text provides a reference resource for all newcomers to the petroleum exploration field.

Chapter 1 is a brief introduction to the exploration oilfield. Chapter 2 discusses Petroleum Geology and Oilfield Fluids; classroom geology is reviewed and related specifically to petroleum geology and formation fluids. Chapter 3, Rig Types and Their Components, describes wellsite equipment and the environment in which the geologist will be working. Chapter 4, Drilling and Completing a Well, discusses what actually happens on an exploratory well. Chapter 5, Formation Evaluation Procedures, describes in detail the various responsibilities of the logging geologist.

The appendixes contain a wide range of information with which the logging geologist must become familiar. While much of the appendix material is not used regularly, some of the test procedures and log-related examples are important in the day-to-day activities of the geologist at the wellsite.

ACKNOWLEDGMENTS

We wish to express our appreciation for the assistance and cooperation of the IADC, AAPG and the Woods Hole Oceanographic Institute for granting us permission to use some of their material in this book.

Exploration Logging's field geologists have provided valuable material gained through years of field experience in the business of mud logging and formation evaluation. Many proven techniques and procedures that were developed by these people are included in this manual to help the trainee field geologist rapidly become familiar with formation evaluation.

The Applications and Training Section wishes to acknowledge their contributions and would like to encourage field geologists to continue to document their discoveries.

Acknowledgments for Illustrations

Figure 2.14 is reprinted by permission from *Geochimica et Cosmochimica Acta*, vol. 29, G. T. Philippi, On the depth, time and mechanism of petroleum generation, fig. 8, p. 1037. Copyright © 1965 Pergamon Press Ltd.

Figures 2.12, 2.18, 2.19 are from *Petroleum Geochemistry and Geology* by John M. Hunt. W. H. Freeman and Company. Copyright © 1979. All rights reserved. Figure 2.12 is adapted from B. P. Yasenev, 1959, Gas sampling in wells and its value in exploration, *Geol. Nefti Gaza*, **3**(2). Figure 2.18 is adapted from A. L. Kidwell and J. M. Hunt, 1958, Migration of oil in Recent sediments of Pedernales Venezuela, in L. G. Weeks, ed., *Habitat of Oil: A Symposium* (Tulsa, Okla.: AAPG), pp. 790–817.

Figure 2.41 is from Alfred G. Fischer and Sheldon Judson, eds., *Petroleum and Global Tectonics*. Copyright © 1975 by Princeton University Press. Figure 2.41, "Distribution and Geological Characteristics of Giant Oil Fields," by J. D. Moody, reprinted by permission of Princeton University Press.

Figures 4.15, 4.17, 5.31, 5.32 are courtesy of Halliburton Services.

The following figures are reprinted by permission of the Petroleum Extension Service, The University of Texas at Austin, in cooperation with the International Association of Drilling Contractors:

Figures 2.4, 2.5, 2.6, 2.7, 2.8, 2.9, 2.10, 2.11, 2.15, 2.16, 2.20, 2.21, 2.22, 2.23, 2.24, 2.25, 2.26, 2.27, 2.28, 2.29, 2.30 are from *Lessons in Well Servicing and Workover, Lesson 2: Petroleum Geology and Reservoirs*. Copyright © 1975 by the University of Texas at Austin.

Figures 3.14 and 3.21 are from *Lessons in Rotary Drilling, Unit I—Lesson 2: The Bit*, rev. Copyright © 1976 by the University of Texas at Austin.

Figure 3.25 is from *Lessons in Rotary Drilling, Unit I—Lesson 4: Rotary, Kelly, and Swivel*, 2nd ed. Copyright © 1981 by the University of Texas at Austin.

Figures 4.1, 4.2, 4.35(a,c,d) are from *A Primer of Oilwell Drilling*, 4th ed., by Ron Baker. Copyright © 1979 by the University of Texas at Austin.

Figure 4.9 is from *Lessons in Rotary Drilling, Unit I—Lesson 12: Mud Pumps and Conditioning Equipment*. Copyright © 1974 by the University of Texas at Austin.

Figure 4.13 is from *Lessons in Rotary Drilling, Unit III—Lesson 4: Subsea Blowout Preventers and Marine Riser Systems*. Copyright © 1976 by the University of Texas at Austin.

Figures 4.14, 4.16, 4.21 are from *Lessons in Rotary Drilling, Unit II—Lesson 4: Casing and Cementing*. Copyright © 1968 by the University of Texas at Austin.

Figure 4.24b is from *Lessons in Rotary Drilling, Unit II—Lesson 3: Drilling a Straight Hole,* 2nd ed., edited by Nancy J. Janicek. Copyright © 1982 by the University of Texas at Austin.

Figures 4.26, 4.27, 4.28, 4.29, 4.30, 4.31 are from *Lessons in Rotary Drilling, Unit III—Lesson 1: Controlled Directional Drilling,* 2nd ed., edited by Nancy J. Janicek. Copyright © 1984 by the University of Texas at Austin.

Figures 4.34 and 4.35(b,e) are from *Lessons in Rotary Drilling, Unit III—Lesson 2: Open-Hole Fishing,* rev. Copyright © 1975 by the University of Texas at Austin.

1
INTRODUCTION

The first oilwells, called "wildcats," were drilled to rather shallow depths of about 1000 feet. Hole depth was limited because of the primitive drilling equipment and the limited technology. The first type of rig, called a "cable tool rig," gave way to the rotary rig which immensely improved drilling techniques and permitted greater depths to be penetrated.

The use of drilling fluids was discovered by accident, but when the advantages were recognized, specially-prepared fluids called "mud" became part of the planned drilling program. At first the mud was used primarily to clean, cool and lubricate the bit as it drilled through formation and to remove the formation cuttings ("returns") from the hole by circulating the mud using surface pumps. In those early years (early 1900s) well analysis in its simplest form consisted primarily of examining the mud and returns for visible signs of oil and by using the sense of smell to detect the odor of hydrocarbons.

Around 1930 it was discovered that mud and cuttings could be correlated to well depth and analyzed to supply more specific data about the formation. The first significant advance in drill returns analysis was made in 1938 when a method was developed to determine the amount of oil in the drilling fluid by centrifuging, and to determine the relative amount of gas in the drilling fluid by using a hot-wire bridge circuit. This early instrumentation was borrowed from the mining industry where it had been used to detect combustible and explosive gases in mine shafts. As these methods pertained to the drilling fluid only, drill returns logging was known originally as "mud logging." (Today, in Exploration Logging, mud logging is referred to as "formation evaluation.")

The predictions of oil and gas accumulations made with these methods were at first very erratic, partly because of the lack of experience and partly because these methods did not furnish sufficient information. Around 1943, a method was developed for systematically analyzing and examining the cuttings for oil and gas. The additional information obtained by this procedure made drill-returns logging an effective and important hydrocarbon detection and evaluation method.

Drill-returns logging is the only evaluation method used as drilling is in progress, and, as such, monitors the formation before any changes (due to mud properties) have occurred. For this reason it is an essential evaluation method. Wireline logging and drillstem testing (DST) evaluate the formation after changes have occurred, and, when used with drill-returns logging, can provide an overall evaluation of the formation properties.

2
PETROLEUM GEOLOGY & OILFIELD FLUIDS

2.1 PETROLEUM GEOLOGY

2.2 INTRODUCTION

Geology is so fundamental to the petroleum industry that a knowledge of its basic principles is desirable for all persons associated in any way with the industry. This section gives a brief outline of the geological processes, the origin and accumulation of petroleum, and how these concepts may be related to the production of oil and gas.

Petroleum geology is based on observation and utilizes many other sciences. The basic principle that "the present is the key to the past" is the concept that processes which acted on the earth in the past are very similar to or the same as those operating today. The geologist's conclusions or deductions are derived by:

- Observing the results of the earth's history and processes

- Reconstructing the events giving rise to certain formations and their arrangement

- Predicting where oil accumulations might occur

The accumulation of oil and gas into a commercial deposit requires the presence of a "trap" consisting of a reservoir rock, a rock seal (or cap rock), and three-dimensional closure. There must also be a rock of petroliferous character structurally and chronologically placed so as to provide a source of petroleum for the trap. The reservoir rock is a container which is usually much more extensive than the hydrocarbon (petroleum) deposit that has been localized by a trap. Below the oil or gas accumulation, the reservoir is almost always filled with water. The qualification of an economic oil pool is that it must contain enough oil or gas to make extraction profitable, must exceed a minimum porosity and permeability and have a minimum thickness, depending on local conditions. Igneous and metamorphic rocks rarely contain oil or gas. Sedimentary rocks are more important to petroleum geology since it is here that most oil and gas accumulations occur because of their greater porosity and permeability.

2.3 Sedimentary Rocks

Sedimentary rocks are deposited by water, wind, or ice. They may be hard and compact or loose sand and clay. Older sedimentary rocks have been compacted by the weight of overlying sediments and cemented by minerals carried by ground water so that they became consolidated sedimentary rocks. Ground water is the water present in pores and cracks in the rocks.

Sedimentary rocks are made up of the following:

- Clastic material, or fragments, composed mainly of broken and worn particles of preexisting minerals, rocks and/or shells, which are carried to the site of deposition by moving streams, waves, wind or glaciers

- Chemical precipitates formed in place by evaporation at the surface or by crystallization of dissolved salts within the sediment

- Organic or biogenic debris such as shells or plant remains accumulated in place; for example, a coral reef or peat

A simple classification of sedimentary rocks is shown in Figure 2-1.

CLASTIC	CHEMICAL		ORGANIC	OTHER
	CARBONATE	EVAPORITE		
CONGLOMERATE SANDSTONE SILTSTONE SHALE LIMESTONE	LIMESTONE DOLOMITE	GYPSUM ANHYDRITE SALT POTASH	PEAT COAL DIATOMITE LIMESTONE	CHERT

Figure 2-1. Sedimentary Rocks

Figure 2-2 illustrates a modern classification based on a tetrahedron at the corners of which are placed carbonate, clay, quartz and chert. This figure also depicts one side of this tetrahedron so that some of the variations between shale, sandstone, and limestone can be seen. For example, starting from shale and going toward limestone, increasing amounts of lime will produce calcareous shale, grading into argillaceous (shaley) limestone, then to pure limestone. Similarly on the other two edges, it is shown how the changes occur from shale to sandstone and from sandstone to limestone. The other three sides show similar variations with chert replacing one of the other constituents.

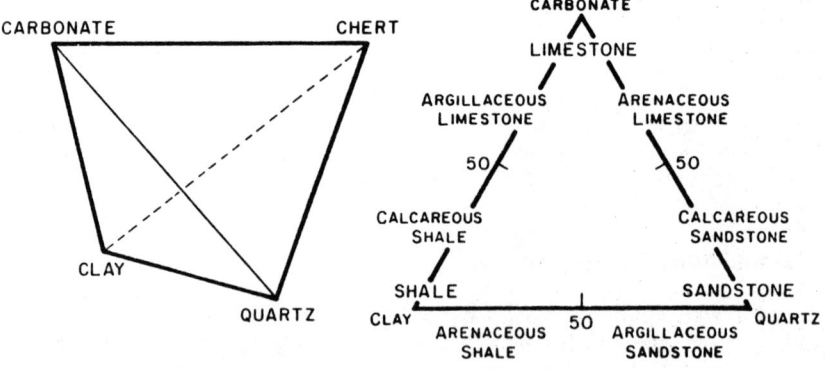

Figure 2-2. Fundamental Tetrahedron for Classifying Sedimentary Rocks

Sediments are deposited under a variety of conditions or environments, both on land and at sea. Clastics (conglomerates, breccia, sandstone, siltstone, limestone and shale) may be deposited under any of the following environments:

- continental aeolian — deposited by the wind on land
- transitional deltaic — deposited in the mouth of a river (delta)
- coastal interdeltaic — deposited on the coast between two deltas
- marine — deposited in the oceans (lagoonal, backreef, estuarine)

A simplified classification of sedimentary environments is shown in Figure 2-3. Note that although a line is shown separating the classifications there is no sharp line of demarcation; they grade one into the other. For example, alluvial grades into upper deltaic plain, or lower deltaic plain into prodeltaic plain which, in turn, grades into normal marine. The map in Figure 2-4 shows the extent of various Recent depositional environments in the area between the Mississippi River and the Rio Grande. The outer part of the deltaic sediments extends seaward of the shoreline and grades into the normal marine deposits. The crosshatched area in the southwest section on the map is an area of sand dunes and is aeolian.

	DELTAIC GROUP		INTERDELTAIC GROUP	
CONTINENTAL	AEOLIAN		AEOLIAN	
CONTINENTAL	ALLUVIAL		ALLUVIAL	
TRANSITIONAL	DELTAIC	DELTAIC PLAIN	COASTAL INTERDELTAIC- MARINE	
TRANSITIONAL	DELTAIC	PRODELTAIC PLAIN	COASTAL INTERDELTAIC- MARINE	
MARINE	NORMAL MARINE	SLOPE	NORMAL MARINE	SHELF
MARINE	NORMAL MARINE	DEEP	NORMAL MARINE	SLOPE
MARINE	NORMAL MARINE	DEEP	NORMAL MARINE	DEEP

Figure 2-3. Depositional Environments

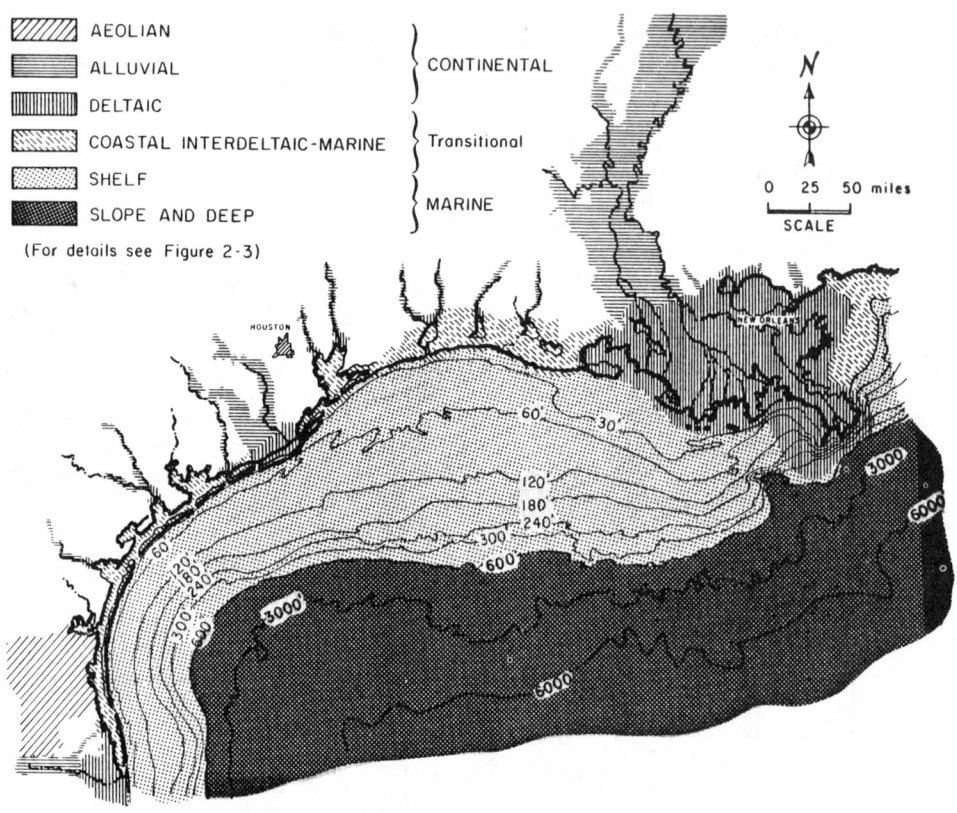

Figure 2-4. Recent Depositional Environments of Northwestern Gulf of Mexico

The distribution of surface sediments in the Bahama Banks is shown in Figure 2-5. Note that the skeletal sands are along the margin, whereas over the bank interiors the sediments are mainly pellet sands or pellet sands and lime muds. Oolitic sands occur only locally where tidal currents are active; for example, at the south end of the deep water tongue of the ocean and at the north end of Exuma Sound. Discontinuous reef barriers occur mainly on the windward northeast of land areas. Many ancient carbonate deposits show the same sedimentation and faunal (animal life) characteristics as Recent sediments.

A large part of the sedimentary rocks in the geologic column were probably deposited in transitional-marine environments in relatively shallow water on the continental shelf. This includes the deltaic-marine sediments and many carbonate (limestone and dolomite) sequences which contain large petroleum accumulations throughout the world.

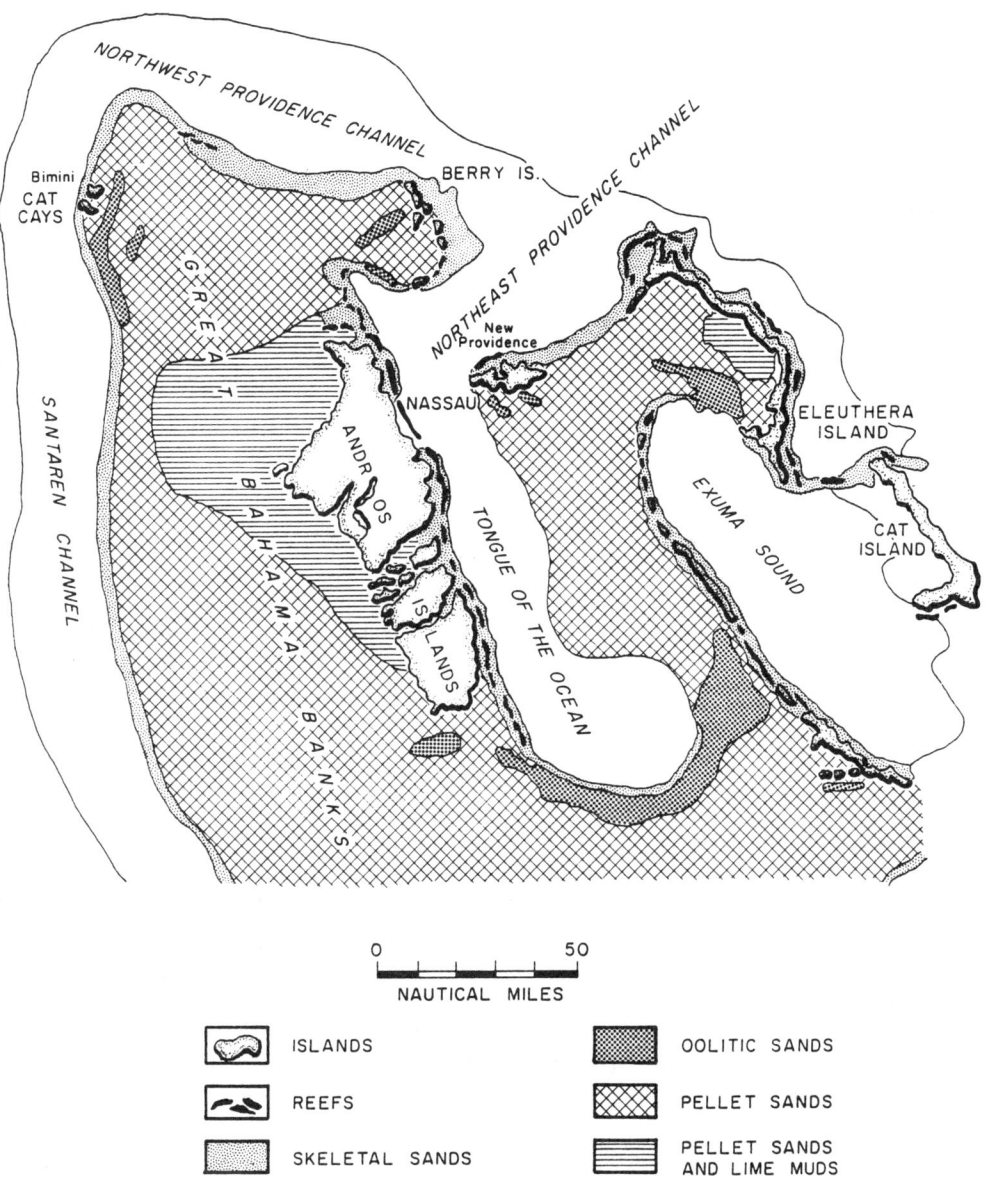

Figure 2-5. Distribution of Surface Sediments of the Bahama Banks

2.4 Geological Structures

Sedimentary rocks are deposited in essentially horizontal layers or shallow slopes called strata or beds. Most rock layers are not strong enough to withstand the forces to which they are subjected, so they are deformed. A common kind of deformation is the buckling of the layers into folds which are the most common structures in present and former mountain chains. They range in size from small wrinkles to great arches and troughs many miles across. The upfolds or arches are called anticlines; the downfolds or troughs are synclines. Folds have many forms, a few of which are shown in Figure 2-6. They may be symmetrical with similar flank dips on both limbs or asymmetrical where one limb is steeper than the other. The ends of anticlines and synclines usually plunge, and a very short anticline — the crest of which plunges in opposite directions from a high point — is called a dome. Many domes are uplifted by an intrusive core, such as the salt domes of the Gulf Coast in the United States, Northern Germany, and elsewhere.

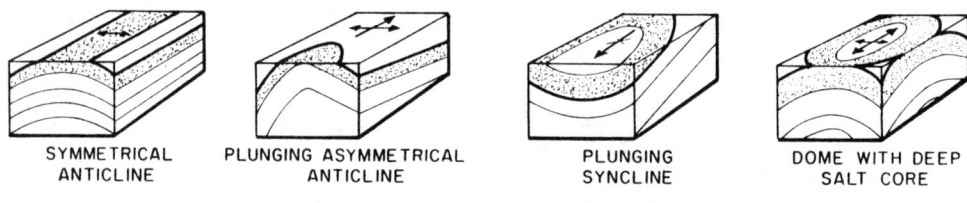

Figure 2-6. Simple Kinds of Folds

2.5 Earth Movements

Nearly all rocks are fractured to some extent during earth movement to form cracks called joints. If the rock layers on one side of a fracture have moved in relation to the other side, the fracture is called a fault. Displacement on a fault may range from only a few inches to many thousands of feet, and even miles in some cases such as along the San Andreas fault in California.

Faults are described according to their present attitude by various names. There are four simple classifications of faults, as shown in Figure 2-7:

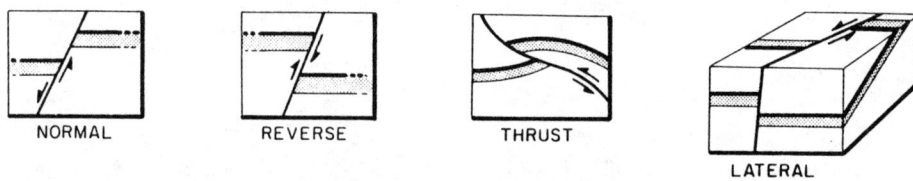

Figure 2-7. Simple Kinds of Faults

- normal
- reverse
- thrust
- lateral

These terms reflect mainly the relative movement of the adjacent blocks with respect to one another. Movement is upward or downward in the case of normal and reverse faults, but is mainly horizontal in thrust and lateral faults. Faults may also have a combination of vertical and horizontal movements.

Rotational faults and upthrusts (Figure 2-8) are variations of normal and reverse faulting. They are most important to the petroleum geologist since they have very important effects upon the location of oil and gas accumulations as compared with the accumulations associated with normal and thrust faults.

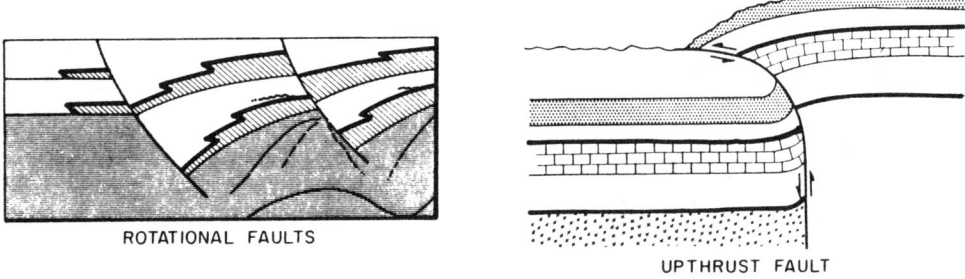

Figure 2-8. Cross-Section Showing Rotational and Upthrust Faults

Another result of earth movement is to erase or to prevent the deposition of part of a series of sediments which are present elsewhere. Such a buried erosion surface is called an unconformity. There are two general kinds as shown in Figure 2-9: A disconformity, where the beds above and below the surface of unconformity are parallel, as shown in the left diagram; and an angular unconformity, where the beds above the unconformity transgress the eroded edges of folded and tilted beds below, as shown in the other two diagrams. Earth movements are most important to the subject of petroleum geology because they produce barriers which contain a large proportion of petroleum accumulations.

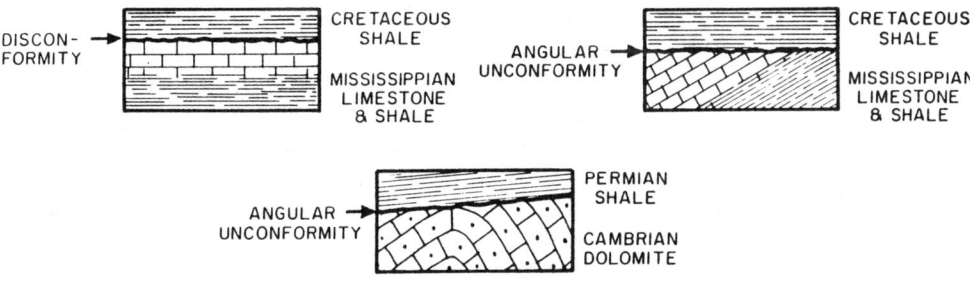

Figure 2-9. Unconformities

2.6 APPLICATION OF GEOLOGICAL CONCEPTS

2.7 Early Exploration

The most successful oil-finding method in the early days of oil exploration was to drill in the vicinity of oil seeps, where oil was actually present on the surface of the ground. In fact, many of the great oilfields of the world owe their discovery in part to the presence of oil seeps. Seeps are of two general kinds: (1) seepage along fractures and (2) seepage updips (Figure 2-10). The diagram on the right shows a seep at the outcrop of a reservoir bed. Such seeps may be active where oil or gas is still flowing slowly. In other cases, the sands near the surface are completely sealed with asphalt from the oil and the seep is no longer active. Seepage from fractures and faults is very common and may be oil or gas or sometimes mud as in the mud volcanos of Trinidad and Russia. Active seeps of this kind are present along the anticlinal crest in the La Paz Field, Venezuela, and the Kirkuk, Iraq.

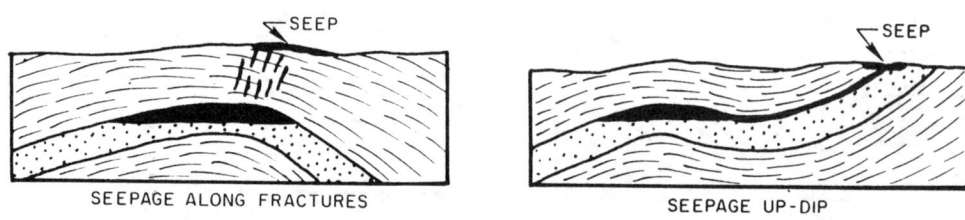

Figure 2-10. Schematic Cross-Section of Seeps

2.8 Geophysical Exploration

By 1920 it was obvious that anticlinal folding was only one of a number of geological factors controlling oil and gas accumulation and that much could not be determined from surface mapping alone. Approximately at this time geophysical methods were developed; first was the torsion balance, and later came the seismograph which enabled subsurface structures to be deduced. The seismic method, the most important in today's predrilling exploration, uses the transit time of sound waves (the time required for a sound pulse to travel a fixed distance between a transmitter and reciever) generated by an explosion. These transit times depend on the nature of the rocks penetrated, particularly their density. The transit time measured is for a wave reflected from a surface in the subsurface (Figure 2-11). Under favorable conditions geologic beds may be mapped quite accurately to create subsurface contour maps of structure and possible reservoir locations.

Other geophysical methods include the gravimeter and magnetometer which make use of the physical properties of the rocks to find favorable structural conditions for petroleum accumulation. In early Gulf Coast exploration, salt domes were located by gravity anomalies. In remote land locations gravity surveying can be done from the air and is therefore much cheaper than seismic surveying.

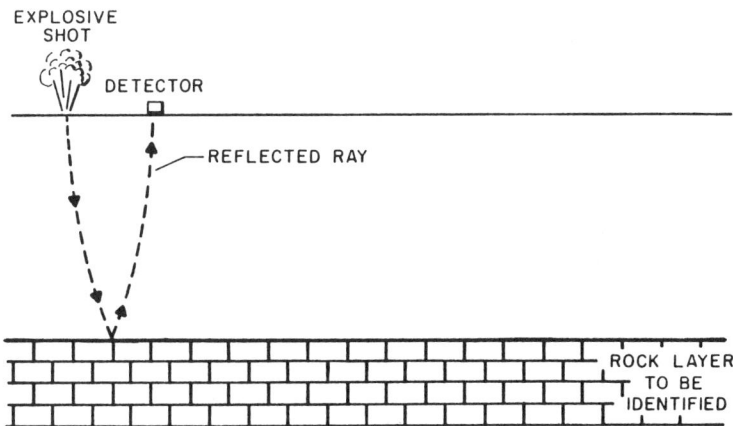

Figure 2-11. Reflected Seismic Waves

2.9 Surface Exploration

Surface mapping and photogeology, i.e., aerial photos and photogrammetric "stereo pairs" using Earth Resources Satellite (ERSAT) scans, are used extensively for lithological and structural determinations. Another type of surface exploration is geochemical prospecting, a recent technique based on the supposition that some hydrocarbons in an oil accumulation migrate vertically to the surface directly over the oilfield. There are at least five general methods of geochemical prospecting:

- Analysis of free hydrocarbon gases in the pore spaces of the soil

- Analysis of gaseous hydrocarbons adsorbed on the soil particles or on subsurface rock samples

- Fluorescence of soil samples presumably due to the presence of high molecular weight hydrocarbons

- Analysis for bacteria that thrive on certain kinds of hydrocarbons

- Radioactive scintillometer surveys

Geochemical prospecting as a direct indicator of oil has had limited success because hydrocarbons do not migrate directly upward from an accumulation. Migration occurs along faulted and fractured zones and through more permeable beds such as glacial drift or continental deposits. Water in sands overlying accumulations can redirect any upward migration. Consequently, surface indications may be useful in defining oil, gas or barren regions, but they cannot pinpoint an accumulation.

Seeps and natural asphalts give surface evidence of oil or gas that has migrated from its original accumulation. Outcrops, if not weathered, contain traces of hydrocarbons that are indicative of the oil potential of the sediment.

2.10 Subsurface Exploration

The methods and techniques used in the study of subsurface geology have developed since the 1920s. Today, more oil and gas discoveries are credited to subsurface geology studies than to any other technique.

Data is determined by the following:

- Distribution of organic carbon. It is generally accepted that shales need more than 0.5% organic carbon to yield commercial oil. The organic carbon content of Devonian shales in the oil-producing region of the Russian platform is 1.6% and in the barren regions 0.5%. In the richest oil region, it ranges from 0.5 to over 5%.

- Gas analysis of mud and cuttings. Sediments in the immediate vicinity of petroleum accumulations have much higher gas yields compared to those away from accumulations. Mapping the vertical and areal distribution of gas and related light hydrocarbons can show the sedimentary sections more apt to contain commercial accumulations (Figure 2-12).

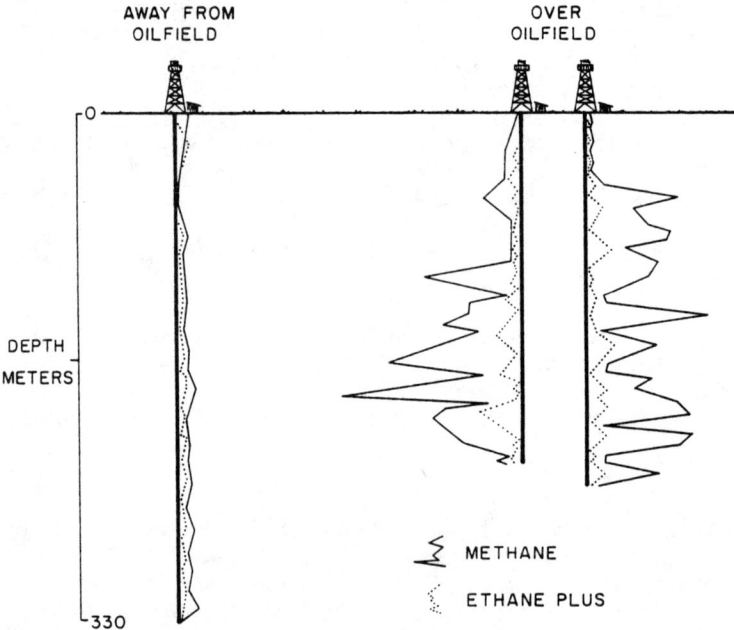

Figure 2-12. Diffusion of Hydrocarbons

- Geochemical temperature facies. The disseminated organic matter of sedimentary rocks, kerogen, is insoluble in acids and organic solvents. The organic matter initially deposited with unconsolidated sediments is not kerogen but a precursor that is converted to kerogen during diagenesis. Kerogen in shales changes color from yellow to orange-brown to black through the subsurface temperature range from about $100°F$ ($38°C$) to $500°F$ ($260°C$). The mature orange-brown range yields oil and wet gas (Figure 2-13). Color changes may be anomalous in carbonates because recrystallization blackens organic matter.

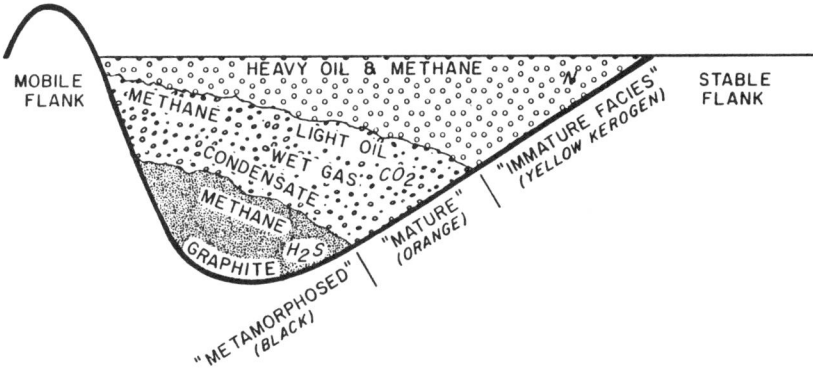

Figure 2-13. Geochemical Temperature Facies in a Typical Basin

- Gasoline and gas in cuttings. The analysis of C1-C4 (gas) and C4-C7 (gasoline) hydrocarbons in cuttings shows the subsurface range in which oil is being generated and wet gas is phasing out. Wet gas is natural gas dissolved in heavier hydrocarbons where the natural gas liquid vapors amount to more than 300 gallons of propane, butane and other liquid hydrocarbons per 1000 cu ft of gas.

- Heavy hydrocarbons, C15+, in outcrops, cores and cuttings. Source rock quality and type (gas, oil or nongenerating) can be evaluated from C15+ analysis on unweathered cuttings, cores, outcrops.

- Kerogen analysis. With increasing depth, analysis of carbon, hydrogen, and oxygen in mineral-free kerogen from cuttings or cores indicates horizons at which hydrocarbons are generated from a reversal in hydrogen content. The threshold of intense oil generation usually occurs below the accumulation zone (Figure 2-14).

- Vitrinite reflectance. Metamorphism of sediments causes disseminated vitrinite particles to harden and reflect light better. Reflectance (Ro) ranges from 0.5 to 1.2 in oil zones, up to 2.8 in gas zones. Recycling of vitrinite particles and oxidation of particles gives anomalous readings.

- Wireline log correlation.

- Lithology (rock type) and environments of deposition from cores.

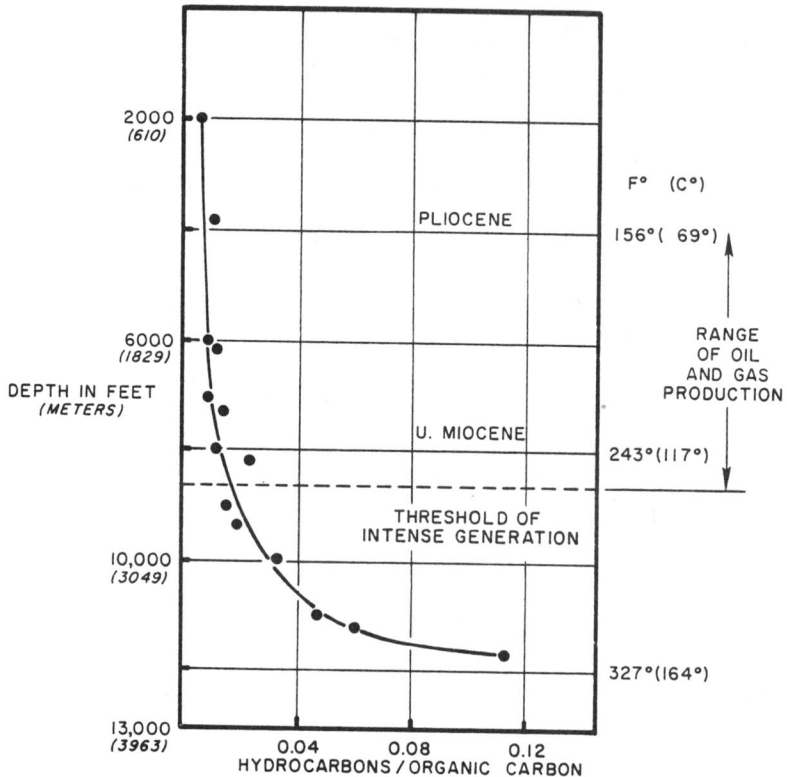

Figure 2-14. Generation of Hydrocarbons in the Los Angeles Basin

2.11 Exploration Applications

Considering the exploration possibilities, the following procedures are conducted whenever possible in the situations listed below:

- Basin evaluation. Determinations are made of organic carbon, C15+ hydrocarbons, kerogen type and color, plus vitrinite reflectance on unweathered outcrops or shallow cores of prospective exposed horizons. In offshore areas, analysis is made on C1-C7 hydrocarbons in surface cores. All seeps onshore and offshore are analyzed.

- Wildcat wells. Analysis is made for two or more of the following on cuttings or cores: gas and gasoline, C15+ hydrocarbons, kerogen type and color, C, H, O elemental analysis, vitrinite reflectance. Any subsurface water samples are analyzed for hydrocarbons. In permafrost, or deep ocean areas, hydrocarbon hydrates (crystalline compounds in which the ice lattice of H20 expands to form cages that contain gas molecules) are monitored.

- Developed areas. Correlation of crude oils with their source rocks and with seeps is done with detailed gas chromatography and mass spectrometry of hydrocarbon groups. Untested horizons are analyzed like wildcat wells.

- Generalizations. A commercial petroleum accumulation requires a hydrocarbon source, a structure or trap with impermeable cover, and good reservoir porosity and permeability. Geochemical techniques define the first requirement by indicating whether there is enough organic matter of the right type for commercial accumulations, and by defining the stage of organic maturation in the sediments penetrated. Well cuttings, core analysis and wireline log interpretation define the other requirements.

This information is used to prepare many kinds of maps and cross-sections. Contour maps are used to show geologic structure on numerous correlation markers. Other contour maps may show fault attitude and intersections with beds and other faults, as well as porosity and permeability variations. Various types of maps may show variations in the characteristics of the rocks and the structural arrangement such as old shorelines, pinchout, or truncation of beds.

Maps give only a plan view, so it is necessary to supplement them with vertical cross-sections. These are of various kinds; they may show fault structure as in Figure 2-15, or they may be designed to show detail of one sort or another for just one small interval. For example, the section in Figure 2-16 shows the effect of pinchout to the North of the darker colored sand body on the thickness between correlation markers A and B above and below it.

2.12 Conceptual Models of Exploration

As applied to petroleum geology, a conceptual model is an idea, developed from available data, as to how a geological area is structured, what it looks like, and where petroleum accumulations are likely to occur.

The prediction of both sand trends and pore space distribution is important in all phases of exploiting sandstone reservoirs. It is quite disconcerting during primary development to drill a well a few hundred feet from a good producer only to find the objective sand absent. The best way to appreciate the value of environmental concepts is to consider a few examples of their applications in oilfield development problems. The following paragraph gives one example.

An environmental study of the Aux Vases sandstone in an oilfield in Illinois showed that the reservoir is a composite of two types of sand bodies (shoreline and tidal channel deposits) having distinctly different trends and distribution as well as different characteristic properties and boundary relations. The tidal channel sandstones were shown to have cut down into the shoreline sandstones. A rework of several pairs of wells indicated that fluid communication was poor or absent between the two types of sand.

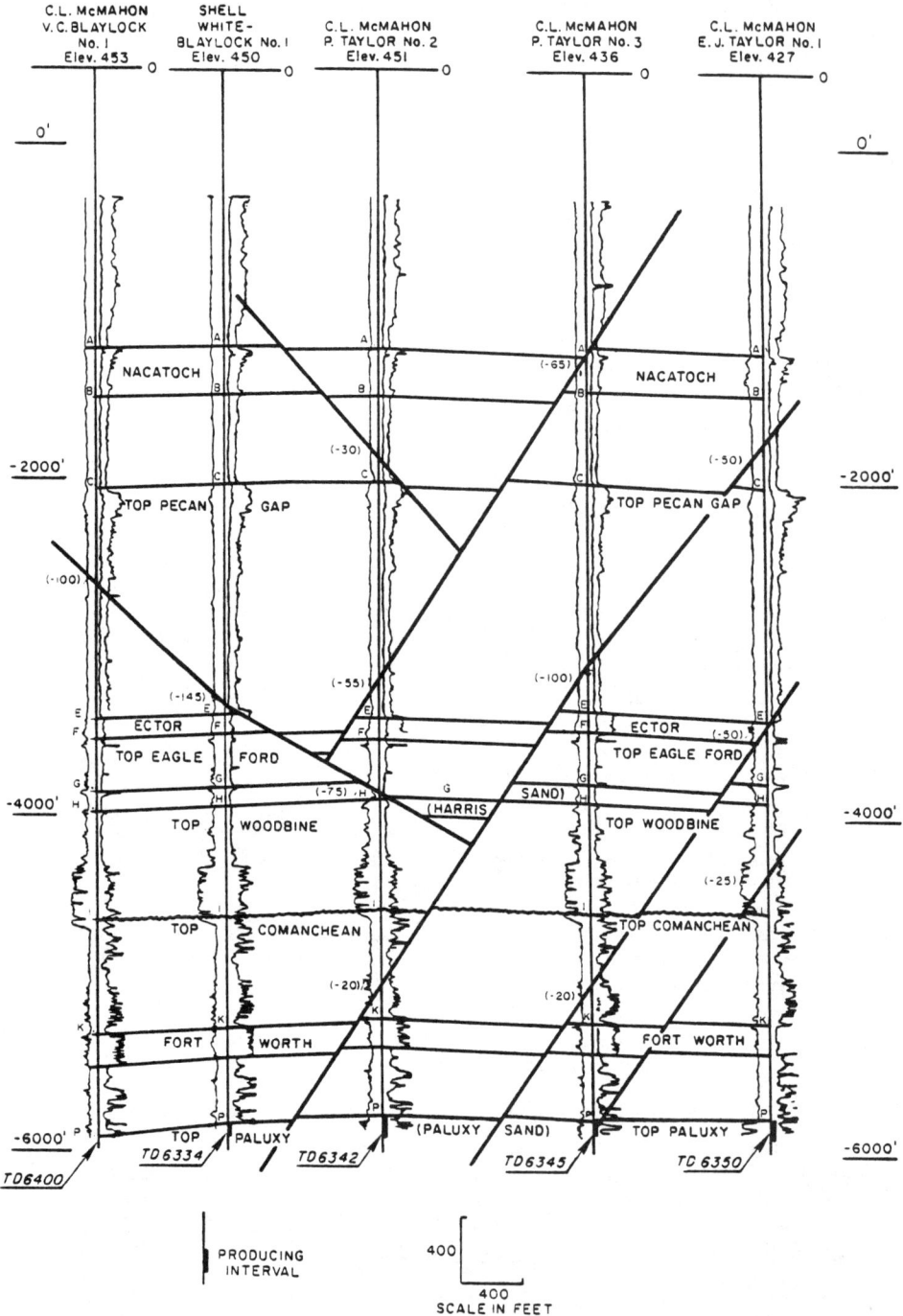

Figure 2-15. Cross-Section Illustrating Fault Pattern, Quitman Field, Texas

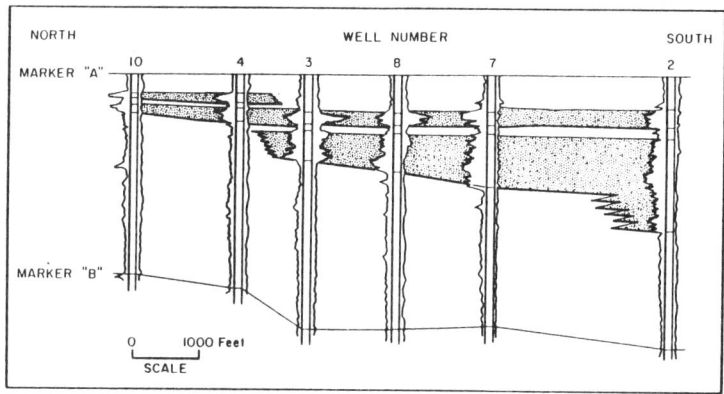

Figure 2-16. Sand Pinchout

2.13 PETROLEUM ACCUMULATIONS

For petroleum to accumulate, there must be (1) a source of oil and gas; (2) a reservoir rock (a porous bed which is permeable enough to permit the oil and gas to flow through it); and (3) a trap (a barrier to fluid flow so that accumulation can occur against it). Most knowledge has been obtained from experience and observations, but certain generalizations can be made:

- Petroleum originates from organic matters

- To become commercial, the hydrocarbons must be concentrated

- Petroleum reservoirs are mostly in sedimentary rocks

2.14 Origin of Petroleum

The following discussion presents very generally one theory for the origin of oil. Oil and gas originate from organic matter in sedimentary rocks. Dead vegetation in the absence of oxygen ceases to decompose. It accumulates in the soil as humus and as desposits of peat in bogs and swamps. Peat buried beneath a cover of clays and sands becomes compacted as the temperature, weight and pressure of the cover increase and water and gases are driven off. The residue, ever richer in carbon, becomes coal.

In the sea, a similar process takes place. Of the marine life that is eternally falling slowly to the bottom of the sea, vast quantities of it are eaten and some is oxidized, but a portion of the microscopic animal and plant life escapes destruction and is entombed in the ooze and mud on the seafloor. The organic debris collects in sunken areas at the bottom and is buried within a growing buildup of sands, clays and more debris until the thickness of sediment attains thousands of feet. Bacteria take oxygen from the trapped organic residues, breaking them down molecule-by-molecule into substances rich in carbon and hydrogen, and the extreme weight and pressure of the mass compacts the clays into hard shales.

The generation of hydrocarbons from the source material depends primarily on the maximum temperature to which the organic material is subjected. It appears to be negligible at temperatures less than about $150°F$ ($65°C$) in the subsurface and reaches a maximum within the range of $225°$ to $350°F$ ($107°$ to $176°C$), the "hydrocarbon window." Increasing temperatures convert the heavy hydrocarbons to lighter ones and ultimately to gas. However, at temperatures above $500°F$ ($260°C$), the organic material is carbonized and destroyed as a source material. Consequently, if source beds become too deeply buried by earth movements, no hydrocarbons will be produced.

2.15 Migration

After generation, the dispersed hydrocarbons in the fine-grained source rocks must be concentrated by migration to a reservoir. Compaction of the source beds by the weight of the overlying rocks provides the driving force necessary to expel the hydrocarbons and to move them through the more porous beds or fractures to regions of lower pressure, which normally means to a shallower depth. Gravity separation of gas, oil and water also takes place in reservoir rocks that are usually water-saturated. Consequently, petroleum is forever trying to rise until it is trapped or escapes at the surface of the earth. Vertical migration via faults and fractures has led to many of the large oil accumulations such as that found at shallow depths in the Bolivar District, Venezuela, and in northern Iraq. In other cases, such as the Khurais field in Arabia, migration over relatively long distances has had to take place by movement up-dip in a porous reservoir bed until a trap was encountered.

2.16 Primary Migration: Primary migration of petroleum from source to reservoir is caused by movement of water, which carries oil out of the compacting sediments. When the source muds are deposited they contain 70 to 80 percent water. The remainder is solids (mostly clay minerals) or carbonate particles or fine-grained silica. As they build up to great thicknesses in a sedimentary basin, water is squeezed out by the weight of the overlying sediments. Under normal hydrostatic pressure (approximately 0.446 psi/ft), the clays lose porosity and the pore diameters shrink approximately as shown in Figure 2-17.

Depth		Clay Porosity	Clay Pore Diameter
(Meters)	(Feet)	(Percent)	(Nanometers)*
610	2,000	27	----
2,000	6,560	15	10.0
3,000	9,840	9	5.0
4,000	13,120	6	2.5
5,000	16,400	4	1.5
($*10^{-9}$m)			

Figure 2-17. Changes During Normal Compaction of Shales

The fluids move toward the lowest potential energy. Initially this is upward, but as compaction progresses there is lateral as well as vertical movement of the fluids. The lateral movement results primarily from the tendency of the flat clay mineral particles to lie horizontally as they are compressed. This reduces the vertical permeability of the compacting muds. In addition, the long continuous sands on the edges of basins orient fluid movement laterally as burial progresses, as illustrated in Figure 2-18. The migration of oil from source to reservoir is as follows:

1. Water flows toward the lowest potential energy.

2. Clay muds often have abnormal pressure because they are slow to release water.

3. Avenues of migration during basin compaction are:
   - sandstones
   - unconformities
   - fracture-fault systems
   - biohermal reefs

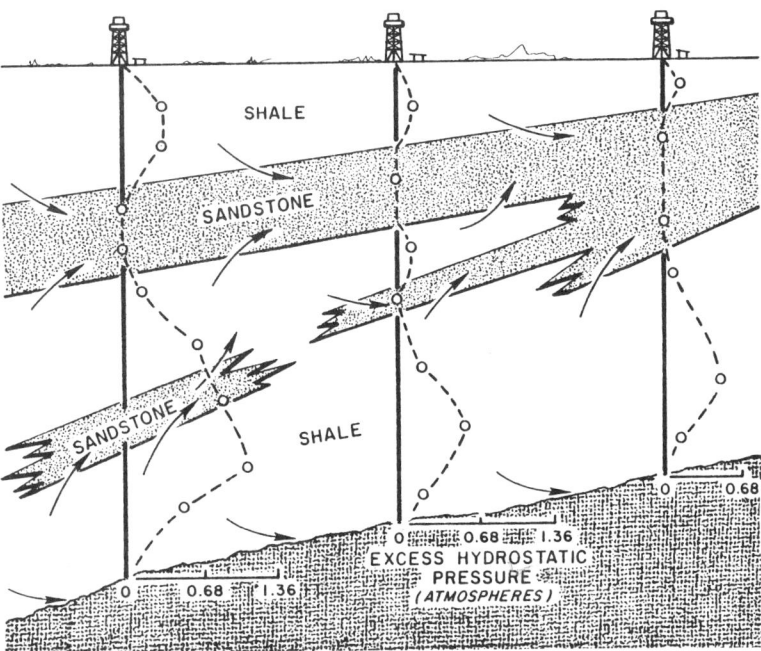

Figure 2-18. Fluid Flow During Compaction

The mechanism by which oil migrates is uncertain, but most likely it is in solution. The solubility of the lighter petroleum hydrocarbons in water is adequate to account for known oil accumulations. When this solution reaches the reservoir, the change in environment may cause coagulation of the hydrocarbons to form discrete oil particles. The heavier hydrocarbons may travel as more soluble nonhydrocarbons and form hydrocarbons in the permeable beds. Another mechanism is simply the squeezing out of oil particles due to the structuring of water. Ordered water exists 2 to 4 nanometers (1 nm = 10^{9} m = 10 angstrom) from a clay mineral surface. Hydrocarbon molecules range up to 2 nm with asphaltenes to 5 nm in size. Since pore openings are less than 10 nm in shales deeper than 2000 m, the hydrocarbons are squeezed out (Figure 2-19). Near-surface fluid migration may be restricted in areas of permafrost or offshore basins where solid methane hydrates may form in sediments.

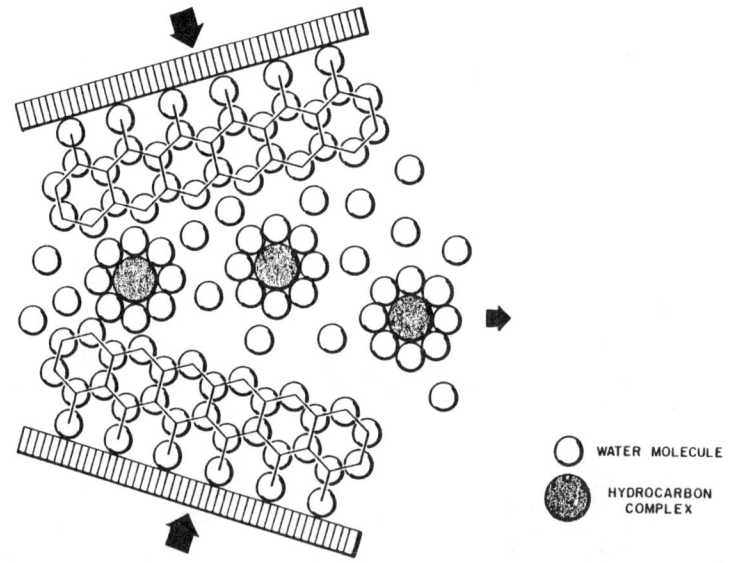

Figure 2-19. Migration Induced by Fixed Water

2.17 Secondary Migration: In secondary migration the oil droplets are moved about within the reservoir to form the pool. Secondary migration includes in some instances a second step during which crustal movements of the earth shift the position of the pool within the reservoir rock.

The position of the accumulating pool is affected by several, sometimes conflicting, factors. Buoyancy causes oil to seek the highest permeable part of the reservoir; capillary forces direct the oil into the coarsest-grained portion first, and into successively finer-grained portions later. Any permeability barriers in the reservoir channel the oil into a somewhat random distribution. Oil accumulations in carbonate rocks are often erratic because part of the original void spaces have been plugged by minerals introduced from water solutions after the rock is formed.

In large sand bodies, barriers formed by thin layers of dense shale may hold the oil at various levels. When crustal movements of the earth occur, oil pools are sometimes shifted away from the place in which they originally accumulated. Faults sometime cut through reservoirs, destroying parts of the pools or shifting them to different depths. Uplift and erosion bring the pools near the surface, where the lighter hydrocarbons evaporate. Fracturing of the cover rock allows oil to migrate vertically to much shallower depth. Wherever differential pressures exist and permeable openings provide a path, petroleum will move.

2.18 Reservoir Rocks

A petroleum reservoir is a rock capable of containing gas, oil, or water. To be commercially productive it must have sufficient thickness, areal extent, and pore space to contain an appreciable volume of hydrocarbons, and must yield the contained fluids at a satisfactory rate when the reservoir is penetrated by a well. Sandstones and carbonates are the most common reservoir rocks. The porosity characteristic of a rock may be primary, such as the intergranular porosity of sandstone, or it may be secondary due to chemical or physical changes such as dolomitization, solution channels, or fracturing. Porosity may be adversely affected by compaction and cementation. The distribution of petroleum reservoirs and the trend of pore space therein are the result of numerous natural processes.

2.19 Sandstones: In sandstones, porosity is controlled primarily by sorting (that is, by mixing the various sizes of grains), cementation and, to a lesser extent, by the way the grains are packed together. Porosity is at a maximum when grains are spherical and all one size, but becomes progressively less as the grains are more angular because such grains pack together more closely. Figure 2-20 shows two ways of packing spherical grains. The one on the left is open (cubic) packing where the porosity is about 48 percent. The close (rhombohedral) packing on the right has a porosity of only about 26 percent because the grains are packed into a smaller space. Artificially mixed clean sand has measured porosities of about 43 percent

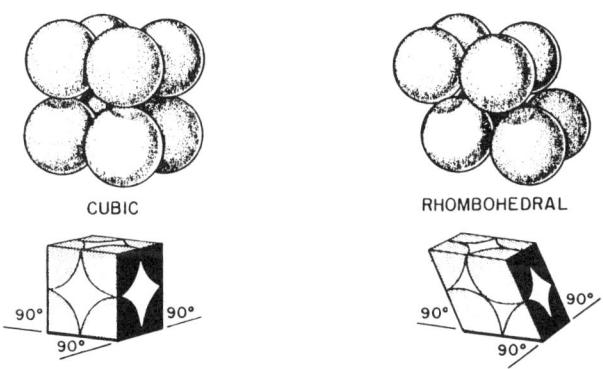

Figure 2-20. Unit Cells of Cubic and Rhombohedral Packing

for extremely well-sorted sands, almost irrespective of grain size, decreasing to about 25 percent for very poorly sorted medium- to coarse-grained sands, while the very fine-grained sands still have over 30 percent porosity. Figure 2-21 summarizes diagrammatically some of the processes controlling pore size distribution in sandstones.

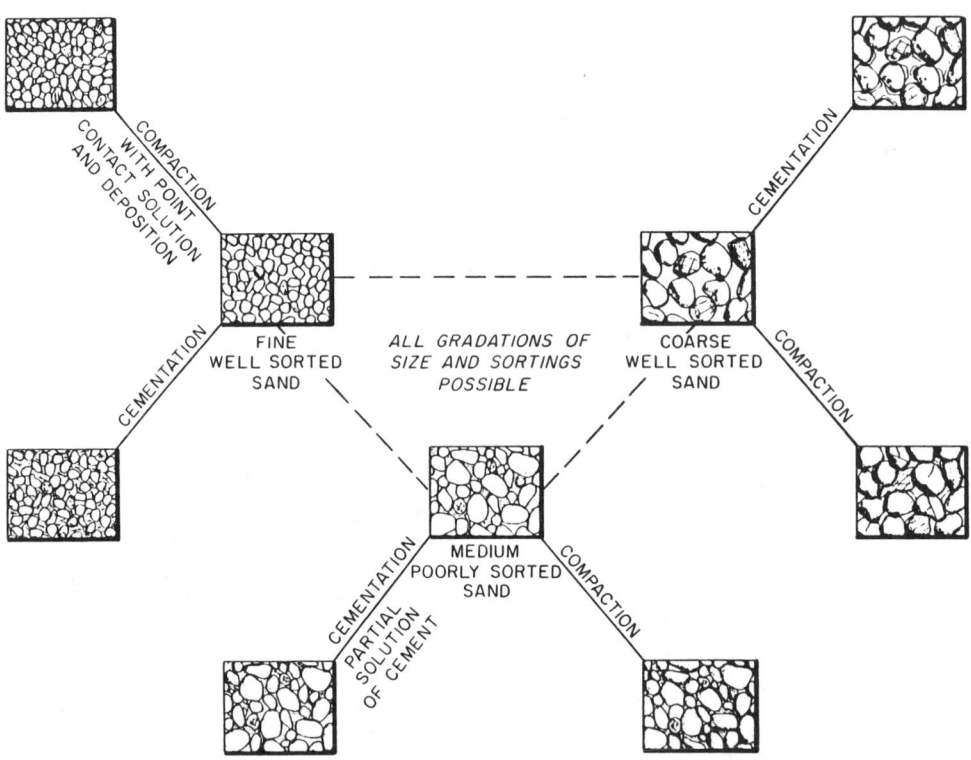

Figure 2-21. Some Processes Controlling Pore Size Distribution in Sandstones

The ease with which fluid moves through the interconnected pore spaces of a rock denotes the degree of permeability. In 1856 Henry d'Arcy, a French engineer, devised a means of measuring the permeability of porous rocks. Numerical expressions of permeability are measured in "darcies" (d). A rock has a permeability of one darcy (1 d) when 1 cc of a fluid of 1-cp (centipoise) viscosity flows through a 1-cm cube (i.e., 1 cm x 1 sq cm) of rock in 1 second under a pressure gradient of 1 atmosphere (N.B. viscosity of water at $68^\circ F$ = 1 cp). Because most reservoir rocks have average permeabilites considerably less than one darcy, the usual measurement is millidarcies (md, thousandths of a darcy). Permeability of a highly porous, well-sorted sand varies from 475 md for a coarse-grained sand to about 5 md for a very fine-grained sand. Permeability may decrease for a coarse-grained sand to about 10 md if it is very poorly sorted.

Compaction by weight of the overburden squeezes the sand grains closer together, and at greater depths may crush and fracture the grains. The result is smaller pores and therefore lower porosity — but more importantly, a drastic decrease in permeability. Thus a sandstone reservoir which could produce petroleum at 10,000 feet might become much too impermeable to be of any economic value at, say 20,000 feet. Cementation, which fills part or all of the pore space, also tends to increase with depth. Figure 2-22 shows the effect of compaction for a poorly sorted sandstone from Ventura, California. The silhouettes are based on photomicrograph, but are easier to see in black and white as shown than as a photograph. Note how the pore space gets less and less from surface conditions to the second, fourth, and so on to the ninth zone and finally to the Miocene sands. There is still quite a bit of porosity in the Miocene, but as can be seen by the very small size of the pores, the permeability must have decreased greatly.

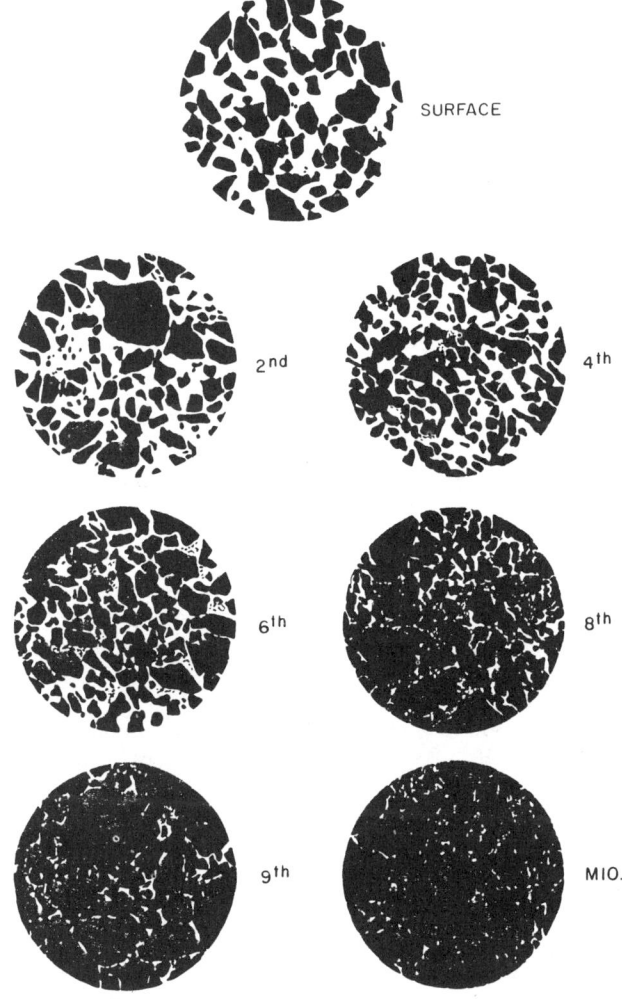

Figure 2-22. General Porosity Decrease with Depth

2.20 Carbonates: Carbonates are a complex group — difficult to study and difficult to interpret. They are very different from sandstones and shales, especially in their susceptibility to post-depositional change, particularly when changed from calcium carbonate to the calcium magnesium carbonate form (dolomite) by the dolomitization process. In carbonates the porosity, permeability, and pore space distribution are related to both the depositional environment of the sediment and the changes that have taken place after deposition. Figure 2-23 is a diagram illustrating the evolution of some dolomite textures and pore types. The original sediments ranging from lime mud to lime sand are depicted in the central column. Diagenetic (or later) processes will change porosity and pore size distribution, as shown to the right and left. Note that lime mud is preferentially dolomitized. The other particles may then be dissolved out, leaving pores or larger holes that may or may not be interconnected. This is shown to the left of the center in the figure. To the right, on the other hand, note the good porosity and permeabilty in the open network of dolomite crystals. This illustrates two basic types of carbonate porosity: <u>interparticle</u> between grains or crystals, and <u>intraparticle</u> due to particle solution; there may be combinations of the two. Going more to the left, it can be seen that cementation and infilling have taken place, destroying most, if not all, of the porosity. When volume loss occurs due to recrystallization, irregular voids are formed called vugs (vuggy porosity).

Permeability is controlled by the size of the passages between the much larger pores and vugs. Consequently, a highly porous rock may have little or no permeability if these interconnections are very small or absent. On the other hand, some very fine-grained carbonate rocks have an extensive network of interconnected pore space with enough permeability to be able to yield economic volumes of oil. Intercrystalline pores tend to be interconnected, and rocks with high intercrystalline porosity are normally permeable as found in many highly productive dolomite reservoir rocks.

Carbonates can be extensively fractured. In this situation, even without porosity and permeability in the main body of the formation, economic amounts of oil can exist if the source and other conditons of accumulation are present.

2.21 Estimations of Porosity and Permeability: Porosity is a measure of the pore space in the body of a reservoir rock, usually expressed as a percent of the unit volume. Permeability is a measure of the ease with which a fluid flows through the connected pore spaces of a reservoir rock. Accurate estimations of porosity and permeability are not possible without the use of core analysis. However, when clean and dry, samples can be generally described in terms of porosity by using the highest possible microscope powers. For this to be meaningful, grain distribution, size, type and abundance must be included in the description.

Porosity is frequently described as follows (this is a guide only — it is not intended to be absolute):

over 15 percent	good
10 - 15 percent	fair
5 - 10 percent	poor
less than 5 percent	trace

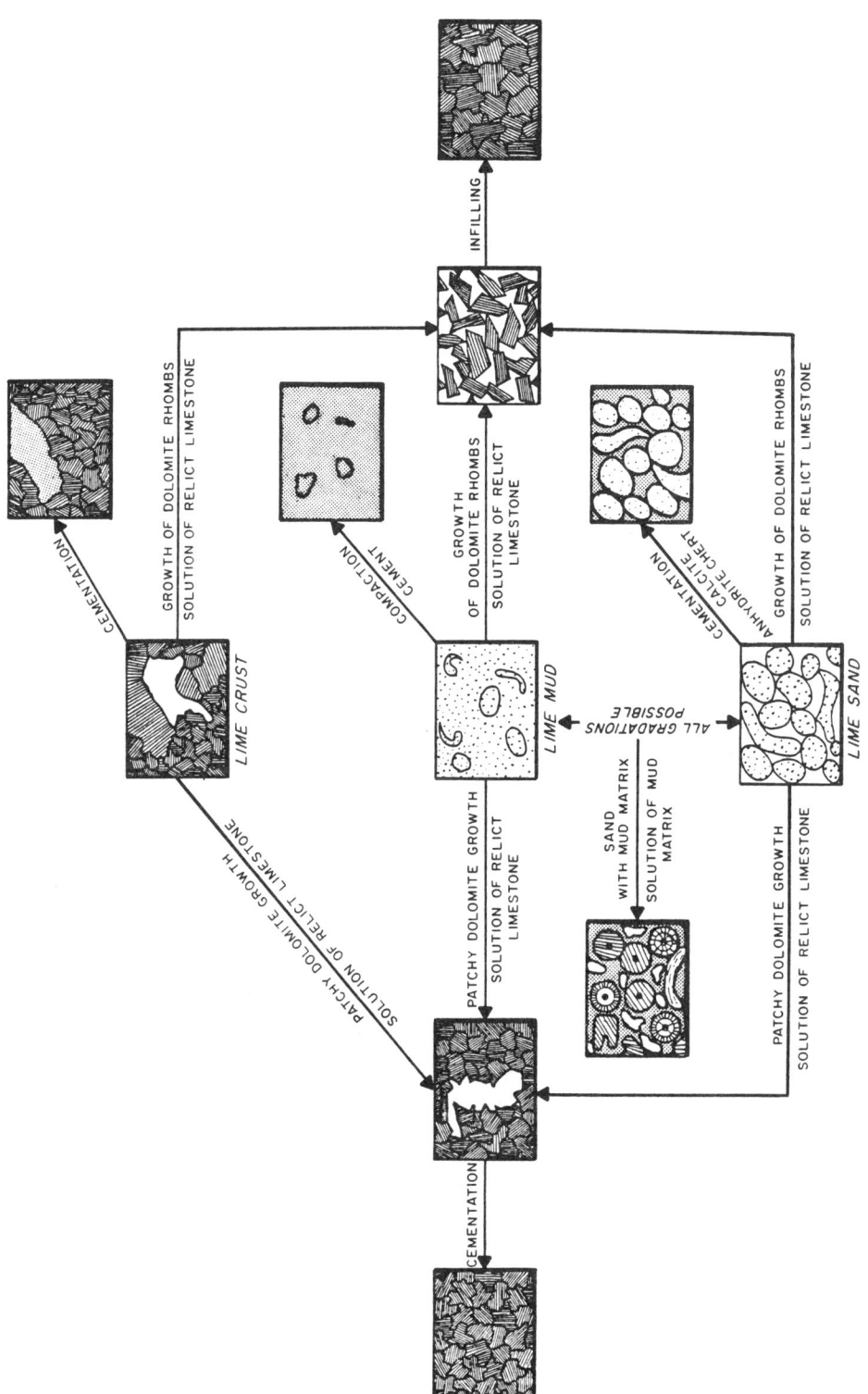

Figure 2-23. Evolution of Some Dolomite Textures and Pore Types

Permeability can be estimated only qualitatively without core analysis, by using the blender. The blender is present in every unit and is used to agitate, break up and mix cuttings. The amount of gas released, which is drawn off and analyzed, is related to permeability and porosity. Cuttings with good permeability have little gas, for it is assumed to have escaped before reaching the blender stage of examination. A more porous but less permeable formation, however, is more likely to contain gas, for the lack of pore space communication produces a trapping effect which is released only when blender-agitated.

It is possible to deduce from this that if a gaseous formation is penetrated (indicated from Total Gas instrument readings), and blender readings (from the Microgas instrument) from the same formation are low, it is likely to be both porous and permeable. A high blender reading would have indicated a low or non-permeable but certainly porous formation.

In a situation where flushing of the formation has occurred (i.e., where jet action of the bit and differential pressure have forced formation fluids and gases away from the borehole), very little or no gas will be detected. Blender gas also will be low or nonexistent, for it follows that the formation is permeable for flushing to have occurred. If the formation is visibly porous from the cuttings and it is possible that the mudweight (hydrostatic head) is overbalanced, flushing should be suspected. Oil and gas shows can and have been overlooked because of flushing, so it should always be borne in mind that this can occur.

2.22 Traps

Once the hydrocarbons have been generated and expelled from the source rock, migration is a continuous process -- regardless of whether they are moving through a reservoir rock or through a fracture system. Obviously, then, a barrier or trap is needed to impede this migration in order to get an accumulation.

A trap is produced by a set of geological conditions which cause oil and gas to be retained in a porous reservoir or at least allowed to escape at a negligible rate. Shales and evaporites make good seals, although any unfractured rock that has a displacement leakage pressure higher than that of the hydrocarbon accumulation will seal a trap.

Most traps are not filled to their structural or stratigraphic spill point. A spill point is illustrated by the successive diagrams in Figure 2-24. Note in Stage 1 the stratification of gas, oil and water above the trap spill point. In Stage 2, hydrocarbons fill the trap to the spill point; oil is spilling out and migrating farther up-dip. In Stage 3, the trap is filled with gas. Gas moving from below enters the trap, but a like volume spills out at the same time; oil bypasses the trap entirely. Incomplete filling of a trap is more likely the result of the seal not sustaining the greater hydrocarbon column pressure rather than being the result of insufficient oil and gas to fill the trap (Figure 2-25). For this reason, traps often may be filled to capacity and yet have water levels far above the spill point.

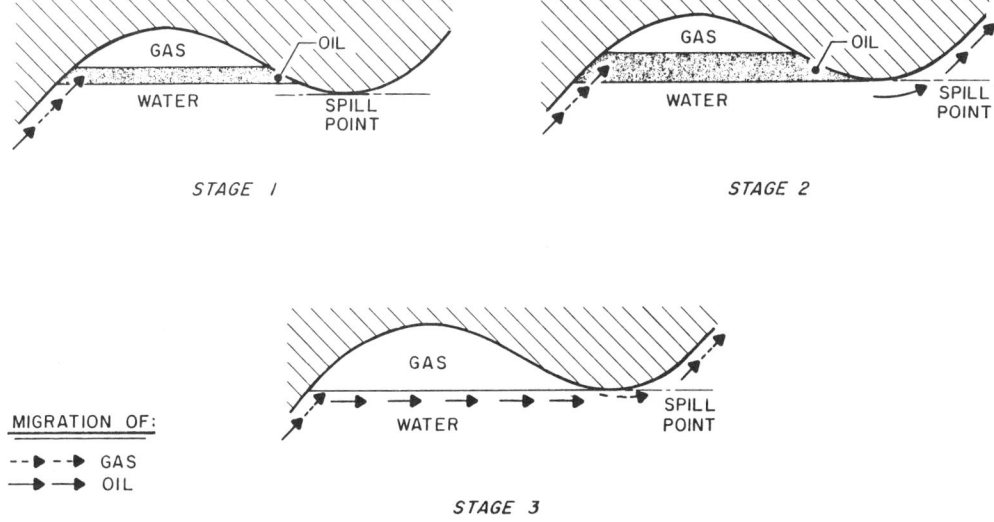

Figure 2-24. Spill Point of a Hydrocarbon Trap

Traps for oil and gas under hydrostatic conditions have two general forms: the trapping factor is either (1) an arched upper surface or (2) an up-dip termination of the reservoir. Figure 2-26 shows some of the simpler forms.

Traps for hydrocarbons under hydrostatic conditions (liquid at rest) are of structural or stratigraphic origin, either alone or in combination, and have horizontal gas-water or oil-water contacts. Hydrodynamic (moving liquid) traps may also occur in different structural environments, but they are characterized by inclined gas- or oil-water contacts.

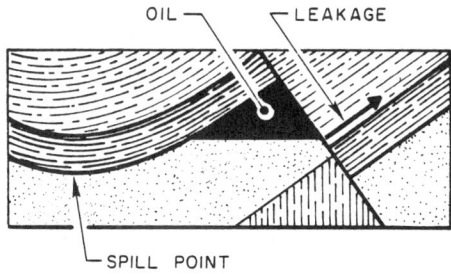

Figure 2-25. Fault Trap Leakage

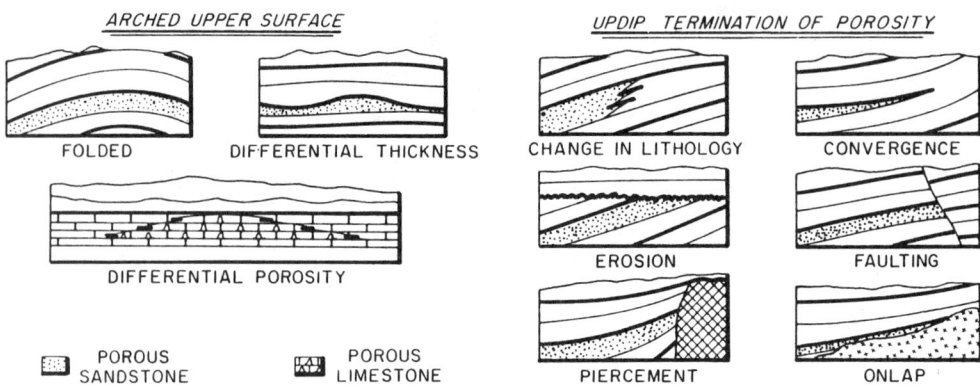

Figure 2-26. Basic Reservoir Traps

2.23 Anticlinal Traps: These vary widely in shape and size. However, they all have a common characteristic in that a gas-water or oil-water contact completely surrounds an accumulation of hydrocarbons. The structure generally extends through a considerable thickness of formation so that traps are formed in all the potential reservoir rocks affected. The culminations of the various hydrocarbon accumulations will be offset if the anticline is asymmetric (not uniformly shaped) so that a shallow accumulation may not overlie a deeper one even though it is on the same structure. An example of this can be seen in the anticlinal traps associated with rotational faults (discussed under Fault Traps).

2.24 Fault Traps: These traps depend upon the effectiveness of the seal at the fault. The seal may be the result of placing different types of formations side by side (for example, shale against sand), or it may be caused by impermeable material called gouge within the fault zone itself. The simple fault trap may occur where structural contours provide closure against a single fault. However, in other structural configurations, such as a monocline, two or even three faults may be required to form a trap. In fault trap accumulations, the oil-water contact closes against the fault or faults and is not continuous as in the case of anticlinal traps. Fault trap accumulations tend to be elongated and parallel to the fault trend.

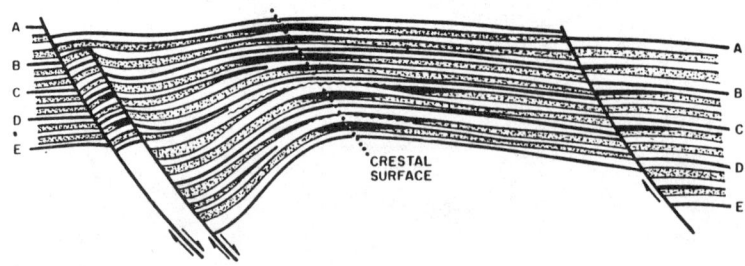

Figure 2-27. Potential Hydrocarbon Traps Associated with Rotational Faults

Many kinds of traps are associated with curved rotational faults; these are especially common in the Texas Gulf Coast area (Figure 2-27). Accumulations tend to be along the faults and are found in fault traps and anticlinal traps in a complex pattern. An understanding of the nature of these traps is most important for their efficient development.

Traps associated with thrust faults may be either fault traps as in the lower sands or anticlinal as in the uppermost sand (as illustrated in Figure 2-28). Accumulations in such traps usually tend to be elongated and parallel to the direction of thrust; they may be quite long but relatively narrow. Thrust traps are often compound.

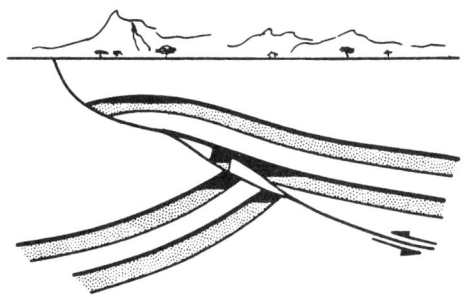

Figure 2-28. Thrust Traps

The intrusion of underlying material (usually salt) into overlying strata often forms a variety of traps, both structural and stratigraphic. Figure 2-29 illustrates three types of traps. Piercement may be more or less circular — typical of the salt dome oilfields in the U.S. Gulf Coast and Northern Germany — or long and narrow as in the oilfields of Romania. The salt and associated material form an efficient up-dip seal. Hydrocarbon accumulations in the peripheral traps around a salt plug may not be continuous. Oil accumulations are usually broken into segments in smaller traps formed by modifying faults or structural closure against the plug. This discontinuous nature of oil accumulations in piercement traps is detrimental to development operations because it cannot be predicted and thus increases the risk of dry holes.

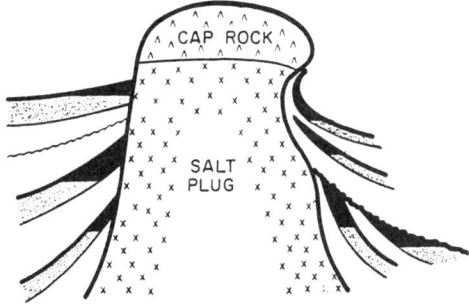

Figure 2-29. Piercement Traps Associated with a Salt Dome

2.25 Stratigraphic Traps: These result from lateral change that prevents continued migration of hydrocarbons in a potential reservoir bed. Many are directly related to their environment of deposition, but others — particularly carbonates — are caused by later changes such as dolomitization. Many large oil and gas fields are associated with this type of trap. The East Texas Field accumulation occurs in the truncated edge of the Woodbine Sand below an unconformity sealed by the Austin Chalk as shown in Figure 2-30.

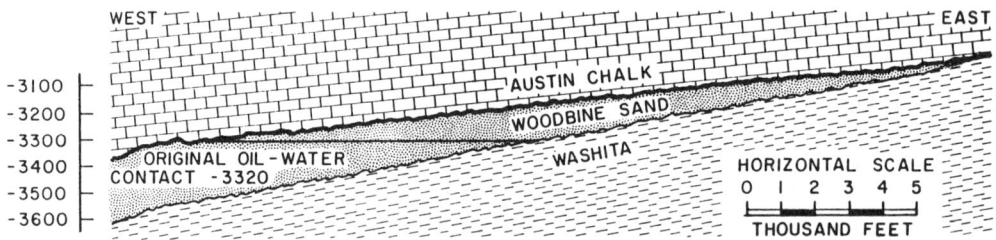

Figure 2-30. Cross-Section of Pinchout Trap Between Unconformities, East Texas Field.

2.26 Lenticular Traps: These pinch out or change permeability on all sides. Lenticular traps are fairly common in carbonate rocks, usually occurring in the upper part of reef carbonate buildups.

2.27 PETROLEUM RESERVOIRS

A petroleum reservoir is a trap containing gas, oil, and water in varying proportions. These fluids are contained in the pore spaces of rock formation among the grains of sandstones or in the cavities of carbonates. The pores spaces are interconnected so that the fluids can move through the reservoir. The porous formations have to be cut off on all sides, above and below, in such a way that the only escape for the fluids will be through a wellbore drilled into the reservoir.

2.28 Physical Characteristics

In order to have a producing oil reservoir, the following conditions must exist:

1) There must be a body of rock having sufficient porosity to contain the reservoir fluids and permeability to permit their movement.

2) The rocks must contain oil or gas in commercial quantities.

3) There must be some natural driving force within the reservoir, usually gas or water.

Special attention should be given to the natural driving force. Oil in itself does not have such force or energy — it cannot move itself. Only the energy stored in the reservoir in the form of gas or water under pressure can move the oil to the well. When this energy has been spent, only the slow method of gravity drainage remains to get the oil to a wellbore. Gravity does not always work to move oil in the right direction to reach a well.

The porosity of a formation is its capacity for reservoir fluids. Porosity may vary from less than 5 percent in a tightly cemented sandstone to more than 30 percent for unconsolidated sands. Accurate determination of formation porosity is an extremely difficult matter. While it is true that laboratory technicians who specialize in this work can make accurate determination on cores taken from a pay section, most reservoirs vary over such wide ranges that it is difficult to arrive at any figure that may be correctly called "average" for a given reservoir (especially in carbonates).

Figure 2-31 shows representative porosity and permeability determinations from some oilfields. The relationship between the porosity and permeability of a given formation is not necessarily a close or direct one. However, high porosity is often accompanied by high permeability. The extreme variations that may be found in a given reservoir are shown in the two permeability determinations from the East Texas Field.

Sand	State	Field	Porosity %	Permeability (Millidarcies)
Woodbine	Texas	East Texas	22.1	3390.0
Woodbine	Texas	East Texas	19.7	192.0
Wilcox	Oklahoma	Oklahoma City	16.9	677.0
Gloyd (Lime)	Texas	Rodessa	20.0	130.0
San Andres Lime	Texas	Goldsmith	12.0	50.0

Figure 2-31. Representative Permeability and Porosity Determinations

2.29 Reservoir Pressure

The fluids in the pores of a reservoir rock are under a certain degree of pressure, generally called reservoir pressure or formation pressure. A normal reservoir pressure at the oil-water contact approximates very closely the hydrostatic pressure of a column of saltwater to that depth. The hydrostatic pressure gradient varies somewhat, depending upon the amount of dissolved salts in the average water for a given area. For fresh water it is 0.433 psi/ft of depth, but for water containing 80,000 ppm of dissolved salts (Gulf Coast) the pressure gradient is approximately 0.465 psi/ft. However, normal fully marine water is about 35,000 ppm dissolved salts, and hence approximately 0.446 psi/ft. Reservoirs can contain fluids under abnormal pressures up to as high as 1.00 psi/ft of depth.

Abnormal pressures may develop in isolated reservoirs as a result of compaction of the surrounding shales by the weight of the overburden. During this process water is expelled from the shale into any zone of lower pressure. This may be into a wholly confined sandstone which does not compact as much as the shale; consequently, its contained water is under a lower pressure than that in the shale. Ultimately, a state of equilibrium can be reached when no further water can be expelled into the sandstone, and its fluid pressure will then approximate that of the shale.

Since compaction of sandstones is related to the pressure of the pore fluid as well as to the pressure exerted by the overburden, it follows that abnormally pressured sandstones are partly supported by the fluid pressure and partly by grain-to-grain contact. Consequently, when the abnormal pressure is reduced by production, compaction of the reservoir bed immediately begins to occur. Subsurface compaction can cause serious problems not only because of collapse of casing in wells, but also because it is reflected at the surface by subsidence. Such occurrences result in very expensive landfill and well repair costs. It has also been demonstrated that there can be a direct relationship between subsidence and the amount of liquid withdrawn. Studies of the Wilmington Field (California) indicated that repressuring by water injection would increase oil recovery and stop compaction. Subsidence was stopped by such a program, and in places the surface regained some of the elevation that was lost. However, careful work on sediments has shown that this compaction is not entirely reversible. Thus some permanent reduction of porosity and permeability results from permitting abnormal reservoir pressures to decline, and this may adversely affect the rate of production and possibly the ultimate recovery.

2.30 OILFIELD FLUIDS

The distribution of fluids will be different for each pool, depending on

- Source rock
- Reservoir rock
- Porosity
- Permeability
- Relative permeability of reservoir fluids
- Relative densities
- Hydrodynamics of the reservoir
- Migration variables (lithology, temperature, etc.)

By definition, a fluid is any substance that will flow. Oil, water and gas are all fluids by this definition. Oil and water are liquids as well as fluids, but gas is a fluid though not a liquid.

2.31 RESERVOIR WATER

Many oil reservoirs are composed of sediments which were deposited on the floor of seas and oceans; consequently, these sedimentary beds were originally saturated with saltwater. Part of this water was displaced in the process of the formation of oil accumulations. That which remained in the formation has been given the name of connate interstitial water — "connate" from the Latin meaning "born with" and "interstitial" because the water is found in the interstices, or pores, of the formation. By common usage this term has been shortened to "connate water" and always means the water in the formation when development of the reservoir was started. Connate water determinations (Sw) using core samples are expressed as a percentage of the volume occurring in the pore spaces of the reservoir. Swi (irreducible connate water saturation) is the fraction of pore space which may be retained as nonmovable wetting phase even though oil and gas may be flowing in the same pore spaces under the influence of relatively large pressure gradients. In addition to the connate water distributed throughout the pay section with the oil and gas, nearly all petroleum reservoirs have water-bearing formations down-dip from the payzones. All the pore spaces of such formations are filled with water. It is this volume of "free" water which supplies the energy for the "water drive" in some reservoirs. With this "water drive mechanism," as some hydrocarbons are liberated via the wellbore the water rushes into the vacated pore spaces, increasing in volume and pushing more hydrocarbons to the surface.

The character of reservoir water is determined by

- Water saturation
- Concentration of dissolved solids
- Compositon of dissolved solids

Water saturation (Sw) is determined directly by core analysis (Section 4, paragraph 4.43) or indirectly from borehole logging tools (Section 5, paragraph 5.57). Concentrations of dissolved solids are analyzed directly by use of a hydrometer and indirectly with a resistivity tool to measure the resistivity of water (Rw). Resistivity of water in the interstitial pore space is a measure of all ions and is therefore an indirect measure of dissolved solids. Density increases with increased dissolved solids (Figure 2-32).

Fresh	Seawater	Heavy Brine
200-300 ppm 8.33 lb/gal	35,000 ppm 8.6 lb/gal	300,000 ppm 10.0 lb/gal

Figure 2-32. Dissolved Solids (Concentrations)

Dissolved solid composition can only be analyzed using water directly from the well. As all brines have similar ionic analyses even though the total concentrations may differ greatly, it suggests they are all diluted forms of the same original water, i.e., seawater. In the oilfield-water analyses in Figure 2-33 it can be seen that, except for the Miocene example, the NaCl ppm readings are very high; in fact, much greater than that of seawater. This is an important fact to be considered when testing a well, for analysis of recovered water may possibly indicate whether it is formation water or water used in the testing procedure.

Pool	Reservoir Rock Age	Parts Per Million (ppm)							Total ppm
		Cl⁻	SO₄⁻	CO₃	HCO₃⁻	Na⁺+K⁺	Ca⁺⁺	Mg⁺⁺	
Seawater Seawater, percent	---	19,350.0 55.3	2,690.0 7.7	150.0 0.2	---	11,000.0 31.7	420.0 1.2	1,300.0 3.8	35,000
Lagunillas (Western Venezuela)	Miocene 2000-3000 ft	89	---	120	5,263	2,003	10	63	7,548
Conroe (Texas)	Conroe sands (Eocene)	47,100	42	288	---	27,620	1,865	553	77,468
East Texas	Woodbine sand (U. Cretaceous)	40,598	259	387	---	24,653	1,432	335	68,961
Burgan (Kuwait)	Sandstone (Cretaceous)	95,275	198	---	360	46,191	10,158	2,206	154,388
Rodessa (Texas-La.)	Oolitic limestone (L. Cretaceous)	140,063	284	---	73	61,538	20,917	2,874	225,749
Davenport (Oklahoma)	Prue sand (Pennsylvanian)	119,855	132	---	122	62,724	9,977	1,926	194,736
Bradford (Pennsylvania)	Bradford sand (Devonian)	77,340	730	---	---	32,600	13,260	1,940	125,870

Figure 2-33. Oilfield-Water Analyses

The oil-water contact is always transitional and may be from two feet to several hundred feet thick. There are three possible definitions and locations:

- Depth above which only irreducible water saturation (S_{wi}) is present
- Depth below which S_w = 100 percent
- Depth below which oil will not be produced

2.32 RESERVOIR OIL AND GAS

The relationship between oil and gas in the reservoir depends upon the degree to which the oil is saturated with gas – i.e., the amount of "dissolved gas" contained in the liquid oil. Natural gas is always associated with oil (yet oil is not always associated with gas), and the energy supplied by gas under formation hydrostatic pressure is probably the most valuable drive in the withdrawal of oil from reservoirs.

Gas is associated with oil and water in reservoirs in two principle ways -- as "solution gas" and "free gas" in gas caps. Given suitable conditions of pressure and temperature, natural gas will "stay in solution" in oil in a reservoir. High pressure and low temperature are favorable conditions for keeping gas in solution. When the oil is brought to the surface and the pressure relieved (as in a separator), the gas comes out of solution.

The volume of gas that remains in solution in the reservoir depends on the reservoir pressure and temperature. When there is less gas in the reservoir than the volume of reservoir oil in place is capable of absorbing, the oil is said to be undersaturated. The East Texas Field with a reservoir pressure of about 1100 psi produces oil with about 325 cu ft of solution gas per barrel. The reservoir temperature is $143°F$. At that temperature and pressure it would require about 400 cu ft of gas per barrel of oil to have saturated conditions. Thus, the East Texas Field produces undersaturated oil. On the other hand, crude oil in the West Pampa Pool of the Texas Panhandle is supersaturated. Oil from this pool carries about 175 cu ft of gas per barrel in solution and produces another 725 cu ft of free gas.

Free gas tends to accumulate in the highest structural part of a reservoir and form a gas cap. As long as there is free gas in a reservoir gas cap, the oil in the reservoir will remain saturated with gas in solution. Having gas in solution lowers the viscosity of the oil, making it easier to move to the wellbore.

2.33 Composition of Petroleum

Petroleum is basically composed of carbon and hydrogen with minor amounts of sulfur, nitrogen and oxygen (Figure 2-34). An increase in minor elements decreases the value of crude.

Hydrocarbons (compounds of carbon and hydrogen only) make up over 90 percent of most crude oils. The hydrocarbons in crude oils vary in molecular size and molecular type (Figure 2-35).

Components	Oil	Asphalt	Kerogen
Carbon	84.0	83	79
Hydrogen	13.0	10	6
Sulfur	2.0	4	5
Nitrogen	0.5	1	2
Oxygen	0.5	2	8
	100.0	100	100

Figure 2-34. Chemical Composition of Oil, Asphalt and Kerogen

Molecular Size	Weight %
Gasoline (C_4-C_{10})	31
Kerosene (C_{11}-C_{12})	10
Gas Oil (C_{13}-C_{20})	15
Lubricating Oil (C_{20}-C_{40})	20
Residuum (>C_{40})	24
	100
Molecular Type	
Paraffins	30
Naphthenes	49
Aromatics	15
Asphaltics	6
	100

Figure 2-35. Composition of a Typical Crude Oil

2.34 Hydrocarbon Size Distribution: Dry gas consists predominately of methane. Wet gas contains methane, ethane, propane, butane (C1 to C4), isobutane, and minor amounts of higher hydrocarbons. Distillation separates crude oil into molecular groups of different sizes, such as gasoline, kerosene, gas-oil, lubricating oil, and residuum. High API gravity crudes have a high gasoline and low residuum content, whereas low gravity oils are low in gasoline and high in residuum. API gravities are used for classifying oils by reference to their densities as defined by the American Petroleum Institute. Numerically, the value is obtained from the formula

$$\frac{141.5}{\text{S.G. at } 16^\circ\text{C } (60^\circ\text{F})} - 131.5 \qquad (\text{S.G.} = \text{Specific Gravity})$$

and it may range from -1 to +101. The larger the number, the lighter the oil; thus a light crude may be 40° API, a medium one 28° API and a heavy crude 20° API.

2.35 Hydrocarbon Type Distribution: Most crude oils contain some of the following different types of structures, which are illustrated in Figure 2-36 (A, B, C and D).

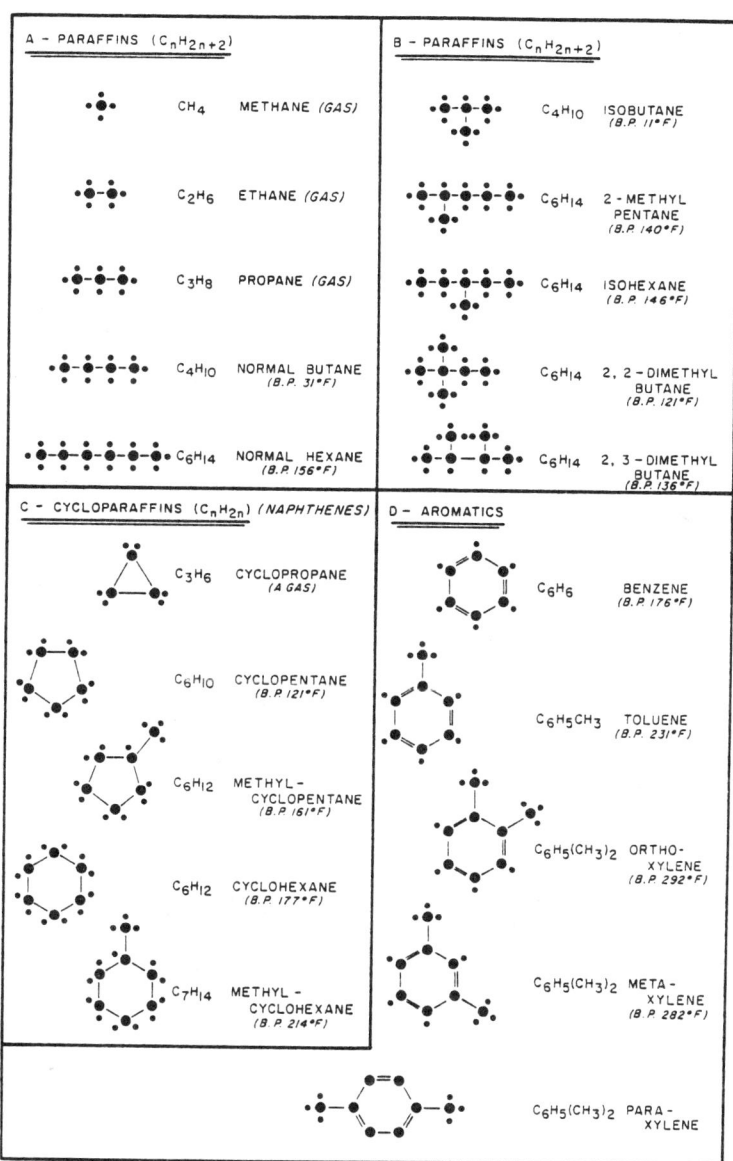

Figure 2-36. Typical Hydrocarbon Structures

- Saturated hydrocarbons (these contain only single bonds between carbon atoms). Normal paraffins are straight chains of carbon atoms. Isoparaffins are branched chains of carbon atoms. Naphthenes (cycloparaffins) are rings of carbon atoms.

- Unsaturated hydrocarbons (these contain double bonds between carbon atoms). Olefins are compounds having one bond between carbon atoms. Aromatics are compounds containing one or more benzene rings. Benzene has a symmetrical ring of six equivalent carbon and hydrogen atoms in a plane.

Typical paraffins are methane, ethane, propane, butane, pentane, hexane, heptane, octane, nonane, decane. Naphthenes, or cycloparaffins, include cyclopropane, cyclopentane, cyclohexane. Olefins include ethylene, propylene, butylene. Aromatics include benzene, toluene, xylene, naphthalene, and anthracene.

Exploration Logging's chromatographs are capable of detecting hydrocarbons up to C5, and under normal conditions detection up to C4, while drilling, is sufficient. If further analysis is necessary, this will usually be done during testing when gas samples are taken downhole using a test tool. Figure 2-37 illustrates hydrocarbons that are commonly detected at the wellsites.

Name	Chemical Formula	Physical State at 60°F and 1465 psi	Molecular Weight	Boiling Point (C) at Normal Conditions
Methane	CH_4	Gas	16.04	-161.4
Ethane	C_2H_6	Gas	30.07	-89.0
Propane	C_3H_8	Gas	44.09	-42.1
n-Butane	C_4H_{10}	Gas	58.12	0.55
iso-Butane	C_4H_{10}	Gas	58.12	-11.72
n-Pentane	C_5H_{12}	Liquid	72.15	36.0

Figure 2-37. Hydrocarbons Commonly Detected at the Wellsite

2.36 Source of Petroleum

Petroleum originates from a small fraction of the organic matter deposited in sedimentary basins. Most of this organic matter is the remains of plants and animals that lived in the sea, and the rest is land-delivered organic matter carried in by rivers and continental runoff, or by winds. Living organisms are composed of carbohydrates, proteins, and lipids (fats) and lignin in varying amounts. These compounds are degraded by micro-organisms into the monomer sugars, fatty acids, etc. These immediately condense into nitrogenous and humus complexes — progenitors of kerogen. Some hydrocarbons are deposited in the sediments, but most form from thermal alteration at depth. Lipids are closest to petroleum in composition among the major life substances (Figure 2-38). Lipids (fats and hydrocarbons) are most concentrated in the lowest forms of life (Figure 2-39).

Substance	Elemental Composition in Weight (%)				
	C	H	O	S	N
Carbohydrates	44	6	50.0	---	---
Lignin	63	5	31.0	0.1	0.3
Proteins	53	7	22.0	2.0	16.0
Lipids	80	10	10.0	---	---
Petroleum	82-87	12-15	0.1-2.0	0.1-5.0	0.2

Figure 2-38. Average Chemical Composition of Natural Substances

Life Form	Weight Percent of Major Constituents			
Plants	Proteins	Carbo-hydrates	Lignin	Lipids
Spruce Wood	1	66	29	4
Oak Leaves	5	44	32	4
Scots Pine Needles	7	41	15	24
Phytoplankton	15	66	--	11
Diatoms	29	63	--	8
Lycopodium Spores	8	42	--	50
Animals				
Zooplankton	53	5	--	15
Copepods	65	22	--	8
Higher Invertebrates	70	20	--	10

Figure 2-39. Composition of Living Matter

Petroleum contains traces of several substances that could have come only from living things. Examples of these are:

- Porphyrins related to hemin and chlorophyll
- Optically active compounds (compounds that will rotate the plane of a ray of polarized light)
- Structures related to cholesterol, carotene and terpenes
- A predominance of odd-numbered paraffin chains

Carbon isotope data suggests that the lipids of plants are an important source of petroleum.

Normal (straight chain) paraffins in crude oil sometimes show a predominance of odd-numbered chain lengths. This odd-numbered chain length has a biochemical origin and tends to predominate in the high molecular weight ranges in oils derived from continental or near-shore organic matter and in the low ranges for marine organic matter.

Coals, kerogen, asphalts, and petroleum all originate from organic matter deposited with sediments. Their differences are due to different source materials, dispersal, and environments of deposition and diagenesis (Figure 2-40). The biological origin of these fossil fuels is clearly proven by their chemical composition and physical structure containing remnants of living organisms.

Coals		Dispersed Organic Matter	Bitumens
Massive land plants ↓ Concentrated as aromatic HMW humus ↓ Peat Lignite Bituminous Coal Anthracite	Algae, plankton, spores, resins ↓ Concentrated as jelly like slime-sapropel ↓ Cannels and Torbanites	Land and marine organic matter deposited with inorganic sediments ↓ Adsorbed on mineral matter as aliphatic LMW humus ↓ Kerogen	Selective fraction (>1%) of organic matter ↓ Dispersed lipids carried by migrating waters to reservoirs ↓ Petroleum Waxes Asphalts Asphaltites Pyrobitumens

Figure 2-40. Origin of Organic Matter Deposited with Sediments

2.37 Origin of Gases

The following discussion presents generally accepted theories for the origin of gases. Methane is formed by bacterial decay of organic material; it is a major product of the diagenesis of coal and is given off by all forms of organic matter during diagenesis. It is the most common hydrocarbon in subsurface waters and is an end product of petroleum metamorphism. When heated, the kerogen in shales from gas-producing areas gives off much greater quantities of methane as compared to the kerogen of shales from oil-producing areas.

Hydrogen sulfide originates from the reduction of sulfate in the sediments and from sulfur compounds in petroleum and kerogen. Carbon dioxide is from the decarboxylation of organic matter, and from HCO_3 and $CaCO_3$. Nitrogen is from the nitrogen in organic matter and from trapped air. Helium is from the radioactive decay of uranium and thorium. During the oil genesis and coalification process, the order of generation is generally nitrogen, CO_2 and methane.

2.38 Occurrence of Petroleum

Only about 2 percent of the organic matter dispersed in fine-grained rocks becomes petroleum, and only about 0.5 percent of that ends up as a commercial reservoir accumulation. This emphasizes the inefficiency of the origin, migration and accumulation process.

The ratio of dispersed hydrocarbons to reservoired hydrocarbons is about 200 to 1 on a worldwide basis, partly because the volume of potential reservoir rock is small compared to total sediments in the earth's crust. Within prospective parts of oil-forming basins, the ratio generally varies between 10 and 100.

Petroleum is found from the Precambrian to the Pleistocene, but it is increasingly abundant in younger sediments for several reasons, such as:

- Older oilfields are increasingly destroyed over geologic time

- An increase in continental margins and restricted basins occurred when the continents split during the Jurassic. This is illustrated in Figure 2-41.

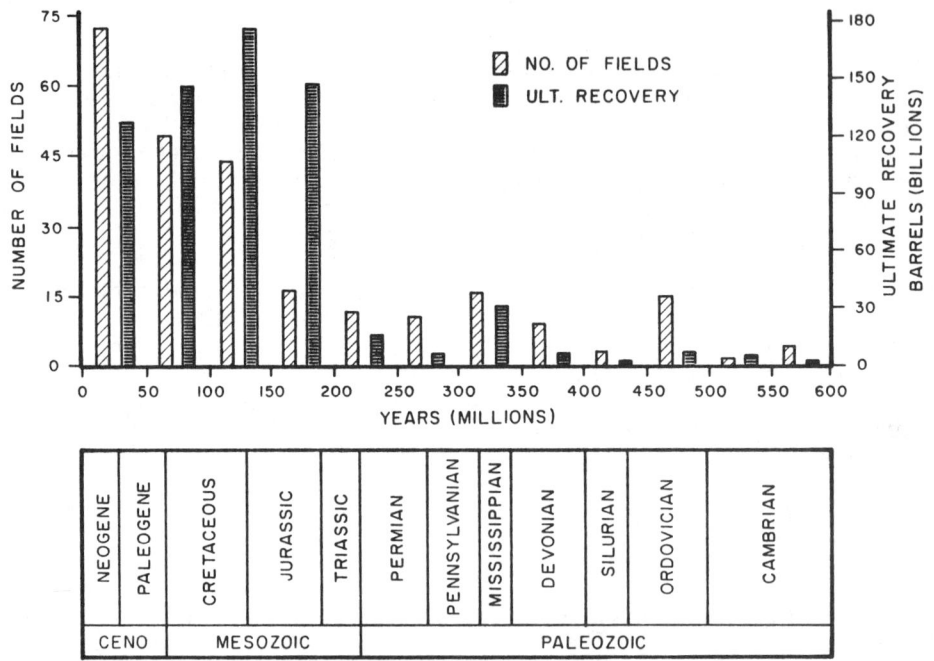

Figure 2-41. Reservoir Age

2.39 Maturation of Petroleum

In most sedimentary basins, the oil in reservoirs becomes lighter (higher API gravity) with increasing depth. Depth has two important effects in altering a crude oil:

- Pressure, which increases with depth, causes diagenesis with resultant clay alteration; hence, salinity variations affect migration which in turn affects "natural filtration." Pressure also inhibits chemical equilibrium where a volume change is concerned.

- Temperature, which increases with depth, causes several changes in oil:

  — Solids with low melting points, or viscous liquids, become mobile liquids.

  — Light hydrocarbons become increasingly soluble in subsurface fluids.

  — Mild cracking of large oil molecules causes small molecules to form.

  — Some large molecules, in turn, polymerize to very large organic structures.

At 100°C (212°F) all hydrocarbons except methane, ethane and propane are unstable with respect to carbon and hydrogen which have a lower free energy. At 200°C (392°F) all hydrocarbons except methane are unstable. Methane, therefore, is the only hydrocarbon found in deep, hot sediments.

The degradation of crude oil may occur in several ways. The two most common are through (1) seepage and (2) contact with circulating meteoric waters. Seepage may take place along unconformities or through fractures or continuous sand beds exposed to the surface. Seepage results in (1) the loss of light hydrocarbons to the air and (2) the formation of a natural asphalt at the seep outlet. Sometimes the formation of an asphalt may seal off a seep and other times a transgressive sedimentation cycle may seal it. This results in the preservation of an oil that is heavy compared with comparable oils from the same formation which have not seeped during their geologic history.

Circulating meteoric waters may degrade an oil by carrying off the light hydrocarbons, and by oxidizing portions of the crude through contact with oxidizing substances such as the sulfate ion. In many areas it is very common to find the heavy degraded oils associated with fresh or brackish water and the light original oils associated with saltwater. Micro-organisms assist in the degradation.

2.40 Factors Which May Affect Crude Oil Gravity

The following factors are all interrelated in too complex a manner to take each as being more than a rough generalization. Formation temperature and pressure seem to be very influential. If they are high, gravity is usually high. Also note the concept of maturity — old oils, which tend to be paraffinic, are more stable and tend to have higher gravity.

- Geological age (maturity). Older rocks tend to have higher gravity, but many tertiary rocks have $API°$ 40 + (e.g., the North Sea); and many Paleozoic rocks have $API°$ 20 -.

- Depth of burial. Deeper reservoir, higher gravity. Deepest wells tend to produce gas.

- Basinal position. Gradient from high at the center to low at the edges.

- Tectonism. High gravities are more common in regions of high stress.

- Lithology. No apparent relationships.

- Salinity. Marine source tends to higher gravity than fresh/brackish; may be due to basinal position.

- Sulfur content. This is high in low-gravity crudes. Main variations are regional, e.g., Middle East crudes are high in sulfur, Nigerian and Libyan crudes are low.

Properties of liquid petroleum reflecting variations in composition and gravity are illustrated in Figure 2-42.

Properties	High API Gravity	Low API Gravity	Remarks
Viscosity	low	high	Temperature sensitive. Inverse measure of ability to flow. Not pressure-sensitive.
Color	light	dark	Yellow, red brown to black by transmitted light (usually geen by reflected light).
Fluorescence	yellows	browns	White-yellow through yellow and brown. Described as color, intensity and hue.
Refractive index	1.39	1.55	Determined with Abe refractometer
Flash and burn points	low	high	$>50°F$
Cloud and pour points	low	high	$-70°F$ to $+110°F$
Coefficient of Expansion	high	low	Noncompressible as far as logging is concerned.
Density	\multicolumn{3}{l}{Dependent on composition. API gravity = $\frac{141.5}{SG(60°F)} - 131.5$. Normal Range = 16 to $50°$ API gravity. SG water = 1 = $10°$ API at $60°F$.}
Odor	\multicolumn{3}{l}{Paraffin and naphthene crudes smell like creosote. Aromatics - unpleasant. Sulfur compounds very unpleasant.}
Optical activity	\multicolumn{3}{l}{Rainbow effect. Due to cholesterol (an alcohol $C_{26}H_{45}OH$). Mainly in intermediate distillation range (boiling point $200°$ to $300°C$).}
Wax content	\multicolumn{3}{l}{Related to pour points and viscosity range 0.5% to 45%; mainly 7% and consists mainly of normal paraffins. No low wax paraffin crudes have been found but there are some high wax - naphthene crudes (NB ceresin). Wax content apparently high in oil from fresh water/brackish source.}

Figure 2-42. Properties of Liquid Hydrocarbons

2.41 Impurities Associated with Hydrocarbons

Impurities occur as free molecules or as atoms attached to the larger hydrocarbon molecules — the so-called nonhydrocarbon compounds. When free, they can in most cases be detected with special equipment (Figure 2-43).

Impurity	Equipment Available to Measure Impurity
Sulfur	H_2S Detector
Nitrogen	All Gas/Nitrogen Detector
CO_2	TC Chromatograph
Helium	TC Chromatograph
Oxygen	All Gas Detector
Water	Salinity Check
Salt	(AgNO3 Titration)
Ash	—

Figure 2-43. Impurities Associated with Hydrocarbons

2.42 Products from Crude Oil

The products from crude oil, their gas ranges and their API gravity are included in Figure 2-44.

Crude	Product	Gas Range	Distillate Type	API Gravity
Gas ↑ ↓ Oil	Natural Gas	C_1, C_2	—	$>110°$
	LNG (Liquid Natural Gas)	C_3, nC_4		
	Natural Gasoline	IC_4, C_5, C_6, C_7	Condensate: Light Distillates BP* < 160°C	$74°$ to $110°$
	Gasoline	C_5^+		$50°$ to $95°$
	Kerosine	—	Middle Distillates: BP 160° to 350°C	$30°$ to $50°$
	Diesel			
	Fuel Oil	—	Heavy Distillates: BP 350° to 520°C	$18°$ to $30°$
	Lube Oil			
	Wax/Asphalts	—	Residuum: BP > 520°C	—

*BP: Boiling Point

Figure 2-44. Crude Oil Products

2.43 FLUID DISTRIBUTION

Practically all reservoirs have water in the lowest portions of the formation, with the oil just above it (Figure 2-45). The oil-water contact line is the prime interest of all concerned in the early development of a field. It would be a mistake to assume that there is a sharp line dividing the oil and water or that the contact line is horizontal throughout a reservoir. There is always a certain amount of water, even at the top of the oil or gas level.

What is true of the oil-water contact is true to a lesser degree of the gas-to-oil contact. There is a great difference in specific gravity of oil and gas; and oil, being much heavier than gas, does not tend to rise very high into the gas zone.

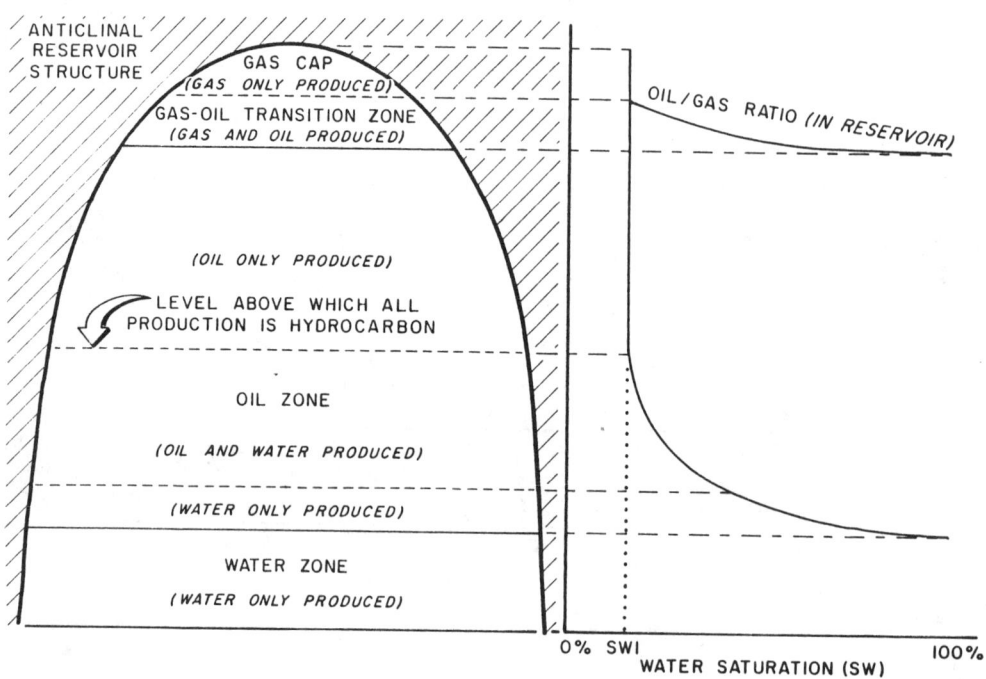

Figure 2-45. Reservoir Fluid Distribution

3
RIG TYPES & THEIR COMPONENTS

3.1   **GENERAL**

The complexity of the drilling operation determines the level of sophistication of the various rig components. However, even with the considerable variety of rig types, the basic components described under 3.11 with only a few exceptions are similar and common to each.

Rigs are generally divided into two categories:

- Onshore
- Offshore

Onshore (land) rigs are all similar, but offshore rigs are of five basic types — each of which is designed to suit a specific offshore environment. Figure 3-1 illustrates the various types of rigs.

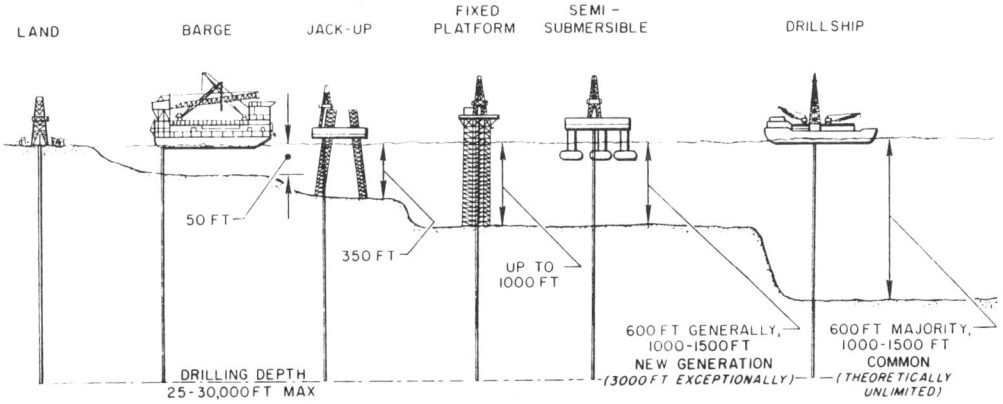

Figure 3-1. Rig Types and Offshore Operating Profiles

3.2   **LAND RIGS**

Before rig equipment is brought in, the land must be cleared and graded, and access roads must be prepared. For jungle locations, however, preparing the drillsite is more difficult, for access roads do not usually exist and cannot be made. The choice of a drilling location is therefore often dependent on its proximity to a

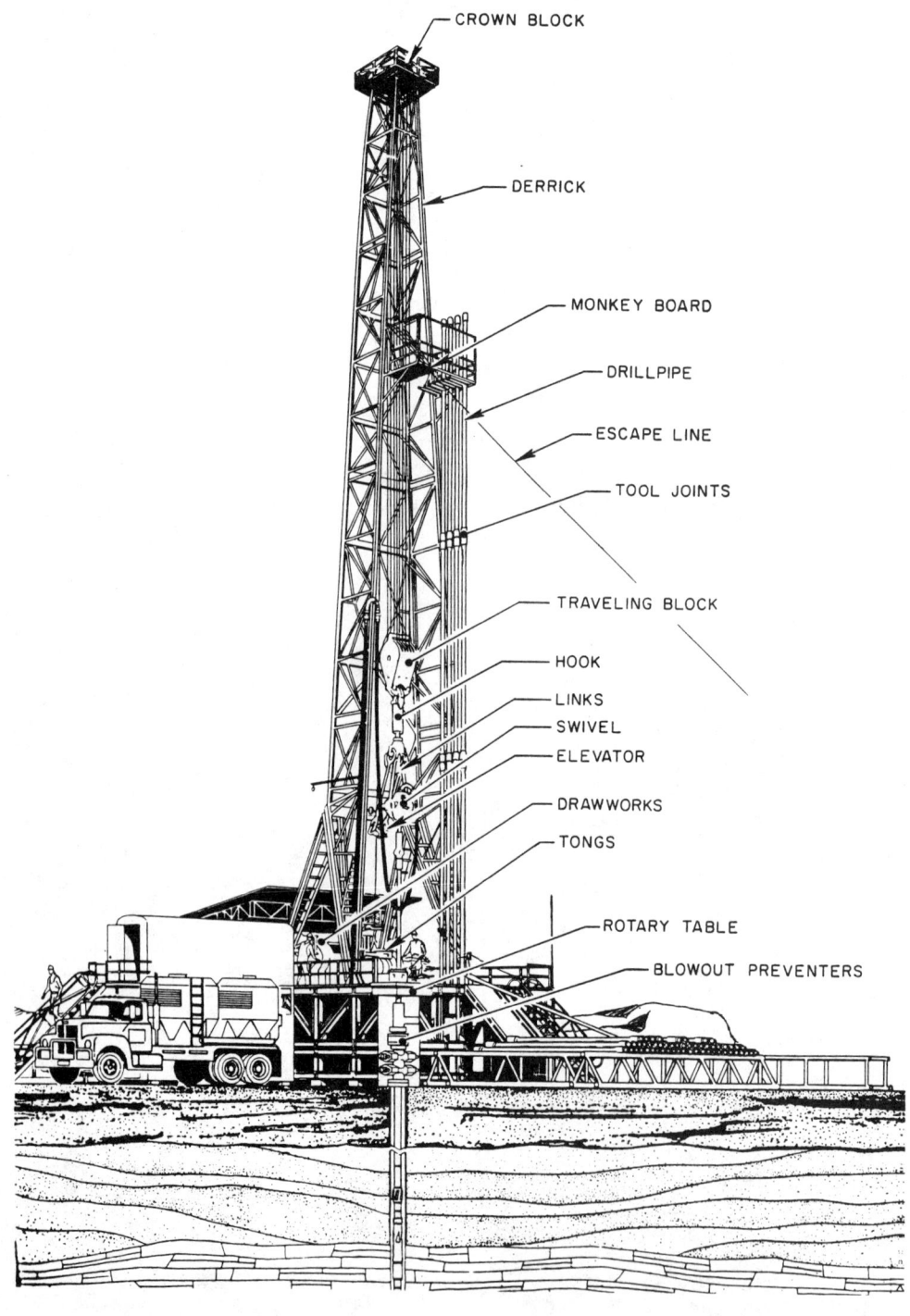

Figure 3-2. Land Rig

navigable river which can be used to transport the rig equipment and supplies for the duration of the well. If there are swampy conditions or if the distance is too great to warrant the construction of a road, helicopters airlift everything to the drillsite from the river.

On the North Slope of Alaska and in other similar areas, site preparation includes an insulating layer to separate the permafrost from the heat in the rig floor area. Wood pilings are set and frozen in place to support the weight of the drilling rig. A thick layer of gravel is added before the substructure is begun. Polyurethane foam may be used to supplement gravel insulation.

The most common arrangement for a land drilling rig is the cantilever mast (sometimes called a jack-knife derrick) which is assembled on the ground, then raised to the vertical position using power from the drawworks (hoisting system). These structures are made up of prefabricated sections which are fastened together by large pins. First, the engine and derrick substructures are placed in proper position and pinned together by the drilling crew, then the drawworks and engines are put in place. The derrick sections are then laid out horizontally, pinned together, and the mast is raised as a unit by the hoisting line, traveling block and drawworks. A working land rig is illustrated in Figure 3-2.

3.3   OFFSHORE RIGS

3.4   BARGE

The barge is a shallow draft, flat-bottom vessel equipped as an offshore drilling unit, used primarily in swampy areas. This type rig can be found operating in the swamps of river deltas in West Africa or in the coastal areas of shallow lakes such as Lake Maracaibo, Venezuela. It can be towed to the location and then ballasted to rest on the bottom.

3.5   JACK-UP

This mobile drilling rig is designed to operate in shallow water, generally less then 350 ft deep. Jack-up rigs, illustrated in Figure 3-3, are very stable drilling platforms because they rest on the seabed and are not subjected to the heaving motion of the sea as are semi-submersibles and drillships. They have a barge-like hull which may be ship-shaped, triangular, rectangular, or irregularly shaped, supported on a number of lattice or tubular legs. When the rig is under tow to a drilling location the legs are raised, projecting only a few feet below the deck, and the structure behaves like a cumbersome floating box; hence it can be towed only in good seas and at a slow speed. Upon reaching its location the legs are lowered by electric or hydraulic jacks until they rest on the seabed and the deck is level, some 60 feet or more above the waves. Most jack-up rigs have three, four or five legs, but a few of the earlier models have eight or ten, and one has fourteen. The legs are either vertical or slightly tilted for better stability. In one design they are fixed to a large steel mat, which gives it the name of mat-supported jack-up. A drilling slot is usually indented into one side of the deck, but on some rigs the derrick is cantilevered over the side.

The chief disadvantage of the jack-up is its vulnerability when being jacked up or relocated, but as a class they are cheaper than other mobile rigs. Nearly half of the world's fleet of offshore rigs in service are the jack-up type, some of which are large, self-propelled units.

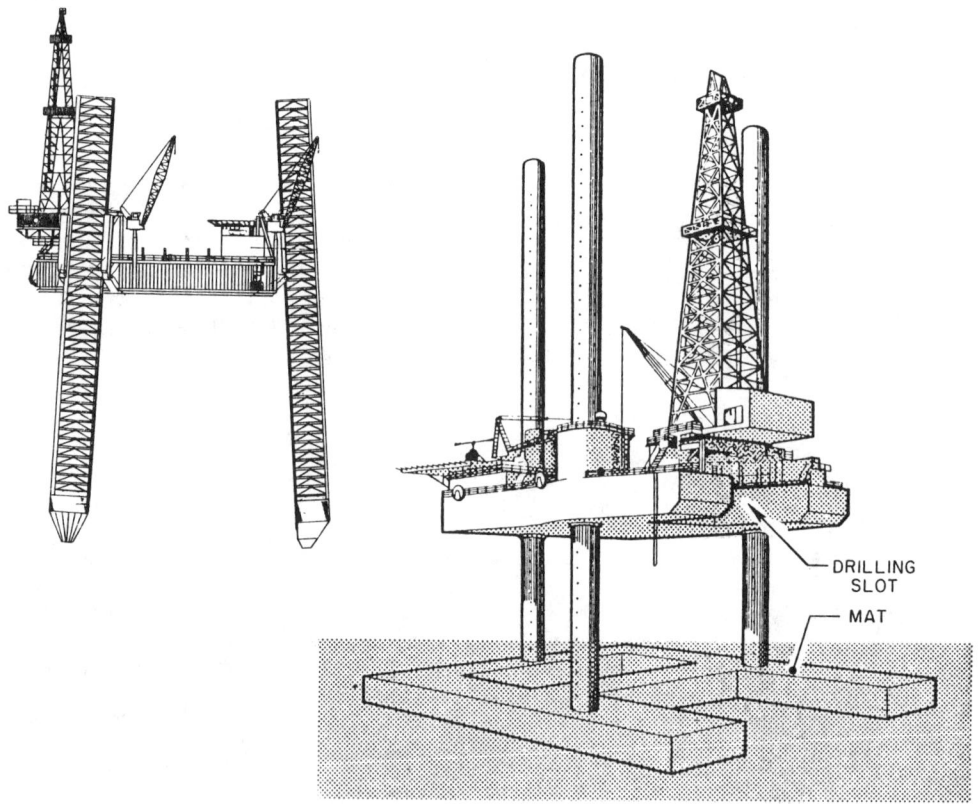

Figure 3-3. Jack-Up Rigs

3.6 FIXED PLATFORM

There are two basic types of fixed platforms: "piled" steel platforms and "gravity structures." Both types are discussed below.

3.7 Piled Steel Platforms

These are conventional drilling and production platforms, and hundreds of them are installed offshore in many parts of the world. The standard configuration (Figure 3-4) consists of a steel jacket pinned to the seabed by long steel piles, surmounted

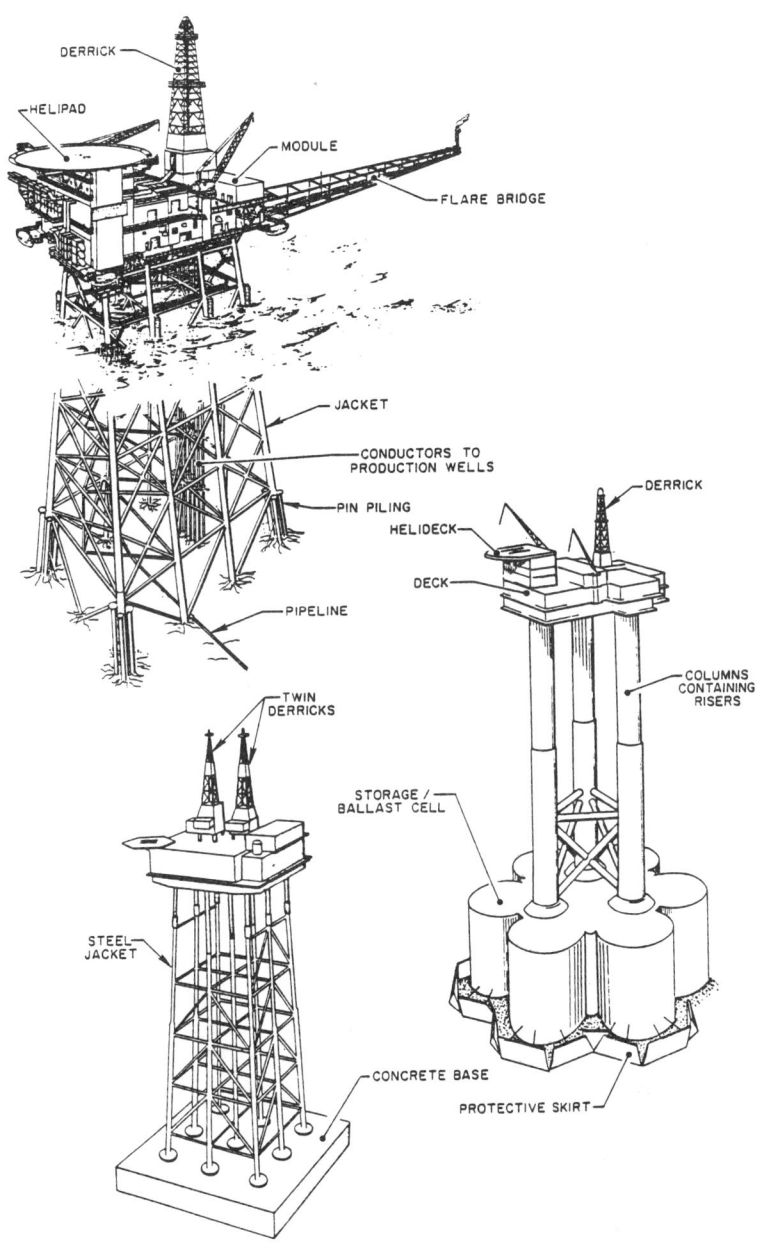

Figure 3-4. Fixed Platforms

by a steel deck which supports equipment and accommodation buildings or modules, one or more drilling rigs, and a helicopter deck. Piled steel platforms have the advantage of being very stable under the worst sea conditions, but they are virtually immobile. In shallow waters the piled platform is probably preferred over all other designs, but they are also practical in very deep water. Construction of the jacket in separate sections usually begins onshore. They are then assembled on a flotation jacket, either onshore or in a graving dock. When completed, the structure is towed carefully to its destination where it is first tilted by adjusting the ballast in the flotation tanks, then uprighted, and finally submerged over the chosen spot on the seabed. The jacket is then pin-piled, the "superstructure" and accommodation modules or buildings erected, and the platform made ready for operations.

3.8   Gravity Structures

This is a family of deep-water structures usually built of reinforced concrete, but may be of steel or a combination of steel and concrete. These structures rely on gravity to keep them stable on the seabed. Unlike piled steel platforms, they are relatively mobile and need no piling to hold them in place. Gravity structures tolerate a wide range of seabed conditions. While they can be used for development drilling and production, they also have the advantage of being able to store oil in their structural cells. A typical gravity structure (illustrated in Figure 3-4) consists of a cellular concrete or steel base for storage or ballast, a number of vertical columns which support a steel deck and give access to the risers, and deck accommodation in the form of detachable modules. Construction of the concrete type begins in a dry-dock basin where the base caisson is partly built. The basin is then flooded and the base towed into deeper water where the caissons are finished, the towers are formed, and the deck installed. Steel structures are assembled in the same manner as piled steel platforms, and all types are towed to their final destinations and settled upright on the seabed by controlled ballasting. Deck modules are then lifted onto the deck and fitted out, after which the platform is ready for operations. There are several configurations of gravity structure, each of which is constructed to client requirements.

3.9   SEMI-SUBMERSIBLE

These are floating drilling rigs consisting of hulls or caissons which carry a number of vertical stabilizing columns and support a deck fitted with a derrick and associated drilling equipment. Semi-submersible drilling rigs differ principally in their displacement, hull configuration, and the number of stabilizing columns. Most modern types have a rectangular deck, a few are cruciform shaped, others pentagon shaped, while some of the smaller rigs have a triangular deck. The most usual hull arrangement consists of a pair of parallel rectangular pontoons which may be blunt or rounded and house thrusters for position-keeping or self-propulsion, although some have individual pontoons or caissons at the foot of each stabilizing column or pair of columns. Eight columns (four stabilizers and four intermediate columns) is a common arrangement as are three and six columns, and both hulls and columns are used for ballasting as well as storing supplies.

The semi-submersible is very stable because its center of gravity is low in the water. It can operate in deeper water than a jack-up rig. Operational depth is limited principally by the mooring equipment and by riser handling problems, so most semi-submersibles have a limit of about 200 meters. However, some units have a capability of drilling in 500 meters of water with the aid of "dynamic positioning." This is a method of maintaining the position of a vessel with respect to a point on the seabed by activating on-board propulsion units in response to signals received from a position-error detector. This method of automatic station-keeping is often computer-controlled.

Figure 3-5a shows a semi-submersible rig with independent pontoons, and Figure 3-5b shows a semi-submersible with ship-shaped hulls.

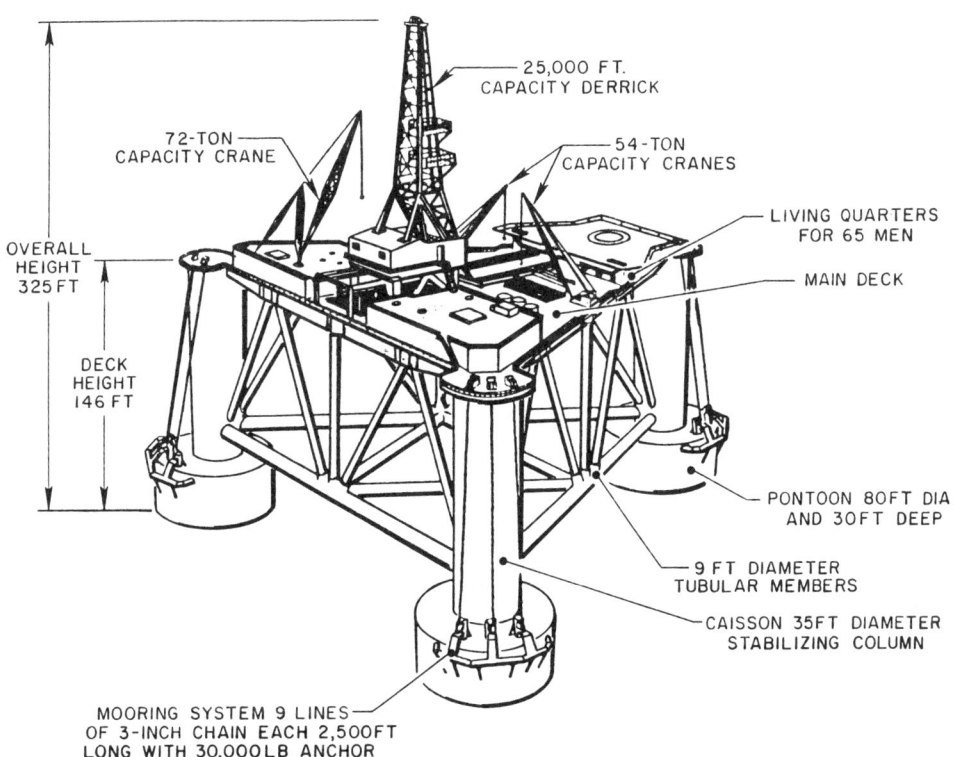

Figure 3-5a. Semi-submersible Rig — Pontoon Type

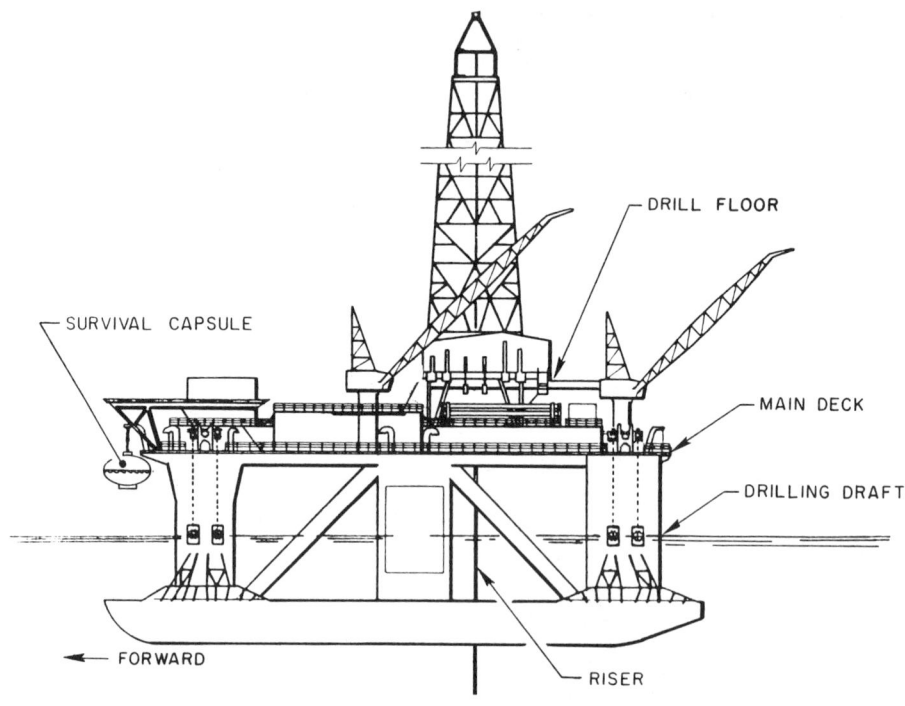

Figure 3-5b. Semi-submersible Rig — Twin Hull Type

3.10 DRILLSHIP

These are ships or "floaters" specially constructed or converted for deep-water drilling. Drillships offer greater mobility than either jack-up or semi-submersible rigs, but are not as stable when drilling. Their main advantage is an ability to drill in almost any depth of water. Many are anchor-moored, but modern ships are fitted with dynamic positioning equipment which enables them to keep on-station above the borehole. Having greater storage capacity then other types of rigs of comparable displacement, drillships are often able to drill deeper wells and operate independent of service and supply ships. A feature of drillships with automated station-keeping facilities is their ability to maneuver accurately with the aid of thrusters fitted with controllable pitch propellers. The drilling slot (moon pool) on a drillship is through the center of gravity, and the derrick mounted above it gives the drillship its distinctive appearance as seen in Figure 3-6.

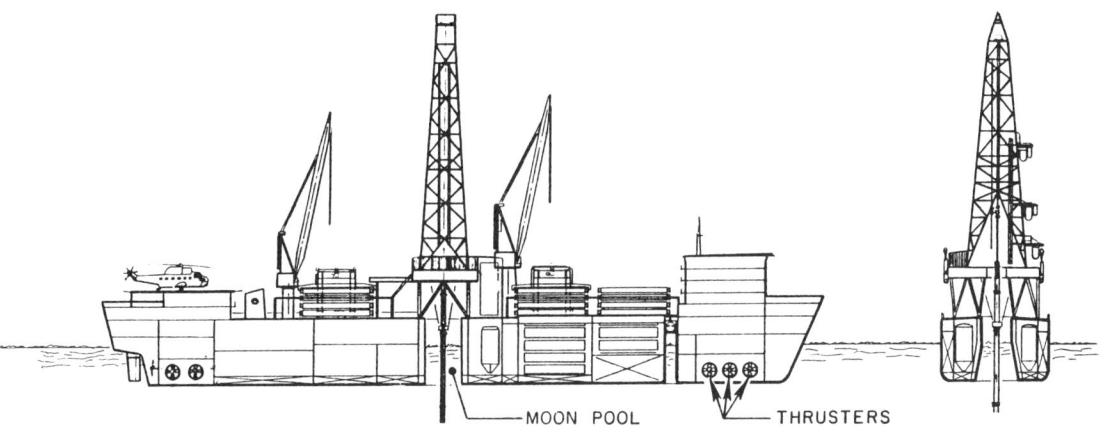

Figure 3-6. Drillship

3.11 RIG COMPONENTS

The principal components of a rig are shown in Figure 3-7. The rig is basically comprised of a derrick, the drawworks with its drilling line, crown block and traveling block, and a drilling fluid circulation system including the standpipe, hose, mud pits and pumps.

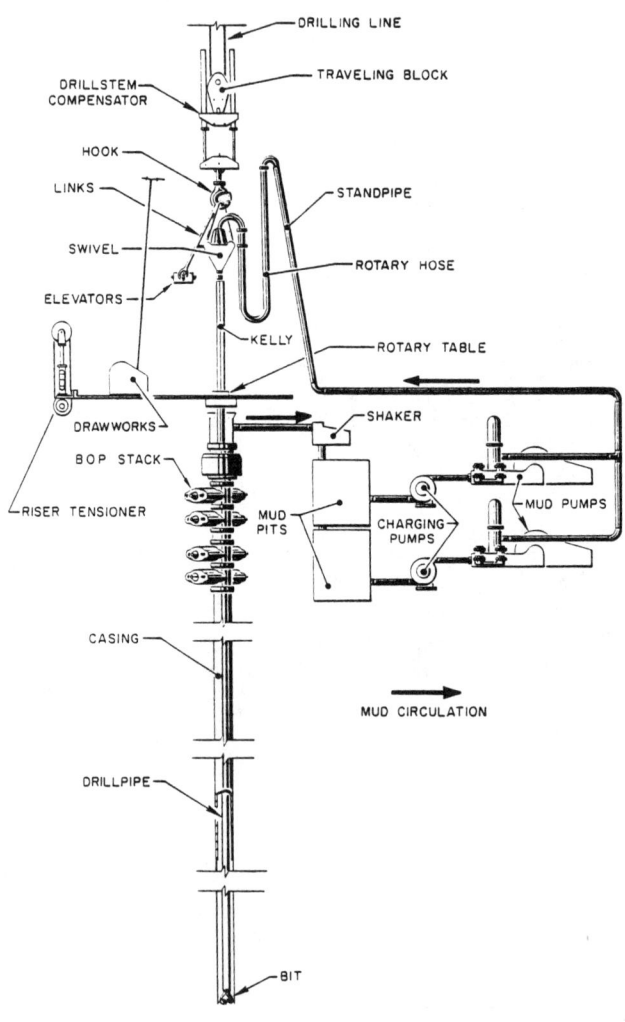

Figure 3-7. Diagrammatical Display of Rig Components, Showing Circulation System

When drilling is progressing the kelly is suspended from the hook beneath the traveling block, and the swivel allows the kelly and drillstring to be rotated in the rotary table while conveying the drilling fluid inside the drillpipe into the hole.

These components work together to accomplish the three main functions of all rotary rigs:

- Hoisting system
- Circulating system
- Rotating system

Two other systems, although not essential in the drilling process, must be mentioned when considering rig components:

- Motion compensation system
- Blowout prevention system

3.12 HOISTING SYSTEM

The mast or derrick supports the hook and elevators by means of the traveling block, wireline, crown block, and drawworks. The drawworks is powered by prime movers — usually two, three or even four engines.

3.13 Derrick (or Mast)

Whenever the drillstem is suspended by the traveling blocks and drill line, the entire load rests on the derrick (Figure 3-8). The standard (or pyramid) derrick is a structure with four supporting legs resting on a square base. In comparison, the mast is much more slender and may be thought of as sitting on one side of the derrick floor or workspace.

The derrick is erected on a substructure which supports the rig floor and rotary table and provides workspace for the equipment on the floor. The derrick and its substructure support the weight of the drillstem at all times, whether the stem is suspended from the crown block or resting in the rotary table. The height of the derrick does not affect its load-bearing capacity, but it is a factor in the length of the sections of drillpipe stands that can be removed. The drillstem must be removed from the hole from time to time, and the length of each drillstem section to be removed is limited by the height of the derrick. This is because the crown block must be sufficiently elevated above the derrick floor to permit the withdrawal and temporary storage in the derrick of the drillstring when it is pulled from the well to change bits or for other reasons.

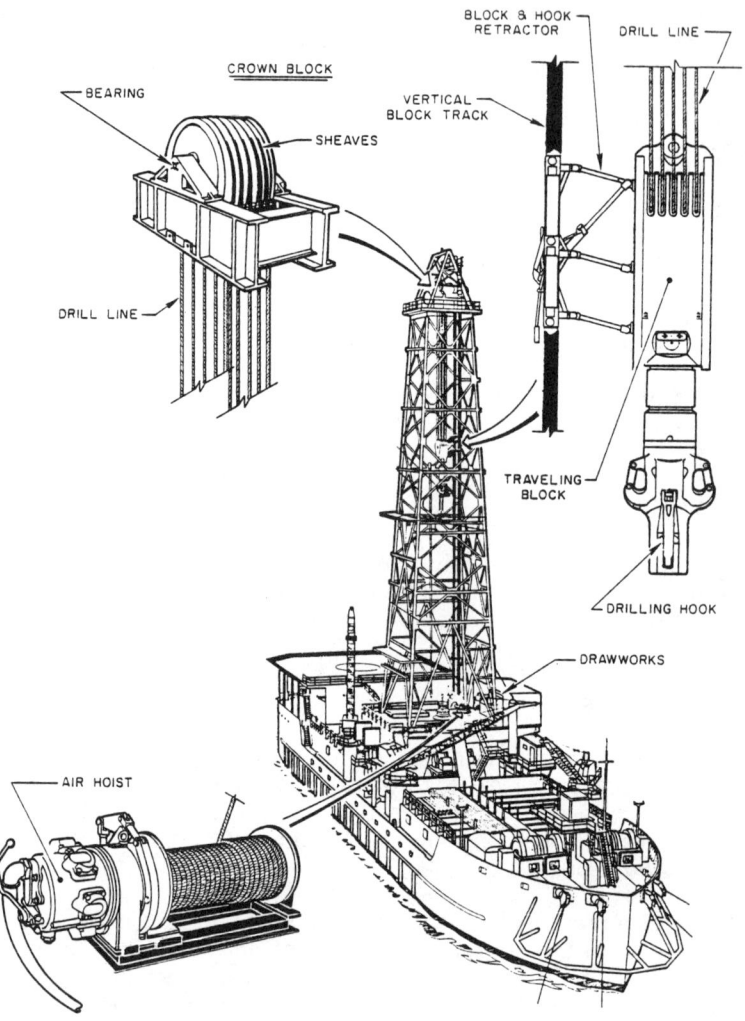

Figure 3-8. Hoisting System Components

3.14 Traveling Block, Crown Block, Drilling Line and Drilling Hook

The traveling block, crown block and drilling line (illustrated in Figures 3-7, -8 and -9) are used to connect the supporting derrick with the load of pipe to be lowered into or withdrawn from the hole. During drilling operations, this load usually consists of the drillpipe and drill collars, with the bit attached to the bottom of the

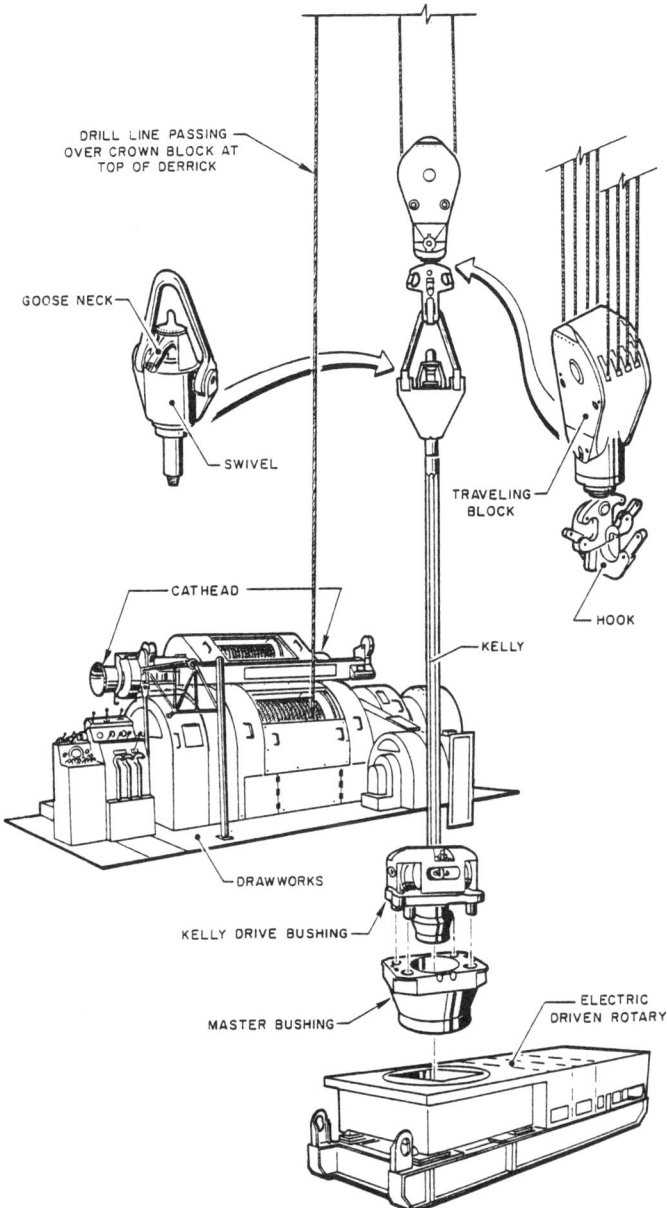

Figure 3-9. Hoisting and Rotary Components

collars. The drilling line passes from the drawworks to the top of the derrick. From here it passes between the crown block and the traveling block to give an eight-, ten- or twelve-line suspension. It is then clamped by the deadline anchor. The drilling line is "slipped" (moved) periodically so that it wears evenly as it is used. Cutoff procedures take into account the amount of usage — that is, the ton-miles of service. (If a wireline has moved a one-ton load a distance of one mile, then the line has received one ton-mile of usage.)

Suspended from the traveling block is the drilling hook which, when drilling, carries the kelly (Figure 3-9); and when tripping (see Section 4), it lifts the drillpipe with the elevators. The elevators are attached to the drilling hook using "links" or "bails" (Figure 3-24).

3.15 Drawworks (the Hoist)

A drawworks is the type of mechanism known in other industries as a hoist. Figure 3-9 shows the drawworks for a heavy-duty rig. The main purpose of the drawworks is to lift the pipe out of and to lower it back into the hole. Wireline is reeled (spooled) on a drum in the hoist. When the drawworks is engaged, the drum turns and either reels in wireline to raise the traveling block or lets out wireline to lower it. Because the drillstem is attached to the block, the stem is thus raised or lowered.

One outstanding feature of the drawworks is the brake system, which enables the driller to easily control a load of thousands of pounds of drillpipe or casing. On most rigs, there are at least two brake systems. One brake is a mechanical friction device and can bring the load to a full stop. The other brake is hydraulic or electric; it can control the speed of the descent of a loaded traveling block, although it is not capable of bringing it to a complete halt. The latter is used to reduce the wear on the primary friction system.

An integral part of the drawworks is the gear (transmission) system. This gives the driller a wide choice of speeds for hoisting the pipe.

Generally, the drawworks has a drive sprocket that drives the rotary table by means of a heavy-duty chain. In other cases, however, the rotary table is driven by an independent engine or electric motor.

Other features of the drawworks are the two catheads. The makeup (or spinning) cathead on the driller's side of the drawworks is used to spin up and tighten the drillpipe joints. The other, located opposite the driller's position on the drawworks, is the breakout cathead; it is used to loosen the drillpipe when the pipe is withdrawn from the hole. An independent air hoist (Figure 3-8) is used on many rigs for handling relatively light loads around the drillfloor.

3.16 CIRCULATING SYSTEM

When drilling is in progress (Figure 3-7), the components of the hoisting system, mud pumps and prime movers are used to circulate drilling fluid from the mud pit through the standpipe, rotary hose, swivel, kelly, drillpipe, and drill collars to the bit. Cuttings are flushed from the bottom of the hole up to the surface, thus cleaning the bottom of the hole and providing the geologist with samples at the surface.

3.17 Mud Pumps

A rig usually has two mud pumps, and these are the heart of a fluid-circulating system for rotary drilling. Their function is to circulate the mud under pressure from the pit, through the drillstem, to the bit (where hydraulic power is used for jetting), return it up the annulus, and back to the pit. Mud pumps for drilling may be either duplex, double-acting, reciprocating pumps; or triplex, single-acting pumps. Each of the two cylinders of the duplex double-acting pump is filled on one side of the piston at the same time that fluid is being discharged on the other side of the piston, as seen in the upper section of Figure 3-10. Each complete cycle of a piston results in the discharge of a mud volume that is twice the volume of the cylinder, minus the volume of the piston rod. The total volume for a duplex pump in one complete cycle is twice this amount because there are two pistons. The volume of fluid pumped per minute is determined by multiplying the volume per complete cycle by the number of stroke-cycles per minute. Strokes per minute (spm) actually means cycles per minute since a duplex, double-acting pump strokes four times during each cycle.

Triplex single-acting pumps (lower section of Figure 3-10) put pressure on only one face of the pistons rather than on both sides as in double-acting pumps. Triplex pumps have three pistons and are much lighter than duplex, double-acting pumps for specific power ratings. More power can be obtained from a relatively small triplex pump because it operates at high speeds of 120 to 160 spm, as compared to only 60 to 70 spm for a duplex pump. Triplex pumps can be operated at higher pressures with a smooth discharge flow than the double-acting units, because equal volumes of fluid are delivered at each 120 degrees of crankshaft rotation.

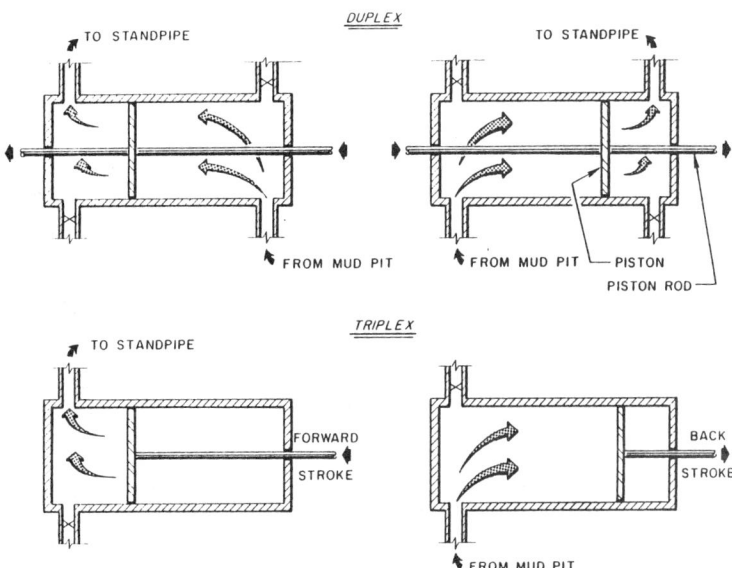

Figure 3-10. Schematic Cross-Sections of Duplex and Triplex Mud Pumps

Because the triplex pump operates at high speed, it usually has a centrifugal pump to charge the suction. Suction-charge pumps may be mechanically or electrically driven. The major cause of poor operation in a triplex pump is improper suction or discharge piping. Fluid knocks develop if the suction arrangement does not permit enough fluid to reach the pistons. With a properly charged suction, triplex pumps operate at nearly 100 percent volumetric efficiency. (The volumetric efficiency of a reciprocating pump is the ratio of the fluid actually pushed out ahead of the piston of a pump, compared to the amount of fluid theoretically displaced by the piston as it moves inside the liner.)

3.18 Standpipe and Rotary Hose

In addition to the pump, the surface portion of a fluid-circulation system consists of high-pressure piping from the pump to the standpipe and rotary hose (Figure 3-7). The standpipe anchors the upper end of the rotary hose and keeps the hose clear of the rig floor when the kelly has been drilled down and the swivel is near the rotary table. The standpipe is firmly clamped to the derrick and topped with a gooseneck fitting. One end of the rotary hose is attached to the gooseneck on the standpipe, and the other end to a gooseneck on the swivel.

3.19 Drillstem

The drillstem consists of three main components:

- Kelly and swivel
- Drillstring (the drillpipe plus the bottomhole assembly: drill collars, crossover subs, shock subs, bumper subs, stabilizers)
- Drill bit

Together they perform the following functions:

— Lower the bit into the hole and withdraw it. However deep the bit may penetrate in the process of drilling, it must be placed at that depth by the drillstem.

— Weight the bit so that the bit can penetrate the formation more effectively. Weight is applied by the drill collars; the drillpipe portion of the drillstem should never be used to put weight on the bit.

— Transmit a rotating action (torque) to the bit. Thus the drillstem is a drive shaft driven by the rotary table. However, when a turbo-drill is being used this is not the case (see Section 4).

— Conduct the drilling fluid under pressure from the surface to the bit. This means the drillstem also serves as a vertical conduit.

3.20 Kelly and Swivel: The upper end of the drillstem terminates where the topmost length of drillpipe screws onto a device called a kelly sub or saver sub. The saver sub saves wear on the threads of the kelly. The kelly is about 40 ft long, square or hexagonal on the outside, and hollow throughout to provide a passage for the drilling fluid. Its outer surface engages corresponding square or hexagonal surfaces in the kelly drive bushing. The rotary activates the kelly drive bushing which, in turn, rotates the kelly. The kelly moves freely up or down through the kelly drive bushing even when the bushing is rotating. Figure 3-9 illustrates the arrangement of the kelly, kelly drive bushing and rotary. At the top of the kelly is the "kelly cock" — a valve which if the well is flowing can be closed to prevent backpressure from damaging the swivel, rotary hose and other surface equipment.

The swivel does not rotate, but supports the kelly, which does rotate. Furthermore, the entire load of the drillstem is carried by the swivel whenever drilling is in progress. In addition, drilling fluid is introduced into the drillstem through the swivel. The fluid enters through the gooseneck connected to the rotary hose.

3.21 Drillstring: The drillstring is made up of the drillpipe, drill collars, crossover subs, shock subs, bumper subs and stabilizers through which fluid and rotational power are transmitted from the kelly to the bit.

    1)   Drillpipe. American Petroleum Institute (API) drillpipe and other tubular sizes are gauged by the nominal outside diameter (O.D.) of the tube. The O.D. of a given pipe size must be a specific measurement in order for threaded fittings and pipe-handling tools such as elevators and slips to fit properly. Although the O.D. of API drillpipe is the same for each size, the inside diameter (I.D.) varies with the nominal weight per foot of length. API drillpipe is also "upset" (made thicker) on the ends for added strength. Here, tool joints are shrunk or welded-on to enable lengths of drillpipe to be screwed together to make up the drillstring. Seamless drillpipe is upset to API dimensions for weld-on tool joints in three types as shown in Figure 3-11. Upsets are

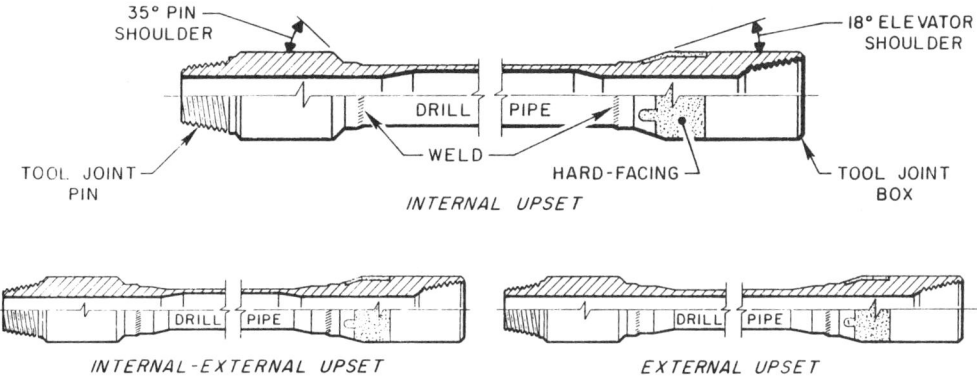

Figure 3-11. API Drillpipe with Weld-On Tool Joints

necessary on drillpipe, both for shrink-on and weld-on tool joints, to provide safety in the weld area for mechanical strength and metallurgical considerations. API seamless drillpipe is offered in five grades of steel varying in strength from D (the weakest) through E, X, G to S (the strongest). High strength drillpipe requires heavier and longer upsets than those used on grades D and E. Lengths of drillpipe are generally about 31 or 45 feet.

2) <u>Drill Collars</u>. These are similar to drillpipe but have larger outside diameters (up to 10 inches) and smaller inside diameters. They are usually about 30 ft long. A string of drill collars has several tasks to perform:

— Provide weight to the bit for drilling.

— Maintain weight to hold the drillstring in tension.

— Provide the pendulum effect to cause the bit to drill a nearly vertical hole.

— Provide rigidity in order to drill new hole aligned with hole previously drilled.

Any drillstring weighs less in mud than in air due to buoyancy of the mud. Actual weight of collars = weight of collars x (1- (0.015 x mud wt)). The denser the mud, the greater the buoyancy effect and the lighter the apparent weight of the collars. This is considered when deciding how many collars are to be run. Total drill-collar weight must exceed that to be applied to the bit when drilling, so that no weight is supplied by the drillpipe. The usual practice is 10 to 30 percent excess drill-collar weight over the amount applied to the bit. In this way the lower portion of the collars is in compression with its weight resting on the bit and the upper portion; plus, the entire drillpipe section remains in tension supported on the hook. It is vital that drillpipe is never subjected to compression as it would bend and "twist off" very easily. Figure 3-12 shows how adequately drill-collar weight keeps the drillpipe in tension and provides weight to the bit for drilling.

The pendulum effect is the tendency of the drillstem to hang in a vertical position due to the force of gravity. The heavier the pendulum, the stronger its tendency to remain vertical and the greater the force needed to cause the drillstem to deviate from vertical.

The length of the pendulum is that section of the drill collar string between the bit and the lowest point tangent to the side of the hole. It is desirable that the pendulum be as long as possible, for it then has a greater tendency to seek a vertical position.

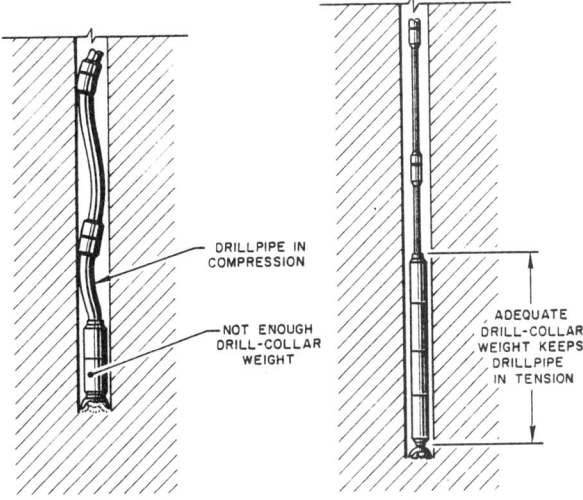

Figure 3-12. Drill Collar Weight

Large, heavy drill collars immediately above the bit stabilize the drillstem in the hole and cause the bit to drill a rifle-bore extension of the well in spite of uncontrollable forces that tend to deviate the hole (such as dip of formation strata, type formation, etc.). Packed-hole assemblies using oversized or square drill collars and frequently with stabilizers guide the bit to drill a true extension of the previously drilled hole. The term "packed hole" refers to the fact that the drill collars in the lower part of the assembly are about 1/2-inch smaller in diameter than the hole. This makes them liable to differential sticking (see Section 4), and various design features (such as spiral grooving, square section, elliptical section, etc.) are incorporated to prevent this.

Bit stabilization enables hole alignment and ensures proper bit performance because the bit is made to rotate on its axis. This prevents the bit from wobbling or walking on the bottom of the hole and uniformly loads the cutting structure of the bit. An unrestrained bit may drill an oversized hole, produce unusual bit wear, and slow the rate of penetration. Bits drill faster and last longer when properly stabilized.

In contrast to drillpipe, the weakest point in drill collars is at the joint, for there is no upset on a drill collar tool joint.

3) Subs. The word "sub" refers to any short length of pipe, collar, casing and so on with a definite job.

- Crossover Subs (Xo Sub). A crossover sub is designed with different threaded ends for changes between different sizes and types of drillpipe or collars.

- Shock sub. This is run behind the bit with a steel spring or rubber packing to absorb the impact of the bit bouncing on hard formation and damaging the rest of the drillstring.

- Bumper Sub. This is a free telescopic sub with 6- to 8-ft closure. Its purpose is to absorb the effects of heave on a floating rig. It is placed in the collars at the point where compression changes to tension — that is, above whatever weight of collars is placed on the bit. In this way any movement of the rig and hence of the drillstring will be taken up in the sub and not transferred to the bit.

These are now largely replaced by the "motion compensator" which is a hydraulic device attached to the traveling block so that the entire drillstring remains stationary as the rig heaves. The motion compensation system is explained in more detail in paragraph 3.30.

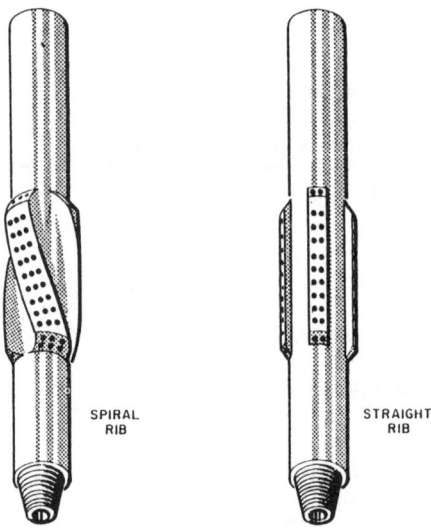

Figure 3-13. Stabilizers

- Stabilizers. These are short subs (Figure 3-13) with "fins" which are full hole size or a fixed amount below gauge. The fins may be aluminum or rubber but more often are steel with tungsten carbide inserts on the edge. They are located between the collars and are intended to maintain a straight hole by keeping the collars centralized. Also, by a scraping action, they maintain a full gauge hole.

- Bit Sub. This is a short sub with a box on both ends so that pipe and collars may always be run pin down.

4) Bottomhole Assembly (BHA). This is the term given to the current arrangement of tools incorporated in the collars. The BHA may consist of any arrangement of bit, drill collars, stabilizers, shock sub, bumber subs and crossover sub, and possibly various other specialized tools.

3.22 Drill Bits: The bit is the most critical item of a rotary rig operation. The most refined of the rotary-rig tools, it is available in more styles and is more highly specialized for every condition of drilling than any other tool on the rig. To select a bit, some information must be known about the nature of the rocks to be drilled.

There are four categories for bits used in rotary drilling:

- Drag bit
- Tri-cone bit
- Diamond bit
- Reamers/hole openers

1) Drag Bit. This bit (illustrated in Figure 3-14) was the earliest type. It came in various forms but is now rarely used except onshore and in extremely soft formations. It is much less expensive than the tri-cone bit.

2) Tri-cone Bit. This is the most common type of bit used today. A variety of modifications and additions is available depending upon the specific conditions involved. Nearly all modern tri-cone bits are provided with jets instead of the older style conventional water-courses (the "regular" bit arrangement). The conventional and jet bits are illustrated in Figure 3-15.

With the conventional bit, fluid passes through its center to lubricate the bit and carry away cuttings. With the jet bit, the nozzles are replaceable; therefore, the openings can be changed to match the pressure and volume requirements of the fluid circulated in the well. Jet-nozzle sizes are described in thirty-seconds of an inch (a #10 nozzle is 10/32-inch in diameter).

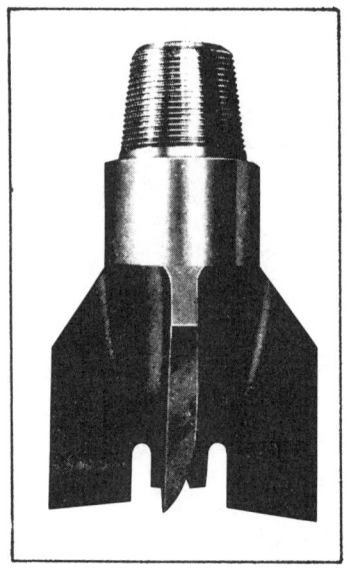
Figure 3-14. Drag Bit

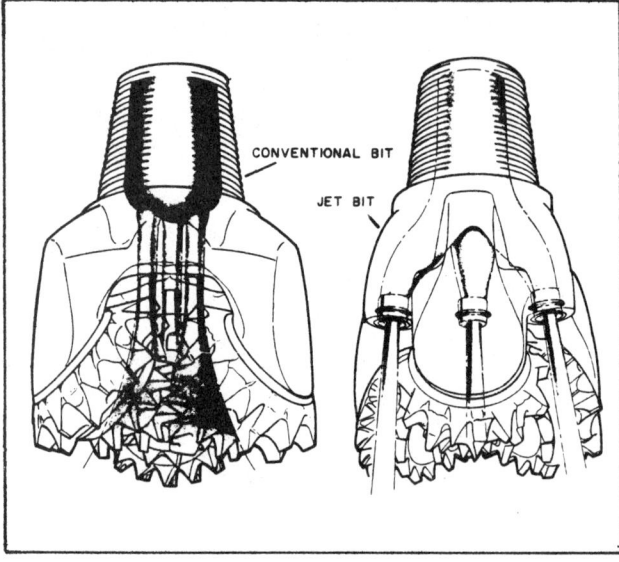
Figure 3-15. Water Courses

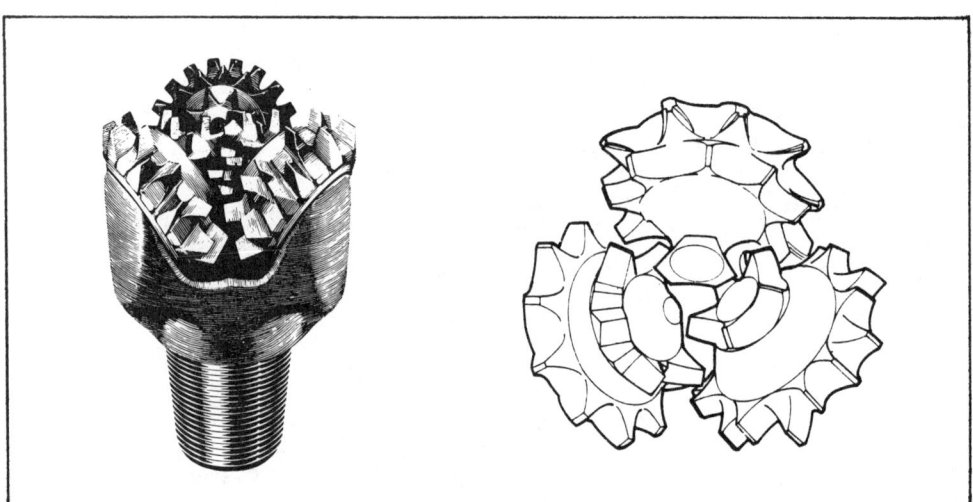
Figure 3-16. Milled Tooth Bit

There are a number of variations of tri-cone bits:

(2a) <u>Milled Tooth Bit</u> (Figure 3-16). Cones have steel teeth of various sizes -- long teeth for soft formation (gouging action) and short, broad teeth for hard formation (chipping and crushing action).

(2b) <u>T-Gauge Bit</u> (Figure 3-17). The outer row of teeth on each cone is T-shaped to give greater surface wear. This maintains the gauge (correct diameter) longer and prevents a worn bit from drilling an undersize hole. It is often treated with "hard facing" material.

(2c) <u>Gauge Insert Bit</u> (Figure 3-18). The outer edge of each cone has flush-set tungsten carbide buttons inserted to maintain the gauge. This would be one step further than T-gauge.

(2d) <u>Insert Bit</u> (Figure 3-19). On these bits the cones do not have steel teeth cut in them; instead, tungsten carbide buttons (teeth) are inserted. These are very much harder and longer lasting for compacted formations (and much more expensive). Inserts can vary in shape — from long chisel shape for firm formations to short round buttons for hard, brittle formations.

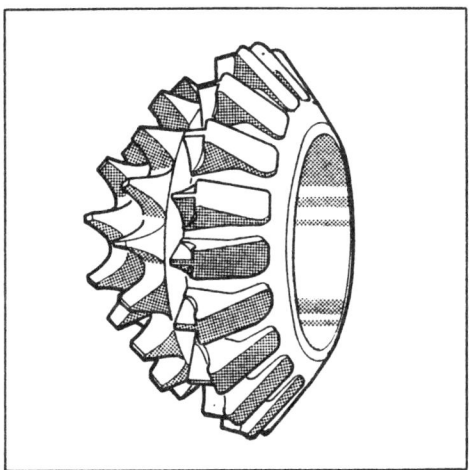

Figure 3-17. T-Gauge Bit (Cone)

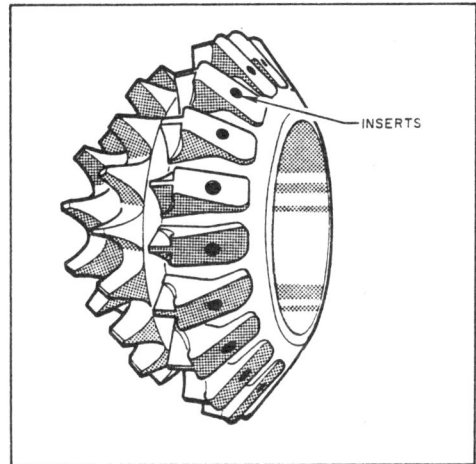

Figure 3-18. T-Gauge Bit Cone with Inserts

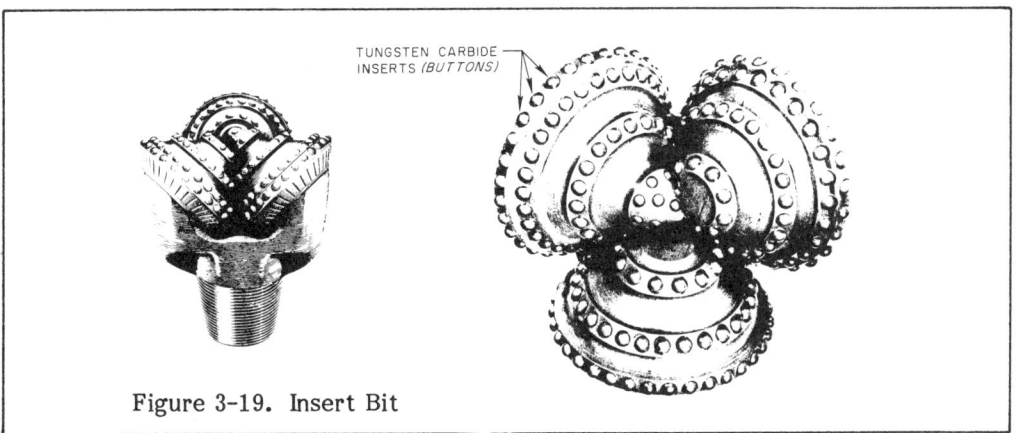

Figure 3-19. Insert Bit

(2e) Bearings. Ordinary bearings for the cone are lubricated by the drilling fluid. Because of the solids content of this fluid, the bearings do not have a particularly long life. However, for long steel-tooth bits which tend to drill rapidly but for only a short time, this type of bearing is sufficient.

- Sealed bearing. This is lubricated by a lubricant in a sealed reservoir rather than by drilling fluid. Sealed bearings have a longer life and may be used for the shorter steel-tooth bits which will last longer.

- Friction bearing. The roller bearing is replaced by a silver journal bearing for insert bits or steel tooth bits which will drill for long periods. They also contain a sealed lubricant chamber.

(2f) Journal. This is part of the rotating shaft seen in Figure 3-20. The journal rests and turns in the bearing and is the weight-bearing segment of the shaft. The journal angle is commonly $33°$ to $34°$ for soft formation bits, $36°$ for medium bits and $39°$ for very hard bits.

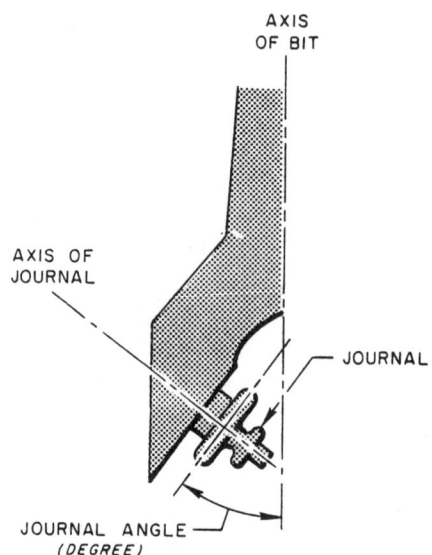

Figure 3-20. Journal Configuration

(2g) <u>Offset</u>. The off-center alignment of the cone makes the teeth scrape and gouge the formation as the cone rolls on the bottom of the hole; the amount of scraping depends on the amount of cone offset. Figure 3-21 shows the cone offset of a soft formation rock bit. For soft bits, 1/4-inch to 3/8-inch is the common offset, while 1/8-inch is typical of medium bits, and none for hard bits.

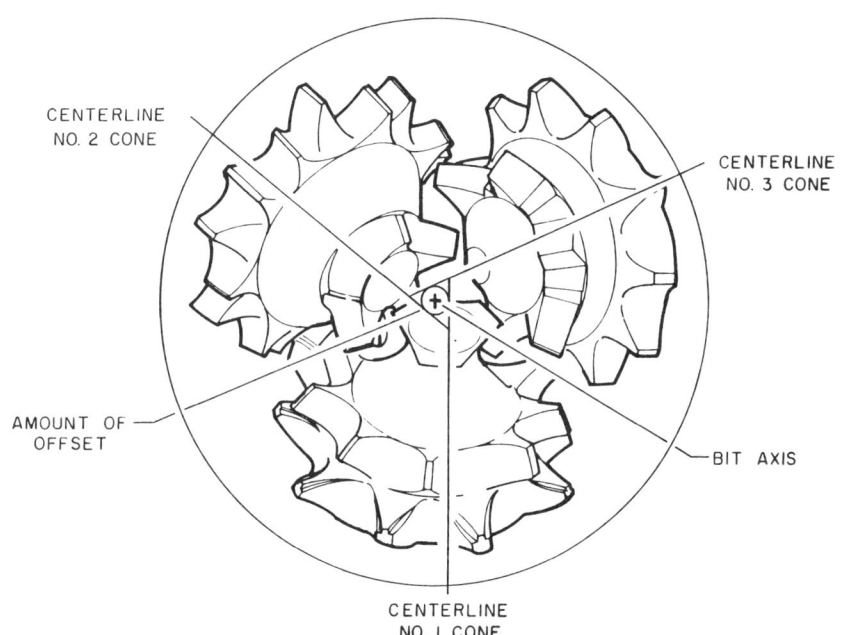

Figure 3-21. Offset on Soft-Formation Cones

(2h) **Bit Classification.** A standard system to classify tri-cone bits has been developed by IADC (International Association of Drilling Contractors). Any tri-cone bit can be described using a four-digit code, as shown below:

1st digit: <u>Series</u> (type of formation for which the bit is suitable).
2nd digit: <u>Type</u> (subdivision of the series).
3rd digit: <u>Feature</u> (mechanical or metallurgical design variations).
4th digit: <u>Vendor</u>. This digit is an Exploration Logging code, <u>not</u> an industry standard, and is used to define the manufacturer of a bit by adding an additional number at the end of the three-digit IADC bit code.

This system of classification permits comparison of the bit types offered by various bit manufacturers. With reference to the IADC classification charts in Appendix D, to find a comparable bit to the Hughes ODG look up its IADC code on the Hughes chart (1331).

```
1333 — Reed Tool Company        Y13G
1335 — Smith Tool Company       DGH
1336 — SMF (Creusot-Loire)      TS5K
```

(2i) **Dull Bit Grading.** This is an aid to bit selection when carried out in a uniform manner on all bit runs and all wells. The degree and type of wear can also be of value to the geologist in lithological evaluation. This information is recorded on the mud log, along with the bit information. An example of this is in Appendix A.

The amount of wear on a used bit is expressed in three ways:

- as a function of the teeth
- bearings
- the gauge

— **Milled Teeth.** Teeth are graded by eighths of total tooth height lost by wear. Hence, T0 indicates a new bit and T8 indicates a bit where the teeth are totally worn away. Teeth may break off instead of wearing down. Therefore, if any <u>one row</u> has the majority of teeth broken, add to the log the letters "BT." Tooth wear will have a different effect on different types of bit. For example, a long-tooth bit may cease to drill efficiently at T4 while a short-tooth bit, although drilling more slowly, may continue to drill to T7. This is due to the different drilling action of the two types.

- Inserts. Grade by eighths of insert height missing, irrespective of how lost. Add the following designators to the log as appropriate:

    BT — broken teeth
    LT — lost teeth
    BG — broken gauge insert

  Grading should be an assessment of overall bit condition, not of the worst row.

- Bearings. Check the cone in worst condition and grade from B1 to B8, as follows:

    B2 — tight
    B4 — medium
    B6 — loose
    B8 — locked or lost

  Bearing wear is not linear with time; it depends upon the weight-on-bit (WOB), rpm and other factors such as the volume of solids in the mud. With a standard bearing, a high sand content rapidly erodes the bearing. If the bearing is sealed, wear is usually minimal until the seal is eroded; wear then becomes rapid. If the condition of the bearing is apparently good or uncertain, grade as a ratio of actual life to expected life.

- Sealed bearings. Grade as normal, and add to the log as appropriate:

    SE — seal effective
    SQ — seal questionable
    SF — seal failed

- Gauge. This is described in the following manner:

    IG — in gauge
    OG — out of gauge (1/2 OG = 1/2-inch out of gauge)

3) Diamond Bits. These are usually smaller by 1/16-inch in the U.S. (1/32-inch in Europe) than the equivalent tri-cone bit, to prevent the diamond bit from being damaged while running into the hole. The design of diamond bits, illustrated in Figure 3-22, varies greatly in the shape of the head and the water courses and in the size and settings for the diamonds. The advantage of diamond bits is that, with no bearings to wear, they can be run for very long periods where their slightly slower drilling rate may be offset by the savings in trip time. A disadvantage is that cuttings will be smaller and may take on a burnt appearance. Diamond bits are now being challenged by the modern friction-bearing insert bits which have reasonably long lives and usually drill more rapidly.

74

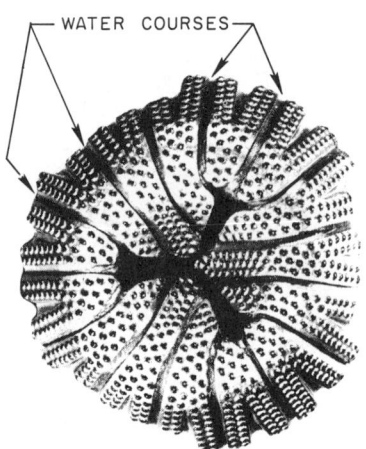

Figure 3-22. Diamond Bit

4) <u>Reamers/Hole Openers</u>. These tools are run immediately above a bit to maintain or enlarge a hole size. Their cutting action is by rotating cones (as with a tri-cone bit) built out from the central stem as seen in Figure 3-23. Both types perform similar functions, but the cones of an under-reamer are built onto collapsible arms which are held out while drilling by the pressure of mud circulating down through the center of the tool. In this way it may be run in or pulled out through a smaller section of hole, such as with a marine riser.

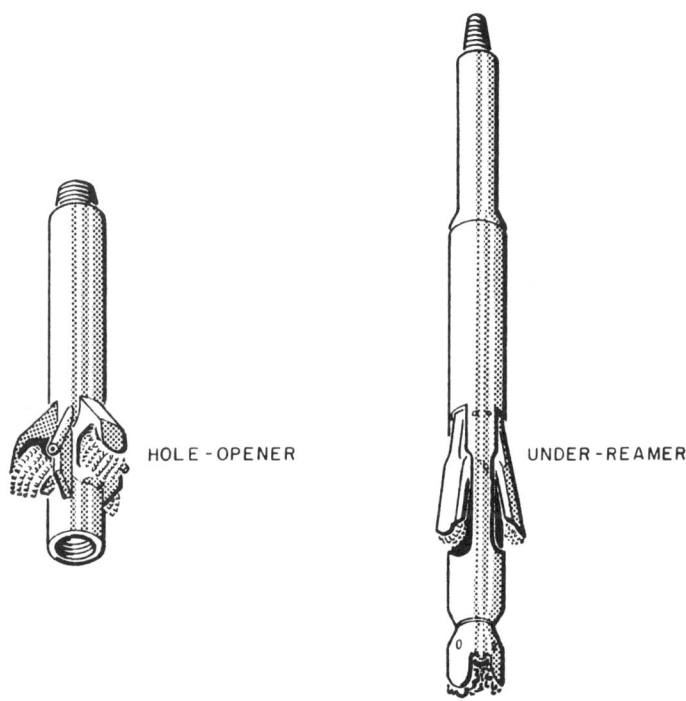

Figure 3-23. Hole Opener and Under-Reamer

3.23 ROTATING SYSTEM

Operating through kelly drive bushings, the rotary table rotates the kelly and through it the drillstring and the bit. Figures 3-9 and 3-24 illustrate the components that make up the rotating system. The kelly drive bushings are driven by four pins which fit into openings in master bushings which, in turn, fit into the rotary table.

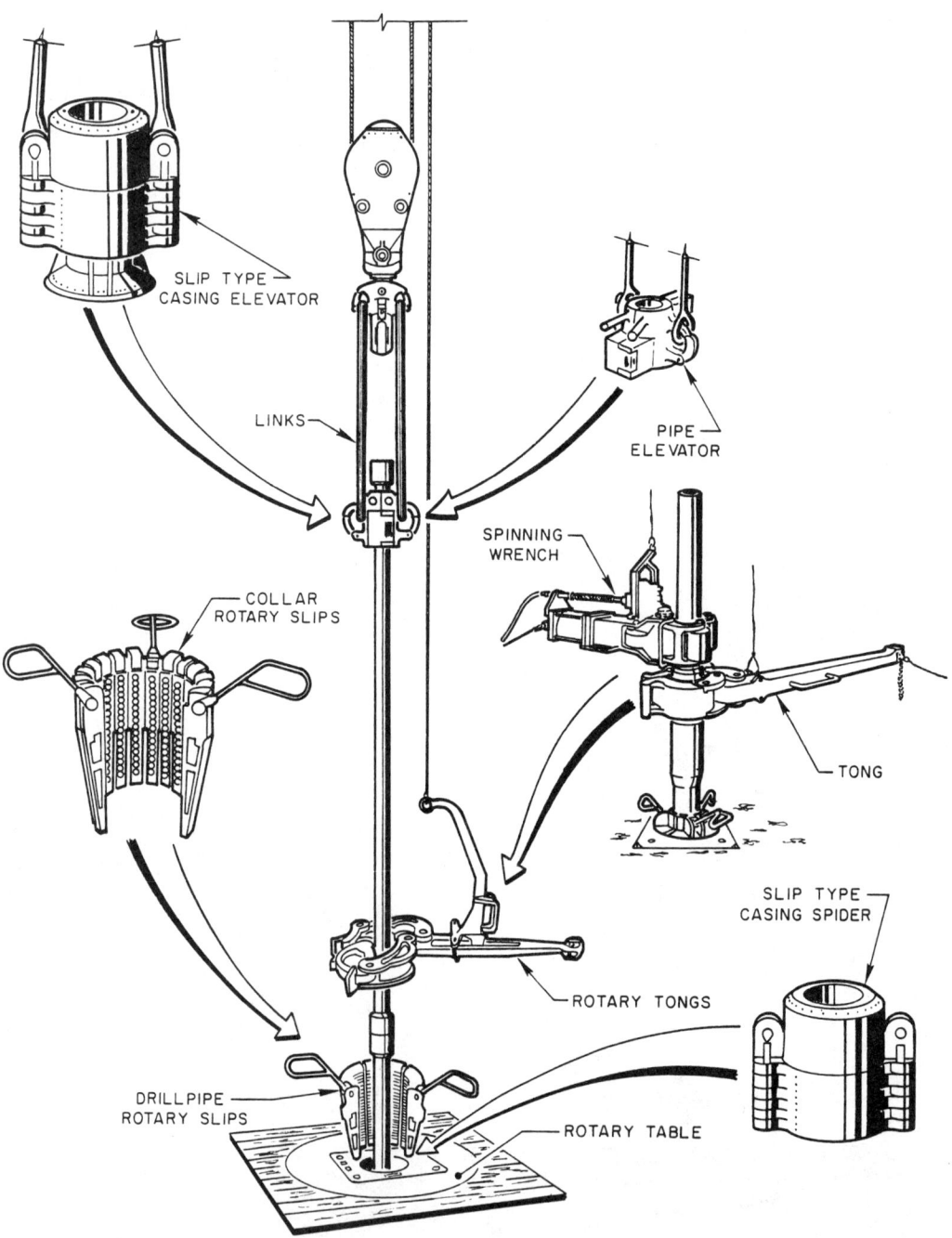

Figure 3-24. Rig Components Used for Pipe Handling

3.24 The Rotary Table

The rotary table serves two main functions:

- It rotates the drillstem.

- It holds devices called slips that support the weight of the drillstem when the latter is not supported by the elevators or hook and kelly.

The rotary drive generally consists of a rotary-drive sprocket and chain, the rotary-drive sprocket being part of the drawworks. However, an independent engine or electric motor with a direct drive to the rotary is also used on many rigs. In such cases, the rotary is driven by a drive shaft rather than by chains and sprockets.

3.25 Master Bushings

Through the master bushings, the rotary table transmits rotary motion to the kelly drive bushings and the kelly. They are also the connecting link between the rotary table and the slips which support the pipe during trips. Master bushings and kelly drive bushings can be seen in Figure 3-25.

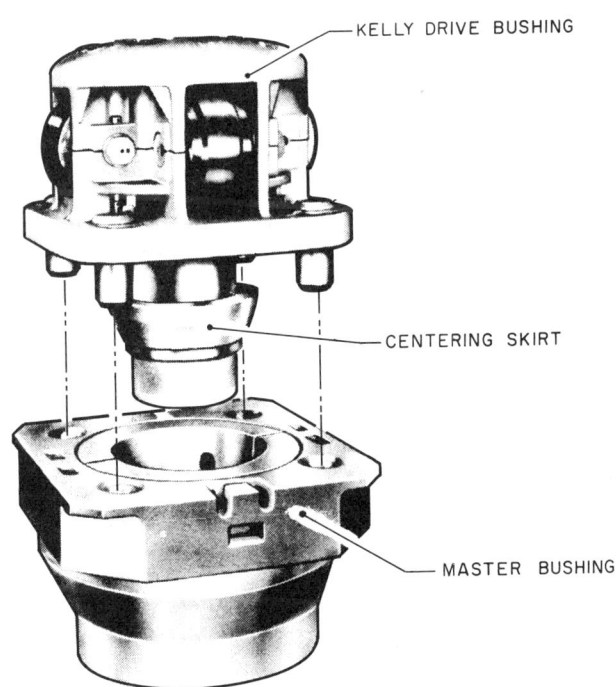

Figure 3-25. Master and Kelly Drive Bushings

3.26 Kelly Drive Bushings

These bushings engage the master bushing which in essence is a part of the rotary. Rollers within the bushing permit the kelly to move freely upward or downward when the rotary is turning or when it is stationary. When the kelly is disconnected and set back, the drive bushing lifts and sets back with it.

3.27 Slips

As can be seen in Figure 3-24 (lower left), slips are wedge-shaped steel dies fitted in a frame with handles, which are placed between the drillpipe and the sides of the master bushing in the rotary table when making a connection or tripping. Their purpose is to support the drillstring and hold it suspended in the hole, and when in this position they are said to be "set."

3.28 Tongs

This is a type of wrench (center, Figure 3-24) used for tightening and loosening drillpipe and drill collars. Two sets of tongs are used, one to hold the drillstring and the other to tighten or loosen the joint. These tongs are called back-up and lead tongs. Power tongs may combine both in a single unit and can be used on both drillpipe and casing.

3.29 Spinning Wrench (Power Tongs, Pipe Spinners)

This is an pneumatically-powered wrench (Figure 3-24, right center) used for rapidly spinning drillpipe or collars when making up or breaking out pipe. Final torque, when making up, and initial loosening of the pipe joint when breaking out, is achieved by using tongs.

3.30 MOTION COMPENSATION SYSTEM

These systems are used entirely on offshore floating rigs. The two basic types of motion compensators are:

- Drillstring compensator
- Riser and guideline tensioner

3.31 Drillstring Compensator

The drillstring motion compensator system (Figure 3-26) is designed to nullify the effects of rig heave on the drillstring or other hook-supported equipment. Mounted between the hook and traveling block, the compensator is connected to deck-mounted air pressure vessels via a hose loop and standpipe and is controlled and monitored from the driller's control console.

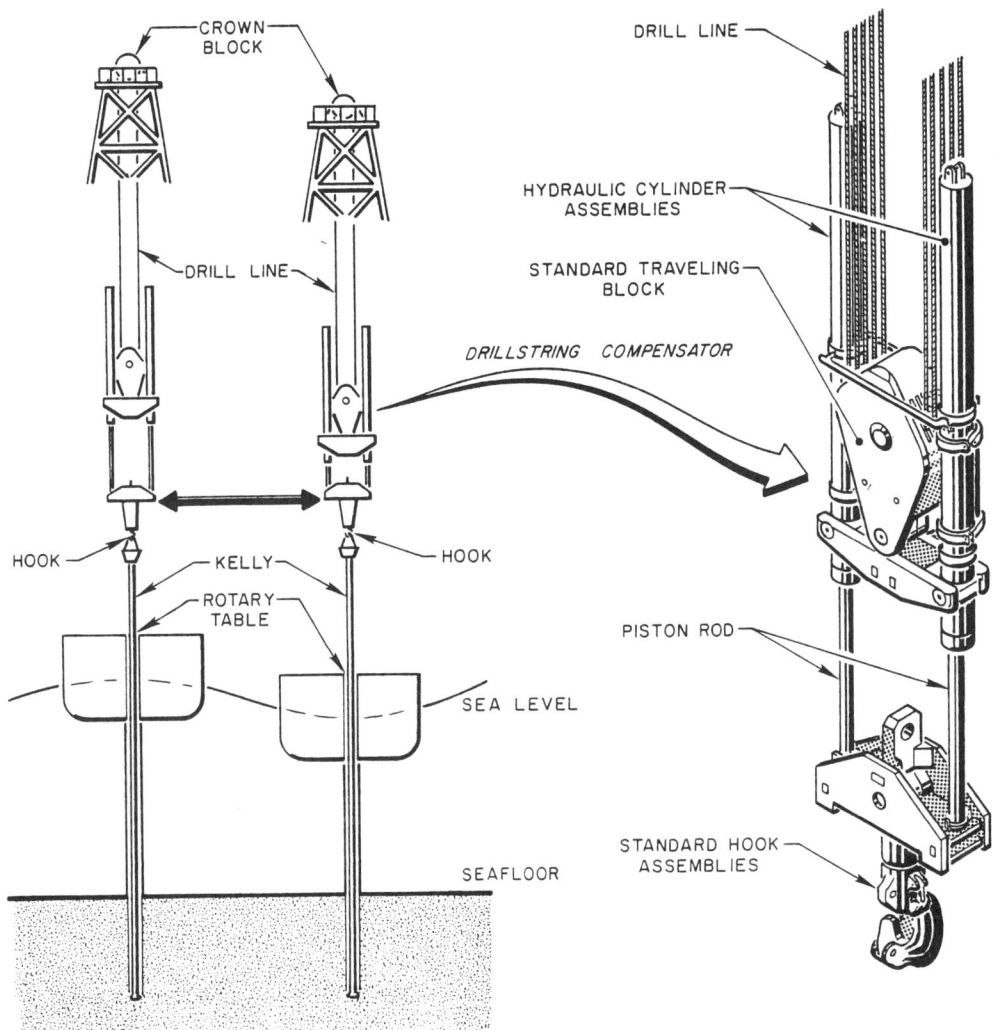

Figure 3-26. Drillstring Compensation System

While drilling, the drillstring compensator controls the weight on the bit. The driller intermittently lowers the traveling block to account for drill-off and to maintain the compensator cylinder within its stroke capacity while the drillstring compensator automatically maintains the selected bit weight.

As the rig heaves upward, the compensator cylinders are retracted and the hook moves downward to maintain the selected loads. Actually, the hook remains fixed relative to the seabed; the rig and compensator move, producing relative motion between the hook and rig. The motion of both the kelly and drillstring is relative to the rotary table.

3.32 Riser and Guideline Tensioners

The marine riser is essentially a conduit but its main purpose is to maintain contact with and give access to the borehole when drilling. Riser tensioners provide tension to the marine riser pipe below the telescopic joint by a system of wires joined via sheaves to a series of pneumatic cylinders, as seen in Figure 3-27. Its purpose is to maintain the riser in tension at all times regardless of the heaving of the rig. It is common practice to install four or six tensioners, each with a load rating of around 60,000 lb. Compensation for vertical movement several times the length of the stroke of the pistons is possible due to multiple cable turns around the pistons.

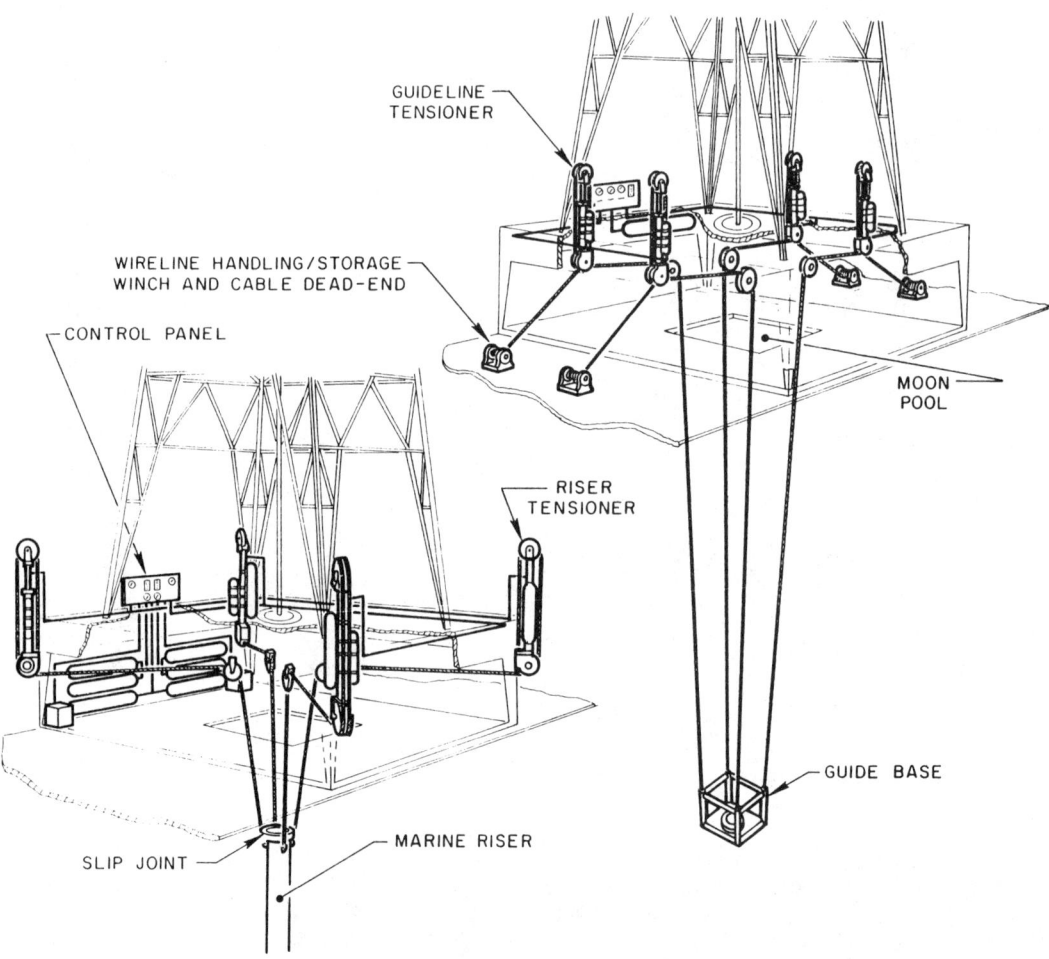

Figure 3-27. Riser and Guideline Tensioner Systems

Guideline tensioners operate on the same principle as the riser tensioners. Their purpose is to maintain the correct tension in the guidelines between the rig and the guide base (which sits on the seafloor), regardless of the heaving of the rig. The guideline tensioner is designed to deal with tensions of less than 10,000 lb; therefore, wires, cylinders and other components are smaller than those used in the riser tensioner system.

3.33 BLOWOUT PREVENTION (B.O.P.) SYSTEM

Normally, hydrostatic pressure of the drilling fluid column is greater than pressure of formation fluids, preventing flow of formation fluids into the wellbore. When hydrostatic pressure drops below formation fluid pressure, formation fluids are able to enter the well. If this flow is small, causing a decrease in density (mudweight) as measured at the surface, the drilling fluid is said to be "gas cut," "saltwater cut," or "oil cut" as the case may be. When a noticeable increase in mud pit volume occurs, the event is known as a "kick." An uncontrolled flow of formation fluids is a "blowout." As long as hydrostatic pressure controls the well, circulation is as indicated by the arrows in Figure 3-28, using the flowline, and the well may be left open.

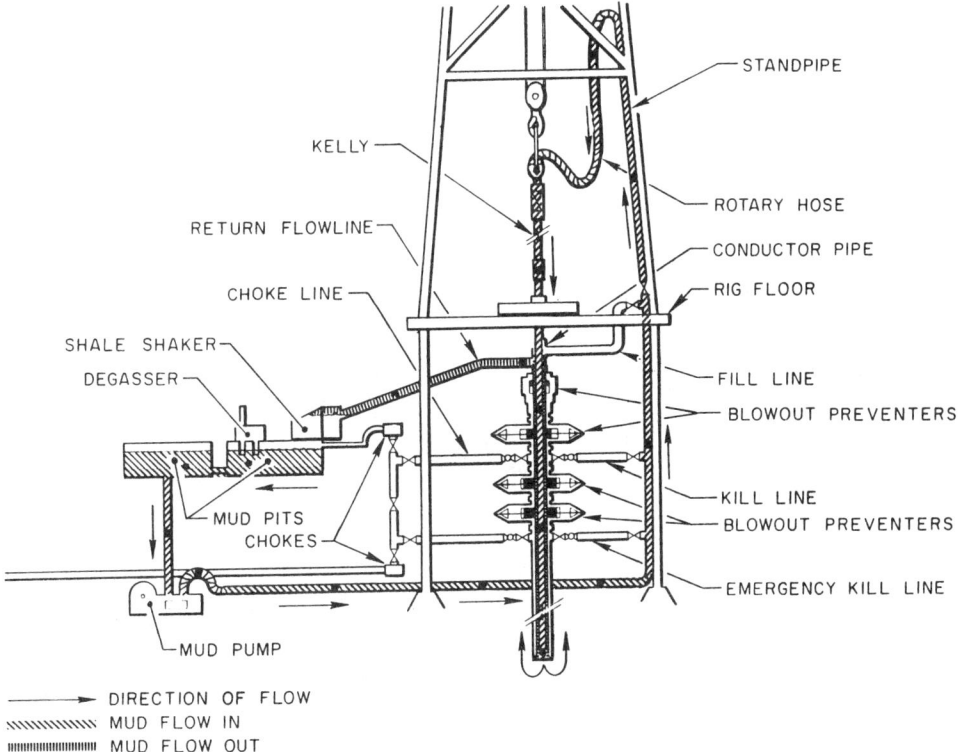

Figure 3-28. Schematic Projection Showing the Relationship Between the Circulation and Blowout Prevention Systems

Should a kick occur, blowout prevention equipment and accessories are needed to close the well. This may be done with an annular preventer (Figure 3-29a), with pipe rams (Figure 3-29b), or when drillpipe is out of the hole using the master (blind) rams (Figure 3-29c). In addition, it is necessary to pump drilling fluid into the well and to allow controlled escape of fluids. Injection is accomplished either down the drillpipe or through one of the kill lines, and flow from the well is controlled by a variable orifice (choke) attached to a choke line. Choke lines continue both to a reserve pit where undesired fluid is discarded and to a choke (splash) box, degasser and mud pit where desired fluid is degassed and saved. Through use of this equipment, the low-density fluids are removed and replaced with a higher density fluid capable of controlling the well.

As seen in Figure 3-29, the B.O.P. stack consists of a number of different blowout preventers. Their arrangement is decided by the degree of protection deemed necessary and the size and type of pipe in the hole. There are three types of blowout preventers:

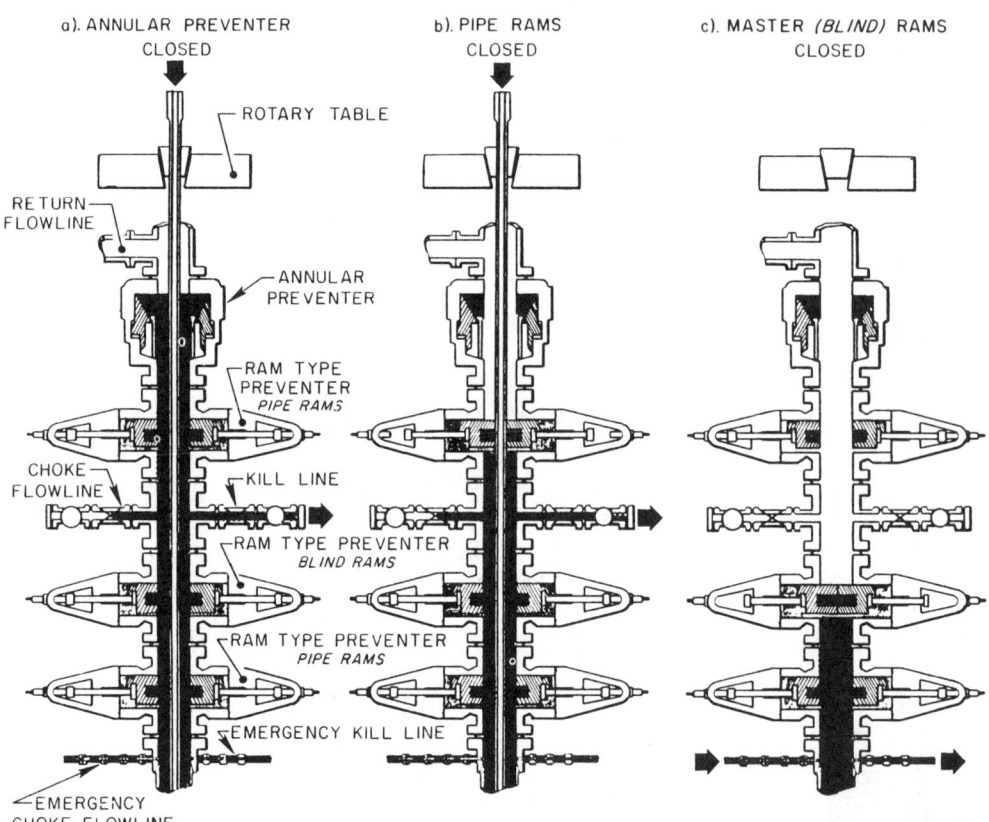

Figure 3-29. Blowout Preventer (B.O.P.) Stack, in Various Operational Modes

- The Annular Preventer (Figure 30). This consists of an annular rubber sealing element which, when pressure is applied from above, closes around the drillpipe or kelly. Since pressure may be applied progressively, the B.O.P. can be made to close on any size or type of pipe. A slight relaxation of pressure may allow a small leakage of fluid and permit the pipe to be rotated or moved within the B.O.P.

- Blind and Shear Rams (Figure 30b and c). Hydraulic rams closing from opposite sides will completely close the borehole when no pipe is in the hole. If closed on drillpipe or casing, shear rams will cut or crush it.

- Pipe Rams (Figure 3-30a). These are similar to blind rams, but the rubber face of the ram is molded to fit a certain size of pipe. If more than one size of drillpipe is in use, there must be one set of pipe rams for each size of pipe.

The hydraulic pressure used to close the blowout preventers is supplied by accumulators containing high-pressure nitrogen. In an emergency, the ram preventers may be closed manually on land jobs and jackups. The B.O.P. stack also includes several other components necessary for controlling the well:

- Conductor pipe. A length of pipe is attached to the top of the B.O.P. to extend the annulus to the drillfloor. The top of this pipe is belled-out (nippled) to prevent tools from "hanging up" when lowered through the rotary table. To one side of the conductor pipe is attached the mud return line (flowline) by which the mud is carried to the shale shaker.

- Choke Line. After the B.O.P. is closed, high-pressure fluid can be released at a carefully controlled rate by use of a hydraulically controlled valve below, while heavier drilling fluid is introduced through the kelly or kill line.

- Kill Line. Heavy mud can be introduced through a check-valve in order to control high formation pressure. It is also used to fill the annulus when pipe is being tripped out, or in the event of lost circulation.

The B.O.P. stack is not always situated directly under the rig floor as indicated in Figure 3-28. On floating offshore rigs, the B.O.P. stack sits on the seabed. The various positions of the B.O.P. stack are governed by the type of rig. This is illustrated in Figure 3-31.

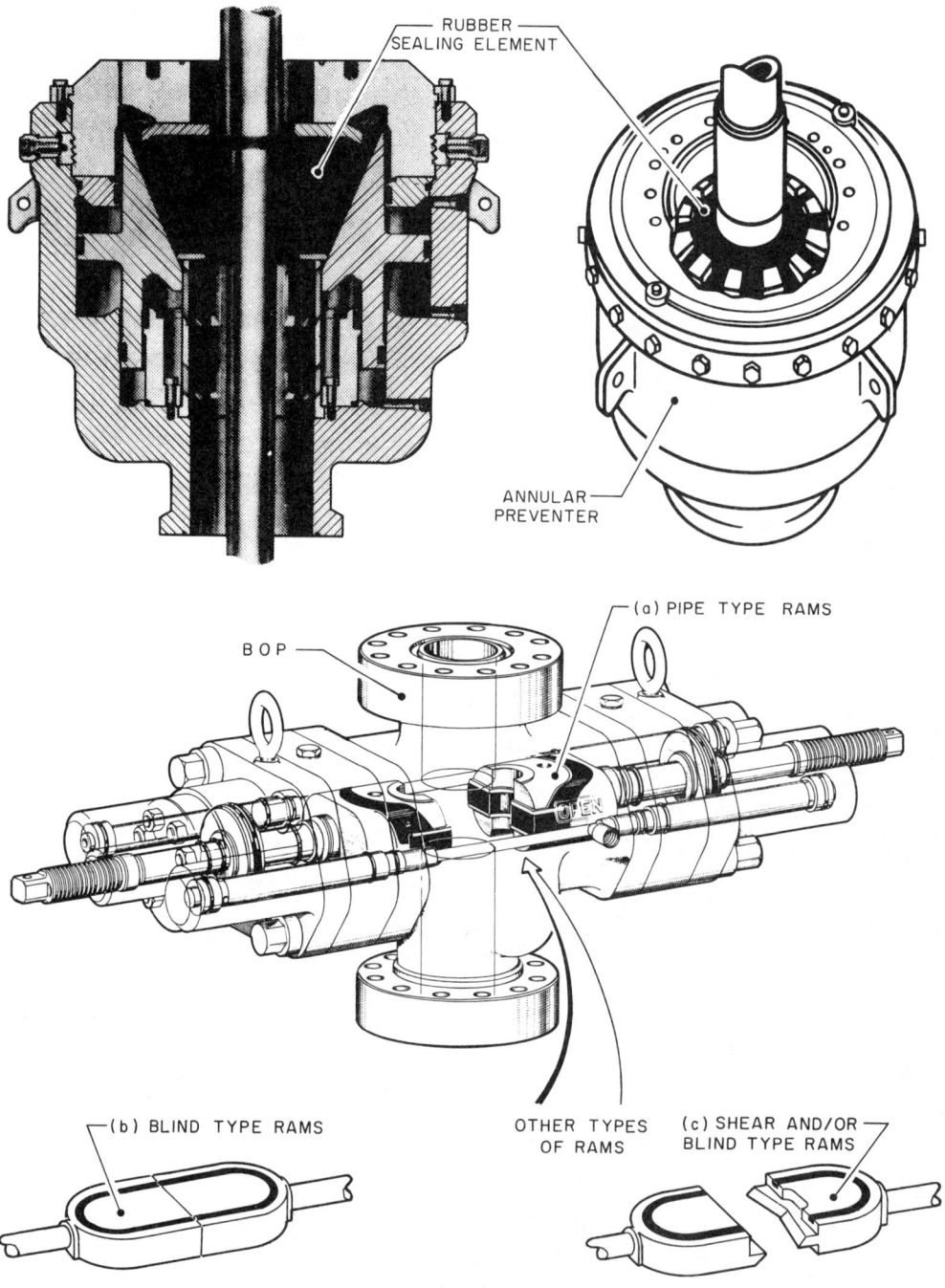

Figure 3-30. Blowout Preventers

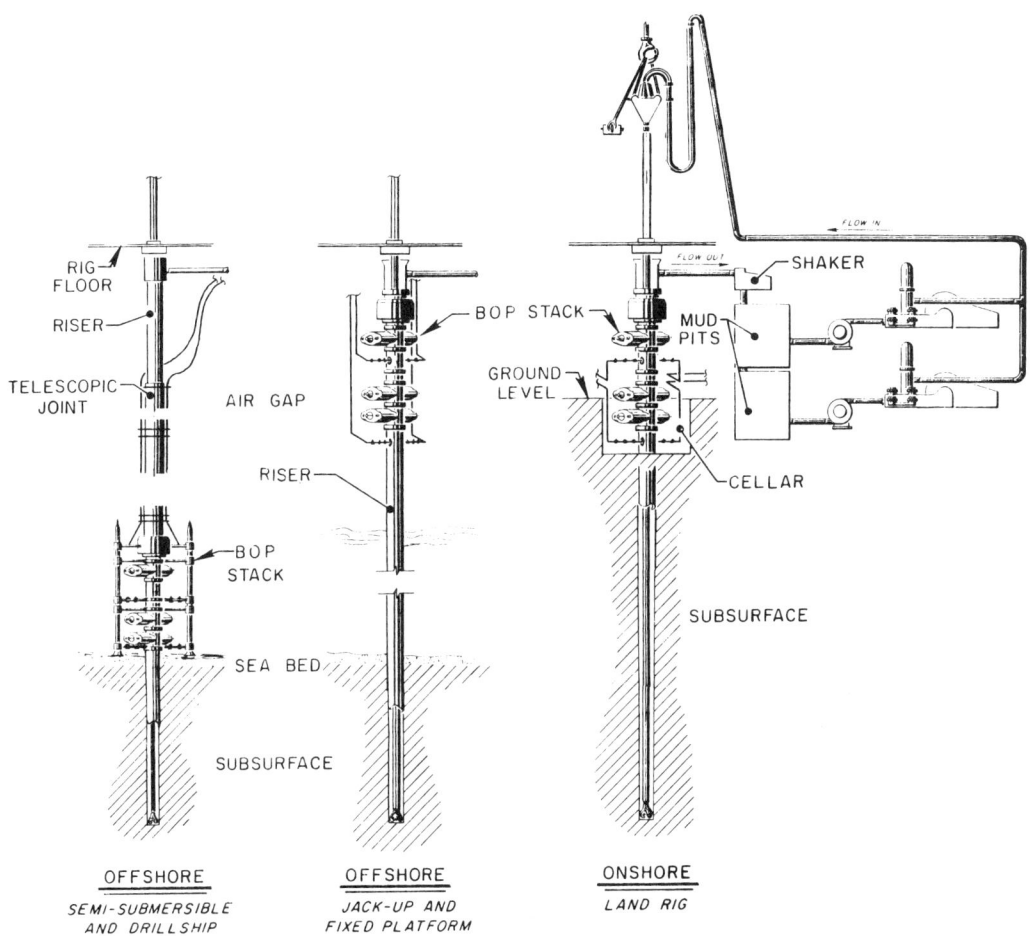

Figure 3-31. B.O.P. Stack Positions Governed by Rig Type

4
DRILLING & COMPLETING A WELL

4.1 ROUTINE DRILLING

The drilling contractor is employed by the oil company to routinely drill ahead (make hole). However, a number of related services must be performed in order to properly drill the hole and complete the well, whether it is a dry hole or a producer.

After the well is spudded, routine drilling consists of continuously drilling increments the length of one joint of pipe, making connections (adding to the drillstring another single joint of drillpipe generally 30 or 45 feet in length), and continuing until it is time to change the bit. The bit must be changed when it is worn or when a formation is encountered for which the particular bit being used is not suitable; see Section 3, paragraph 3.22. Changing the bit is accomplished in an operation called "making a trip." A round-trip includes coming out of the hole, changing the bit, and going back into the hole.

4.2 CONNECTIONS

When the kelly has been drilled all the way down, it is withdrawn and a new length (or joint) of drillpipe is added. Refer to Figure 4-1 (a through d) for the following text description.

In (a) the kelly is nearing the "kelly down" position, where another joint of pipe must be added. The new length of drillpipe has been placed in the mousehole, ready to be connected. (The mousehole is used to stow the next joint of pipe until it is required for threading into the drillstring.) The crew breaks out the kelly so that it can be swung over to the joint of pipe in the mousehole as in (b), made up on the joint, and tightened with tongs. In (c) and (d) you can see the new joint of pipe being picked up, "stabbed" and spun into the pipe hanging in the rotary. This is tonged-up tightly before being lowered into the hole to drill another joint length.

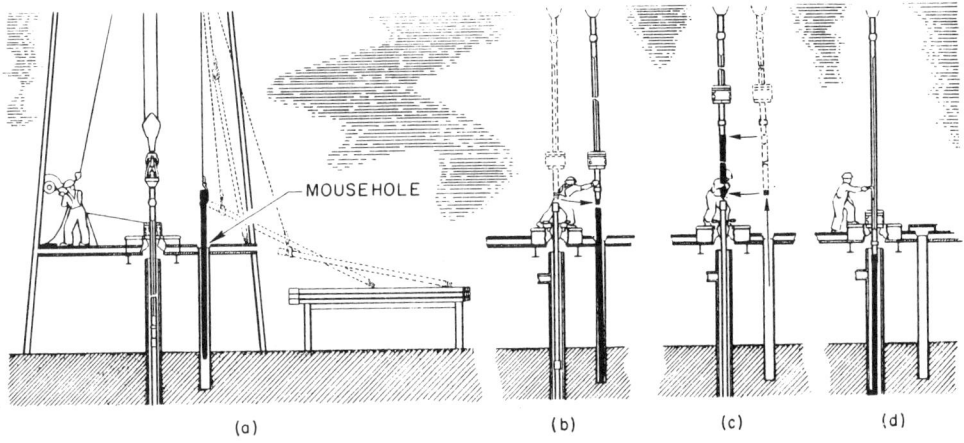

Figure 4-1. Making a Connection

4.3 TRIPS

When making a trip, drillpipe is handled in "stands" of usually two or three joints each (i.e, approximately 90 ft). Smaller rigs used for drilling to moderate depths (less than 5000 ft) will generally handle the drillpipe as two 30-ft-joint stands. Pipe is removed from the hole and stood back on the floor. Refer to Figure 4-2 and follow the procedure for pulling out of the hole (tripping out). The kelly, rotary bushings and swivel are placed in the rathole (used to stow the kelly when not in use) during the trip, as seen in (a). With the kelly out of the way, the elevators (which are part of the hook and block assembly) are latched around the pipe just below the tool joint box.

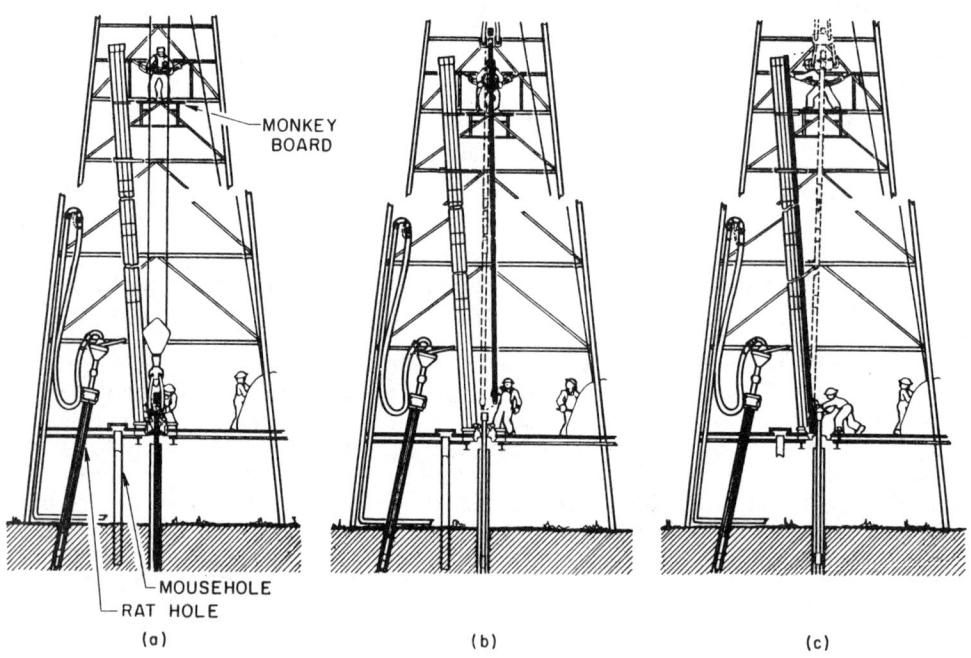

Figure 4-2. Pulling Out of the Hole (Tripping Out)

The tool joint box provides a shoulder for the elevators to pull against. The pipe is then pulled from the hole, and, after being secured in the rotary table (using slips), the connnection is loosened with the breakout tongs. In some instances the pipe in the rotary is spun in reverse until the connection separates, but commonly a spinning wrench (power tongs) is used. This is illustrated in Figure 4-3.

The top of the stand which has been pulled past the derrickman (standing on the monkey board, Figure 4-2c) has a rope around it. The bottom of the stand is swung to one side of the drillfloor where it is set down (as in Figure 4-2c), and the derrickman racks the top of the stand in fingers in the derrick to secure it. The

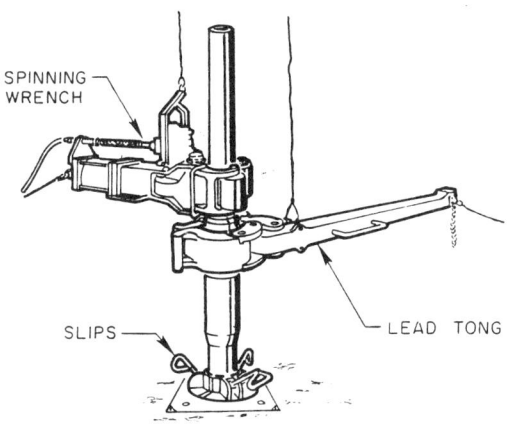

Figure 4-3. Separating a Connection

drill collars and bit are the last to come out of the hole. A bit breaker is placed in the rotary table, and, by use of the breakout tongs, the bit is loosened and removed from the bottom collar. "Tripping in" the hole is just the reverse procedure of tripping out.

Some rigs have a pipe handling system (Figure 4-4). This system provides three racker arms mounted at different elevations in the derrick. Each racker consists of a carriage which travels across the width of the derrick on tracks, an arm which is powered toward and away from the centerline of the well through the carriage, and a racker head which is fitted to the end of the arm. The head closes around the drillpipe or collar to stabilize it; and in the case of the intermediate racker, the head is actually used to lift the stand for setback.

4.4 RELATED SERVICES

Services required at the wellsite include those of specialty companies whose contracts are separate from that made with the drilling contractor. Some of these are:

- Drilling Fluid (Mud) Engineering
- Drill Returns (Mud) Logging
- Wireline Logging
- Casing
- Cementing
- Fishing
- Directional Drilling
- Formation Testing

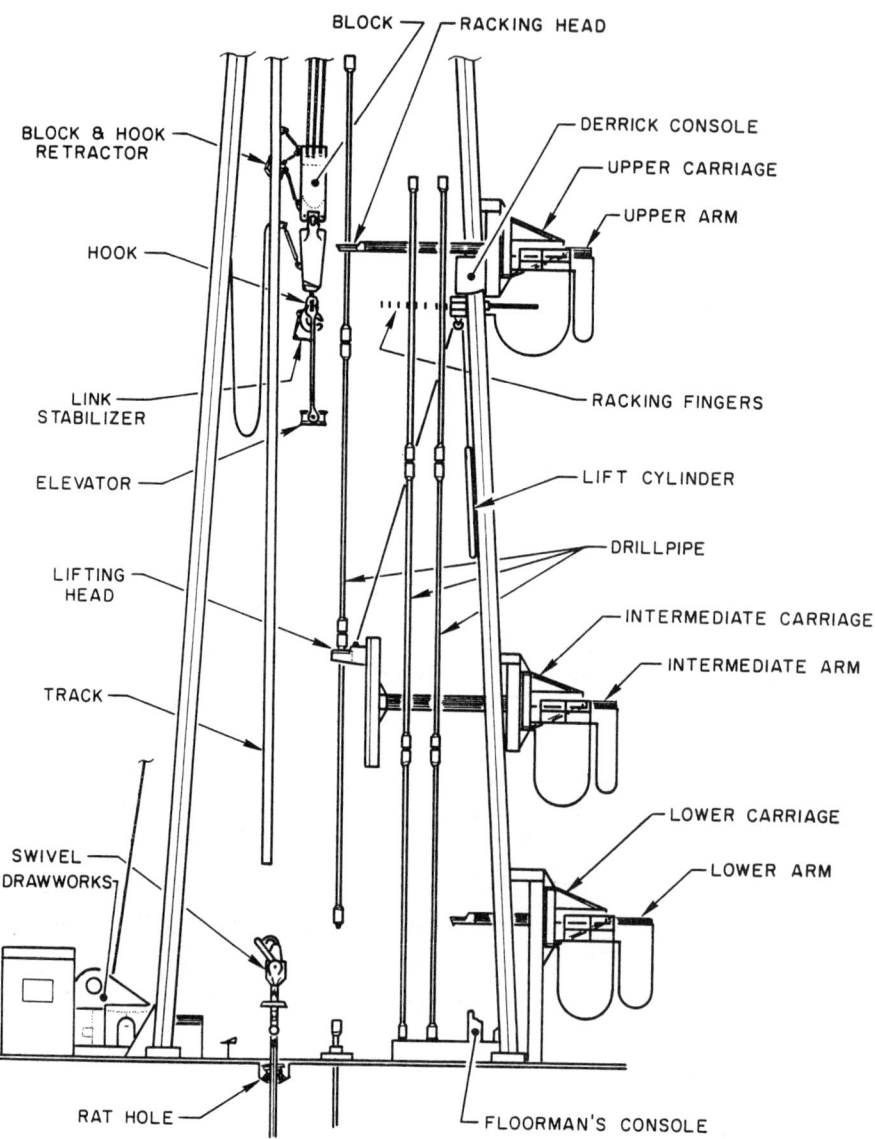

Figure 4-4. Pipe Handling System

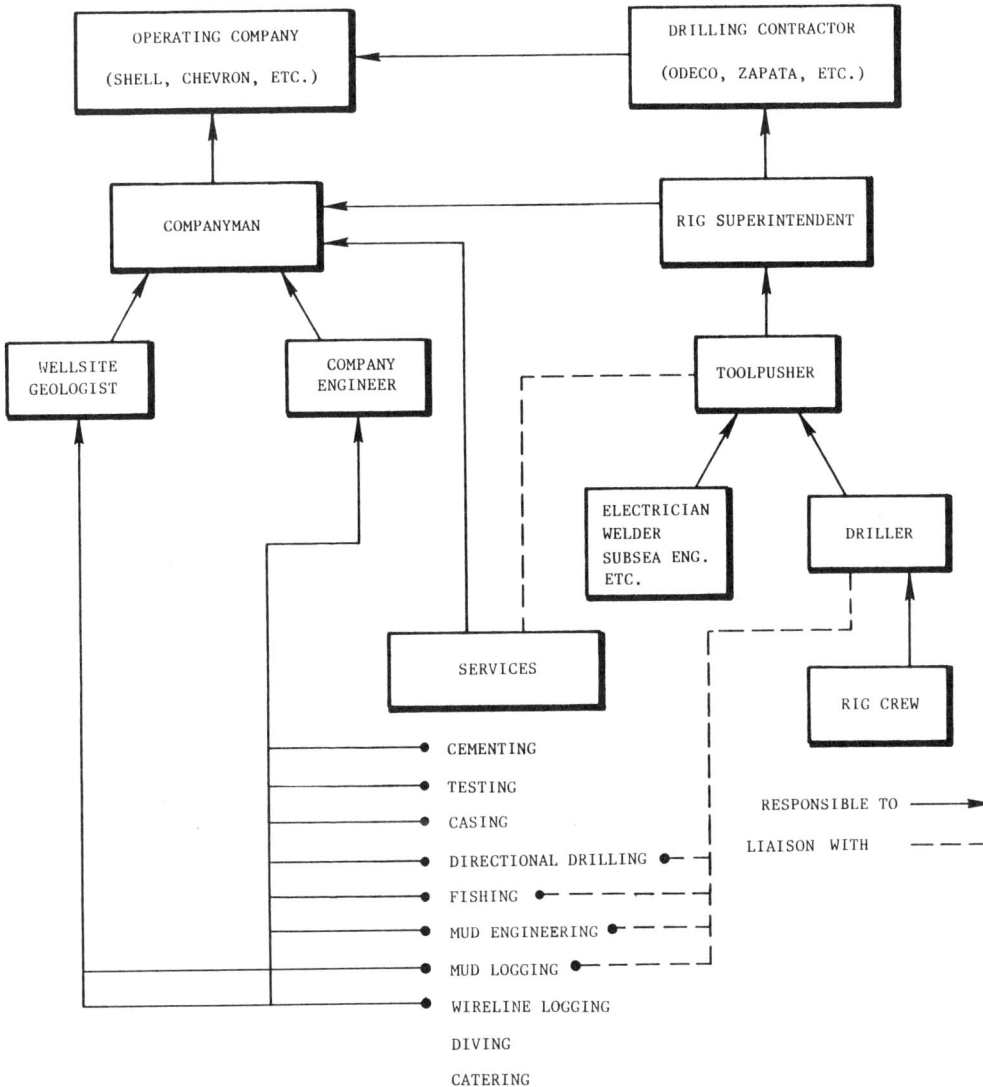

Figure 4-5. Interrelationship of Operator, Drilling Contractor and Service Companies

These services are coordinated by the operating company (the oil company client). The interrelationship of the client, drilling contractor and service companies is illustrated in Figure 4-5. Drill returns logging, wireline logging and formation testing are explained in Section 5, Formation Evaluation Procedures. The other services are discussed in this section.

4.5 DRILLING FLUIDS ENGINEERING

Initially, the primary purpose of drilling fluid was to clean, cool and lubricate the bit and continuously remove cuttings from the hole. But with progress came sophistication, and more was expected from the drilling mud. Many additives for almost any conceivable purpose were introduced, so what started out as a simple fluid has become a complicated mixture of liquids, solids and chemicals for which "mud engineers" are contracted by the operating company to maintain. Today, the drilling fluid must permit the securing of all information necessary for evaluating the productive possibilities of the formations penetrated. The fluid's characteristics must be such that good cores, electric logs and drill returns logs can be obtained. Drilling fluid is discussed here from the standpoints of

- Technology (4.6)
- Chemistry (4.11)
- Mud conditioning equipment (4.15)

4.6 Drilling Fluid Technology

Numerous types of mud are available due to the varied hole conditions. Factors such as depth of the well, type of formation encountered, local structural conditions, etc., all enter into the choice of a particular mud. The functions and corresponding properties of a drilling fluid are to

- Control subsurface pressures and prevent caving (density; mudweight)
- Remove cuttings from the hole (viscosity)
- Suspend cuttings when circulation stops (gel strength)
- Cool and lubricate the bit and drillstring (oil and additive content)
- Wall the hole with an impermeable filter cake (filtrate; water loss)
- Release the cuttings at the surface (viscosity/gel strength)
- Help support the weight of the drillstring and casing (density)
- Ensure maximum information from the formation
- Do all the above without damage to the system (low sand, corrosive inhibition, etc).

4.7 Controlling Subsurface Pressures: The pressure of the mud column at the bottom of the hole is a function of the mud density and column height. This pressure must be adequate at all times to prevent the flow of formation fluid into the mud column. Should mud density fall below that which is necessary to hold back formation pressures, then formation fluids can enter the well. <u>This is termed a "kick."</u> If this condition is allowed to continue unchecked for even a short time, the mud density may reduce (cut) severely and an uncontrollable flow will result. <u>This is termed a "blowout."</u> On the other hand, it is not practical or economical to have the mudweight too high. Excessive mudweights result in low rates of penetration and in the fracturing of "weak" formations and cause loss of drilling fluid into them (lost circulation). Density is also important in preventing unconsolidated formations from caving into the hole.

1) The effect of mudweight on drill returns logging

    - Hydrostatic pressure in excess of formation pressure will cause gas or oil to be flushed back into the formation being penetrated either at the bit or just ahead of it. This flushing occurs at all times whether marginally or greatly overbalanced. If circulation is lost, the cuttings and drilling fluid and any oil or gas they may contain will be lost.

    The way in which the lost circulation zone behaves generally indicates the type of porosity of the formation into which the fluid is being lost. Three types of formation can be recognized:

    (a) <u>Coarse, permeable unconsolidated formations</u>: There is normally some loss by filtration into these formations until an impermeable filter cake is formed. If pore openings are large enough, then loss of mud occurs. Other than in extreme cases this is a slow, regular seepage loss, while partial returns are maintained.

    (b) <u>Cavernous and vugular formations</u>: Loss is usually sudden and of finite amount, after which full returns are maintained. Continuous loss indicates that the formation is fissured.

    (c) <u>Fissured or fractured formations</u>: Fractures may be natural or initiated and opened by hydrostatic pressure. Losses are large and continuous.

    - Formation pressure that approximates to or is greater than the hydrostatic pressure may allow entry of formation fluids, depending upon permeability. In low permeability formations (e.g., shales), caving may occur and make cuttings analysis difficult.

The following are some definitions of pressure-control terminology.

2) Hydrostatic pressure

    This is the pressure which exists due to the unit weight and vertical height of a column of fluid. Size and shape of the fluid have no effect.

    $$P = 0.0519 \times W \times D \tag{1}$$

    where

    P = hydrostatic pressure (psi)
    W = mudweight (ppg, pounds per gallon)
    D = vertical depth (ft)

3) Pressure gradient

    This is the rate of change of hydrostatic pressure with depth for any given unit of fluid weight. That is,

$$\text{Pressure gradient} = \frac{P}{D} = 0.0519 \times W$$

For fresh water: W = 8.33 ppg
Gradient = 8.33 × 0.0519 = 0.432 psi/ft

Typical formation water: W = 8.6 ppg
Gradient = 8.6 × 0.0519 = 0.446 psi/ft

The value 8.6 ppg is an average used worldwide, and therefore may not fit local conditions. However, this value should be used until a local value is determined.

4) <u>Equivalent Mudweight</u>

From Eq. (1), $$W = \frac{P}{0.0519 \times D} \tag{2}$$

In pressure control it may be necessary to calculate total pressure exerted by a column of fluid and a pressure imposed on the top of the column. The equivalent mudweight is the mudweight which would exert a hydrostatic pressure equal to the sum of the imposed pressure and the hydrostatic pressure, and can be found by substituting the sum of the two pressures in Eq. (2) or by converting the imposed pressure to mudweight and adding it to the actual mudweight. Thus,

$$W_e = W_o + \frac{P}{0.0519 \times D} \tag{3}$$

where

W_e = equivalent mudweight (ppg)
W_o = actual mudweight (ppg)
P = imposed pressure (psi)
D = vertical depth (ft)

Figure 4-6 illustrates two situations. In the upper diagram, pressure is felt at the surface due to a difference in fluid column height (artesian situation). In the lower diagram the column heights are equal, but pressure is felt because of a difference in column density. In practice these two may often be combined.

5) <u>Apparent and effective mudweight</u>

Mudweight measured at the pits is the apparent mudweight going into the hole, and will exert a hydrostatic pressure indicated by Eq. (1). However, this is a static pressure — and if the mud is circulated, additional pressure is placed against the formation due to frictional effects in the mud, causing a "backup pressure". This additional pressure can be estimated by calculating the pressure loss in the annulus using Bingham's formula for laminar flow.

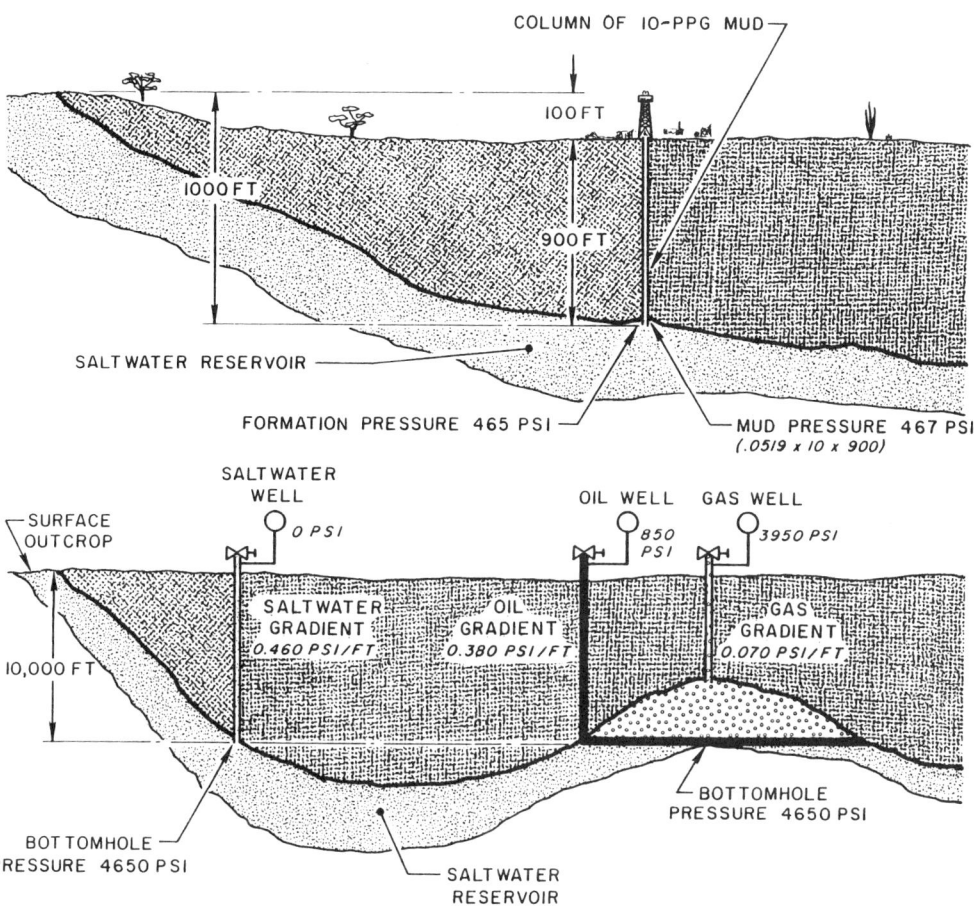

Figure 4-6. Formation Pressures

$$\text{Pressure loss (psi)} = \frac{PV \times V \times L}{60,000\,(d_h-d_p)^2} + \frac{YP \times L}{200\,(d_h-d_p)} \quad (4)$$

where

- L = section length (ft)
- YP = yield point (lb/100 ft^2)
- (d_h-d_p) = hole diameter minus pipe outside diameter (inches)
- PV = plastic viscosity
- V = annular velocity in laminar flow (ft/min)

Each section of annulus should be considered separately and losses summed to total loss.

The effective mudweight (or effective circulating density, ECD) is the equivalent mudweight for the sum of the hydrostatic pressure plus the pressure loss in the annulus. At any point in the hole, the total pressure is the sum of the hydrostatic pressure at that point plus the pressure lost in the annulus above that point. Thus in Eq. (3),

$$ECD = W_o + \frac{\text{pressure loss}}{0.0519 \times D}$$

and in Eq. (1),

$$\text{Bottomhole circulating pressure (BHCP)} = ECD \times 0.0519 \times D$$

Similarly, when pipe is pulled, the pressure on the formation is reduced by an amount of similar magnitude to the pressure loss in the annulus. That is,

$$W_e = W_o - \frac{\text{pressure loss}}{0.0519 \times D}$$

A safe rule of thumb is to use an actual mudweight required to balance the formation pressure and add to this a mudweight equivalent to twice the annular pressure loss. That is,

$$\text{Weight for safe trip} = W_o + 2 \frac{(\text{pressure loss})}{0.0519 \times D}$$

6) Pore pressure

Pore pressure is that exerted by fluids contained in the pore space of the rock and is the strict meaning of what is generally referred to as formation pressure. All rocks have porosity to some extent. If permeability also exists and formation pressure is greater than mud hydrostatic pressure, pore pressure will tend to flush fluids out of the pore space and into the wellbore. In impermeable formations (such as shales), only the fluids in the formation immediately adjacent to the borehole will enter; pore pressure, therefore, will be confined and will produce sloughing or caving.

7) Overburden pressure

The combined weight of formation solids and fluids in the formation exerts a total pressure. This overburden pressure increases with depth.

$$S = \partial + G$$

where

S = overburden pressure gradient (psi/ft) (commonly called O.B.G.)
∂ = rock grain pressure gradient (psi/ft)
G = pore pressure gradient (psi/ft)

8) Normal formation pressure

Normal formation pressure is equal to the hydrostatic pressure exerted by all the fluids above the depth of interest. Since water is the most common fluid, normal formation pressure is equal to the hydrostatic pressure due to a column of water equal to the depth of the formation. Similarly, since density of water varies with salinity, normal formation pressure gradient varies. For example,

Fresh water gradient = 0.432 psi/ft
Marine area gradient = 0.446 psi/ft

Since salinity varies with depth, the average value may not be valid for all depths; and in <u>any</u> area, a log-data-derived pressure/depth profile must be found.

9) Abnormal formation pressure (overpressure)

Formation pressures greater than normal can exist for a variety of reasons, some of which are listed below.

- Formation may extend to the surface at an elevation higher than ground elevation at the well, as in Figure 4-7. This shows the relation of the piezometric surface (the surface to which the water from a given aquifer will rise under its full head) to the surface of the ground as a cause of either apparent "excess" or deficient pressure for the depth of the reservoir below the surface of the ground.

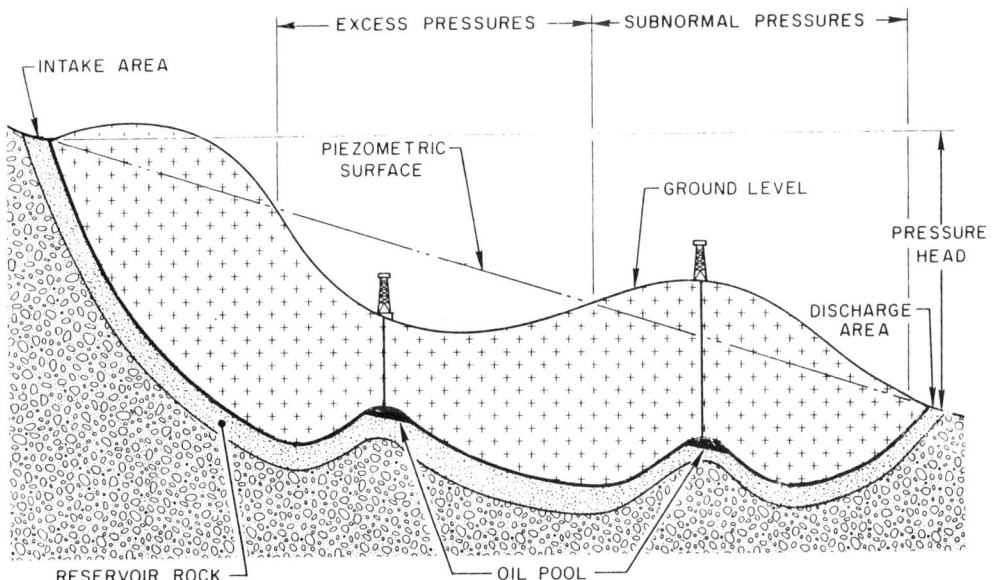

Figure 4-7. Relationship Between Piezometric Surface and Ground Level

- Formation may be abnormally pressured from another higher pressured formation due to interconnection or due to fluid injected during secondary recovery operations. This is the most dangerous situation since, frequently, formation pressure is determined only after the formation has been penetrated.

- Rapid burial of sediments of low permeability can result in "contained" abnormal pressure. Under normal conditions, water is squeezed out of shales as successive layers are deposited so that fluid content decreases with depth of burial; but if rate of burial exceeds rate of extrusion of water, the overburden is in part supported by water, with a consequent increase in pore pressure. Permeable formation interbedded with shales tends to assume the same pressure existing in the shale. Several methods of evaluation of this type of abnormal pressure have been developed and are explained in the Pressure Log manual, MS-156.

- A chemical seal formed by migration and precipitation of dissolved minerals may prevent the normal release of water and cause an abnormally pressured situation.

- Faulting, folding or diapiric salt intrusion may cause burial before complete dewatering has occurred and hence cause an abnormally pressured situation.

- Sealed, deep formations may be uplifted into a lower pressure environment, yet still retain their fossil-pore pressure.

- Diagenesis or metamorphism in a closed system can produce fluid that becomes trapped and therefore overpressured. For example:

  — volcanic ash (methane and CO_2/clay minerals)
  — rehydration of anhydrite
  — diagenesis of clay minerals
  (montmorillonite + heat → illite + water)
  — metamorphism of limestone

- Water drawn into evaporite sections by osmosis.

10) Subnormal formation pressures

Pressures lower than normal are encountered where

- Partial depletion of the formation has occurred
- Surface exposure is lower than the wellsite elevation (Figure 4-7)
- A shallow, low-pressured formation has been buried at depth

4.8 Removing and Suspending the Cuttings: The drilling fluid must carry the cuttings up the hole and suspend them when circulation is stopped. The most important factors involved are the speed at which the mud travels up the hole (annular velocity) and the viscosity and gel strength of the drilling fluid. Annular velocity is discussed in Section 5, paragraph 5.7

1) Viscosity

   As applied to drilling fluids, viscosity may be regarded basically as meaning the resistance that the drilling fluid offers to flow when pumped. The viscosity affects the ability of the drilling fluid to lift the rock cuttings out of the hole. The viscosity is dependent on the amount and character of the suspended solids. Viscosity is ordinarily measured in the field by means of a Marsh funnel. The funnel is filled with one quart of mud, and the elapsed time taken to empty itself is noted and recorded in seconds — as "funnel viscosity." This value can range from 20 to 80 but is normally maintained between 40 and 50.

2) Gel strength

   Gel strength refers to the ability of the drilling fluid to develop a gel as soon as it stops moving. Its purpose is to hold the cuttings and the material used to weight the mud in suspension, while they are in the hole, and not permit them to settle around the bit when circulation is halted.

   In general, the gel strength should be low enough to

   - Allow the cuttings to be removed at the shakers and settling pits
   - Permit entrained gas to escape at the surface
   - Minimize swabbing when pipe is pulled from the hole
   - Permit starting of circulation without the use of high pump pressure

   The gel strength is most commonly determined with a Fann viscometer or VG meter and expressed in pounds per 100 sq ft. Drilling muds ordinarily have gel strengths between 5 and 30 pounds per 100 sq ft.

3) Effect of viscosity and gel strength on drill returns logging

   If the viscosity or gel strength (or both) is too high, the drilling fluid tends to retain any entrained gas as it passes through the surface mud system, with the effect that the gas may be recycled several times. Swabbing of the hole may also introduce extraneous gas anomalies (see Section 5, 5.16, third paragraph).

   Fine cuttings may be held in suspension so they cannot be removed at the shakers or settling pits, thus recycling and contaminating the cuttings samples. Also, cuttings consisting of clays or other dispersible material may be dissolved.

4.9 Cooling and Lubricating the Bit and Drillstring: Practically any fluid that can be circulated through the drillstring will serve to cool the bit and drillstring; however, lubrication commonly requires special mud characteristics that are gained by adding oil, chemicals and other material.

4.10 Walling the Hole with an Impermeable Filter Cake: The hydrostatic pressure of the column of drilling fluid exerted against the walls of the borehole helps prevent the caving of unconsolidated formations. A "plastering" effect, or the ability to line permeable portions of the hole with a thin, tough filter cake, is also produced.

Control of the filtration rate (water loss) is necessary for two reasons:

1) A poor quality filter cake may cause excessive water loss and produce an excessively thick filter cake thereby reducing the diameter of the hole and increasing the possibility of sticking the drillpipe or swabbing the hole when pulling the pipe.

2) High water loss can cause deep invasion, making it difficult to interpret electric logs. See Section 5, paragraph 5.57 on wireline logs.

4.11 Drilling Fluid Chemistry

Drilling fluids are intended to fulfill the functions described in paragraphs 4.8 through 4.10. While the list of chemical additives used to develop these functions is extensive, there are only three basic mud types (they are described in paragraphs 4.12 through 4.14):

- Water/clay muds
- Water/oil/clay muds
- Compressed gases

4.12 Water/Clay Muds: This is the major type of mud used. It consists of a continuous liquid phase of water in which clay materials are suspended. A number of reactive and nonreactive solids are added to obtain special properties. Water-based mud is a three-component system consisting of water, reactive solids and inert solids.

1) Water component

   This may be fresh or salt. Seawater is commonly used in offshore drilling, and saturated saltwater may be used for drilling thick evaporite sequences to prevent these sequences from dissolving and causing oversized hole (washouts). Saturated saltwater is also used for shale inhibition.

2) Reactive solid component

   - Clays: basic material of mud commonly referred to as "gel"; affects viscosity, gel strength and water loss. Some of the clays are:

- bentonite for fresh-water mud
- attapulgite for saltwater mud
- natural formation clays which hydrate and enter the mud system

- Dispersants: reduce viscosity by adsorption <u>onto</u> clay particles and reduce atttraction between particles (for example, tannins, quebracho, phospates, lignites, and lignosulphonates).

- Filtration control agents: control the amount of water loss into permeable formations due to pressure differential by ensuring the development of a firm impermeable filter cake. Some are listed below:

  - starch: pregelatinizèd to prevent fermentation.

  - sodium carboxy-methyl cellulose (CMC): organic colloid, long chain molecule which can be polymerized into different lengths or "grades." The grades used depend upon the desired viscosity.

  - polymers: for example, cypan, drispac; used under special conditions.

- Detergents, emulsifiers and lubricants: assist in cooling and lubricating. Also used for spotting in order to free differentially stuck pipe.

- Defoamers: prevent mud foaming at the surface in treatment equipment.

- Sodium compounds: precipitate or suppress calcium or magnesium which decrease the yield of the clays.

- Calcium compounds: inhibit formation clays to prevent hydration and swelling.

3) <u>Inert solid component</u>

- Weight material: finely ground high-density minerals held in suspension to control mud density (e.g., barites, galena).

- Lost circulation material (LCM): added to the mud in order to bridge over or plug the point of loss. It is available in many sizes and types in order to suit the particular circulation loss:

  - Fibrous
     wood fiber
     leather fiber
  - Granular
     walnut shells (fine, medium, coarse)
  - Flakes
     cellophane
     mica (fine, coarse)

— Reinforcing plugs
bentonite with diesel oil
time setting clay
attalpulgite and granular material (squeeze)

If none of these materials successfully plug the point of loss, the zone must be cemented off.

- Antifriction material: added to the mud to reduce rotary torque and decrease the possibility of differential sticking. The most frequently used material is inert silica spheres (e.g. Lubraglide). More frequently it is used on high angle directional holes where torque and differential sticking are a problem. Mud loggers must be careful with sand descriptions when this material is being used.

4.13 Water/Oil/Clay Muds: Two types of water/oil muds are used:

1) emulsion (oil/water) system in which diesel or crude oil is dispersed in a continuous phase of water.

2) invert emulsion (water/oil) in which oil is the continuous phase and the water is dispersed.

These muds have desirable properties as completion fluids or when drilling production wells (e.g., nonreactive with clays, filtrate will not damage formation, etc.), but their high cost, difficulty of running and complication of geological evaluation preclude their use on exploration wells other than in certain very troublesome evaporite and clay sections. Apart from these emulsions containing roughly equal portions of oil and water, there are true oil-based muds which may contain only 5 percent water. When invert emulsion or true oil-based muds are in use, special consideration has to be made regarding formation evaluation. These considerations are included in "Mud Logging Techniques with Oil-Based Drilling Fluids" in Appendix C.

4.14 Compressed Gases: Compressed air or natural gas is occasionally used as a drilling fluid (often with a foaming agent to improve carrying capacity), but its use is applicable only in areas where there is little underground water. The compressed air or gas is circulated the same as conventional drilling mud, but occasionally may be circulated in reverse direction.

4.15 Drilling Fluid Conditioning Equipment

Drilling fluid returning from the wellbore contains drilled cuttings sand, other particles from the hole, and sometimes gas — all of which must be removed before the mud is suitable for recirculating in the well. Also, mud treatment clays and chemicals must be added to the mud system from time to time to maintain the required properties. The equipment necessary to perform these functions is presented and listed in Figure 4-8.

- Shale shaker
- Settling pit (sand trap)
- Desander and desilter
- Centrifuge
- Degasser
- Mixing hopper
- Suction pit

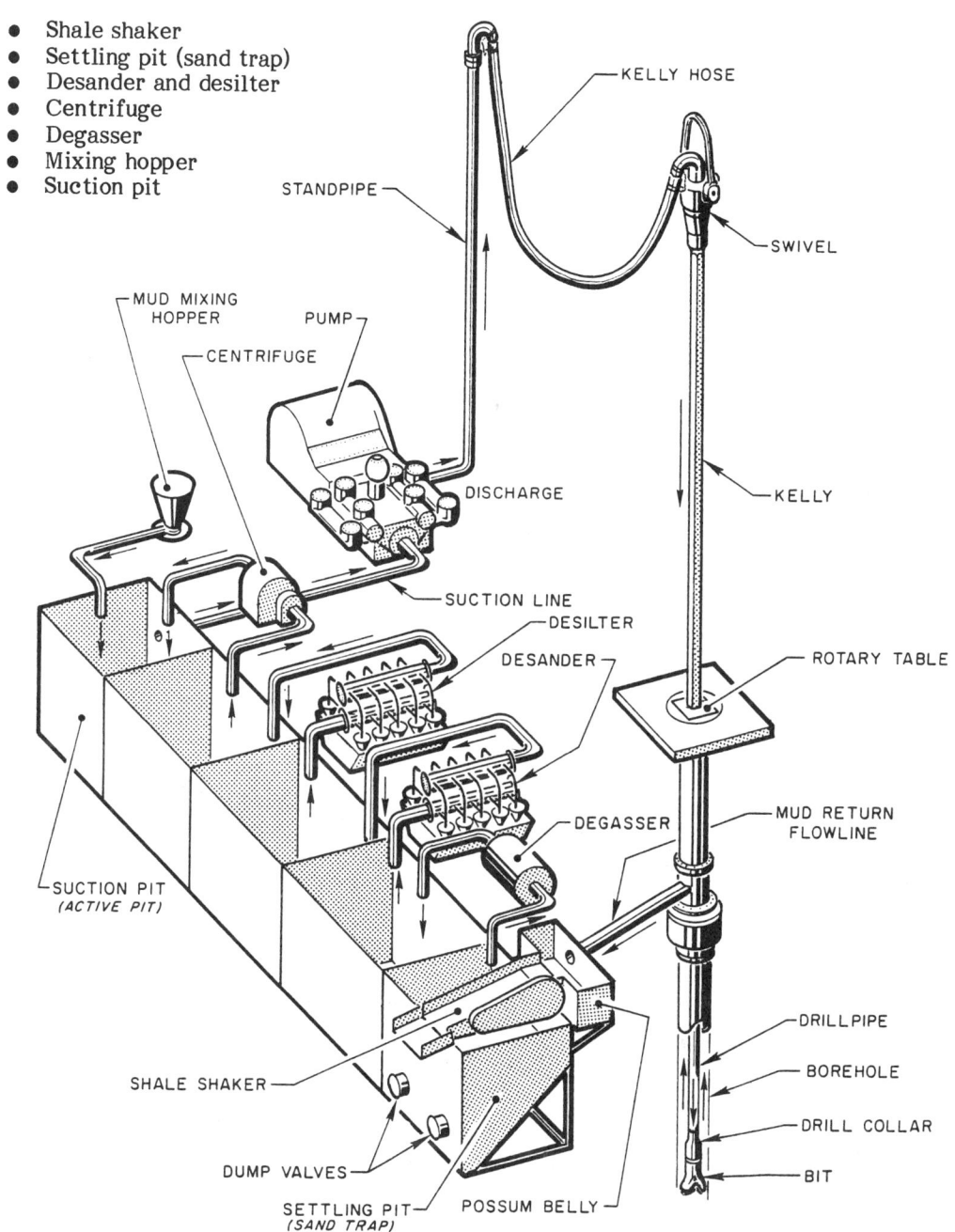

Figure 4-8. Mud Conditioning Equipment

4.16 Shale Shaker: Fluid returning from the hole immediately passes over the shale shaker which contains a sloping, vibrating screen. The mud falls through the screen and returns to the mud tanks, but large solids travel to the bottom edge of the screen where they are "dumped." Some, however, are collected for geological examination.

4.17 Settling Pit (Sand Trap): The first pit to receive the drilling fluid after it leaves the shakers is the sand trap. The bottom of a sand trap is usually sloped so that particles segregated by gravity settle toward cleanout valves which are opened periodically (usually during trips) so that the solids can be dumped.

4.18 Desander, Desilter and Centrifuge: The desander and desilter separate solids in a hydroclone in which the fluid rotates and the solid content is caused to separate by centrifugal force, as illustrated in Figure 4-9. A hydroclone (which has

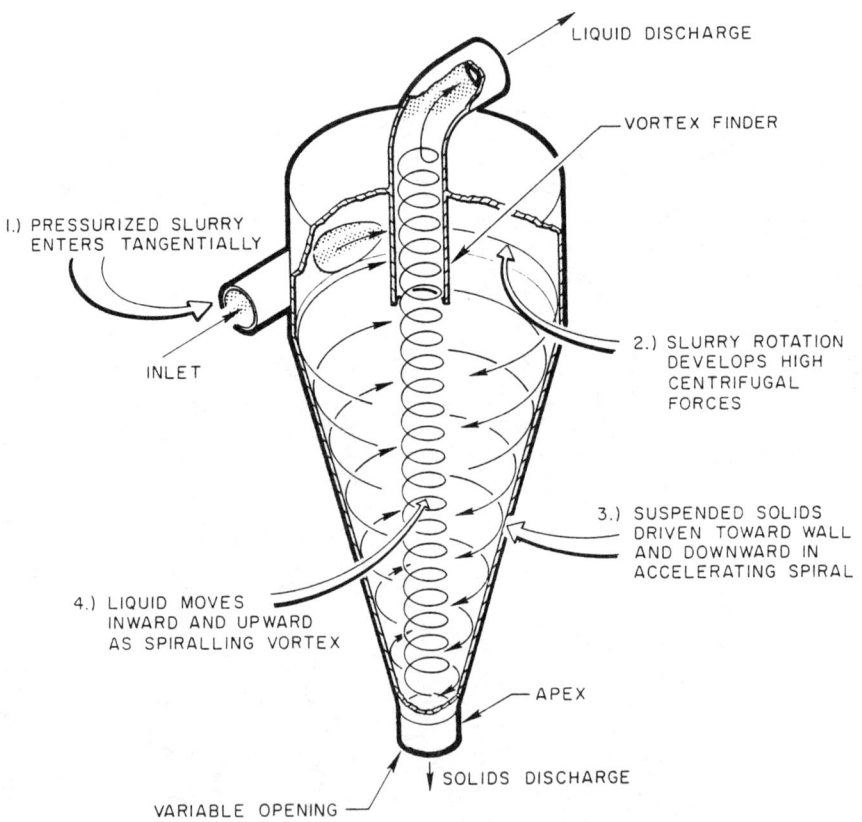

Figure 4-9. Hydroclone

no moving parts within itself) imparts a whirling motion to the fluid, thereby achieving sufficient centrifugal force to separate various particle sizes. A pump is used to feed mud through a tangential opening into the large end of the cone-shaped housing. A hydroclone operates in a similar manner, whether used as a desander, desilter or for the recovery of weighting materials. When being used as a desander or desilter, the underflow from the apex contains the coarse solids (which are discarded) while overflow, or effluent, is returned to the active mud system. Conversely, when the hydroclone is used on weighted muds, the underflow contains the barite which is saved and returned to the active mud system, while the effluent contains the clays and colloidal particles to be discarded. Hydroclones are used most often on low-weight water-based muds to remove coarse drilled solids. Individual cones are manifolded in parallel to provide any desirable throughput and are sized to fit the capacity of the pump provided for circulation through the cone units.

The centrifuge (Figure 4-10), like the barite salvage hydroclone, is used for salvaging materials that are to be retained in the mud system. The centrifuge, however, is much more efficient. It consists of a rotating cone-shaped drum (or bowl) that turns at a high rate of speed and a screw conveyor within the bowl that moves the coarse particles to the discharge port and back into the active system.

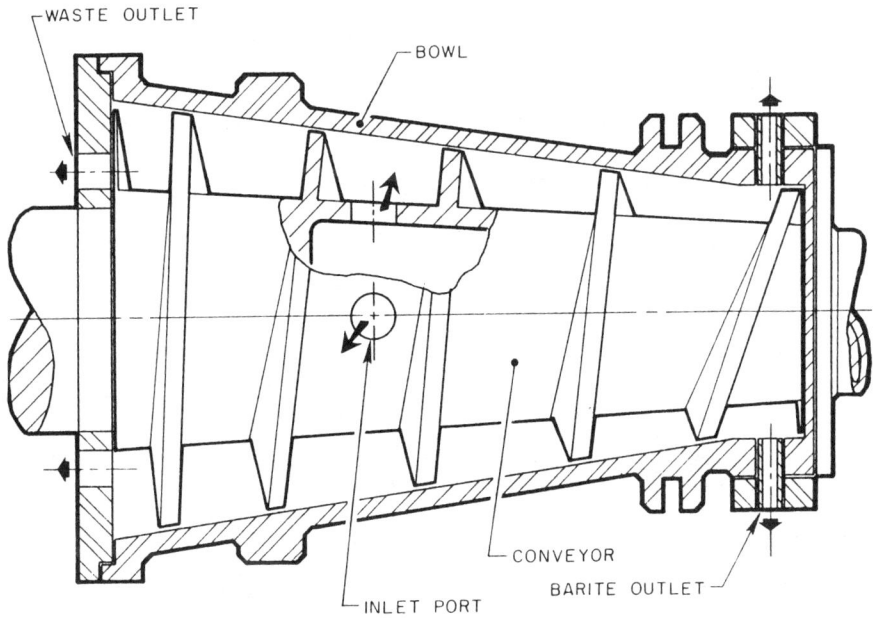

Figure 4-10. Centrifuge

4.19 Degasser: Recirculation of gas-cut mud may be hazardous and can result in reduced pumping efficiency and less hydrostatic pressure to contain formation pressure. The usual practices of running the returns mud across a shale shaker, using settling action in the pits and stirring the mud by paddle mixers or mud guns, may not completely release entrained gas from the mud. In this case, it may be necessary to run the mud into a degasser. Two general types of degassers are commonly employed:

- Mud-gas separators
- Vacuum degassers

A mud-gas separator is desirable to safely handle high-pressure gas and mudflows from a well when a "kick" takes place. The vacuum degasser is more appropriate for separating entrained gas, which resembles foam on the surface of the mud.

1) Mud-gas separators

   Various types of mud-gas separators are used. Most, however, consist of a vertical vessel arranged to vent free gas from the upper end and discharge relatively gas-free mud from the bottom. To operate a mud-gas separator, the hole must be shut in and mud circulated through the choke manifold; well flow is diverted from the flowline or choke manifold to the mud-gas separator. The separator releases the gas which is then carried by the vent line at the top to a remote flare.

2) Vacuum degasser

   This type of degasser (Figure 4-11) is mounted over a mud tank from which it takes suction. The mud enters near the top of a horizontal barrel and flows along a section of large pipe that is closed at its far end. The top of the pipe is sliced away in a horizontal plane so that the mud can spill over the sides and down an inclined plane extending the full length of the feed pipe and sloping downward. As the mud streams down the inclined plane, a vacuum in the vapor space causes the gases to to leave the mud and be withdrawn from the tank by a vacuum pump. The degassed mud, back to its normal weight, flows to the bottom of the barrel for exit.

4.20 Mixing Hopper: The most common hopper in use is the "jet hopper" (Figure 4-12); it was originally developed for mixing cement and water slurry for oilwell cementing. It is now used for adding material to mud to achieve the desired physical and chemical properties. In operation, a mixing pump or centrifugal pump circulates mud from the pit through the jet and back to the pit.

The fluid velocity through the jet nozzle lowers the static pressure in the housing to below atmospheric. A vacuum is created, and material placed in the hopper is sucked into the stream of fluid where it becomes mixed with the fluid. Powdered mud materials (such as bentonite, barites and chemicals, as well as solids such as cellophane, nut hulls and other pulverized material) can be added to the mud through a jet hopper.

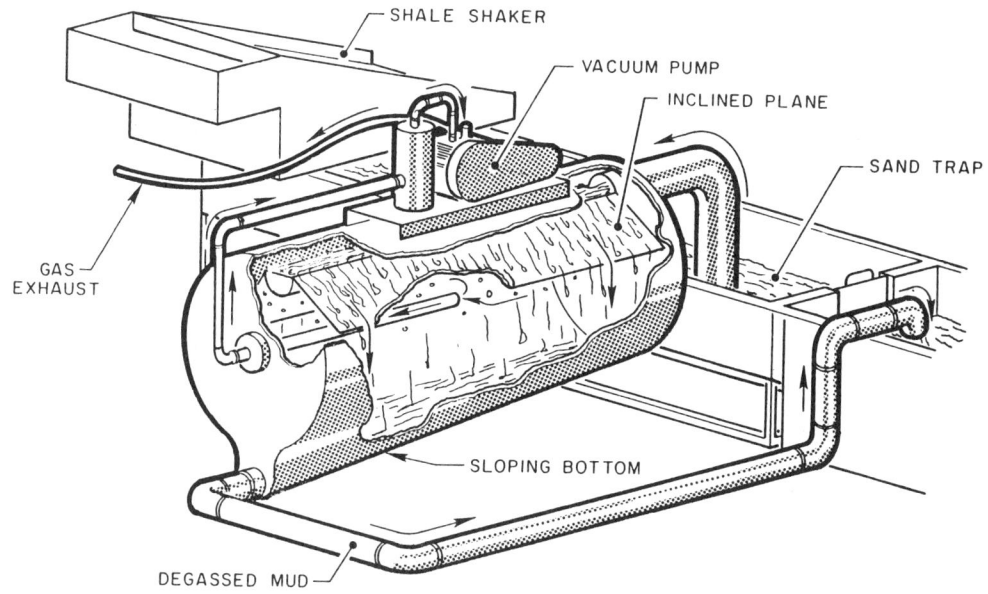

Figure 4-11. Vacuum Degasser

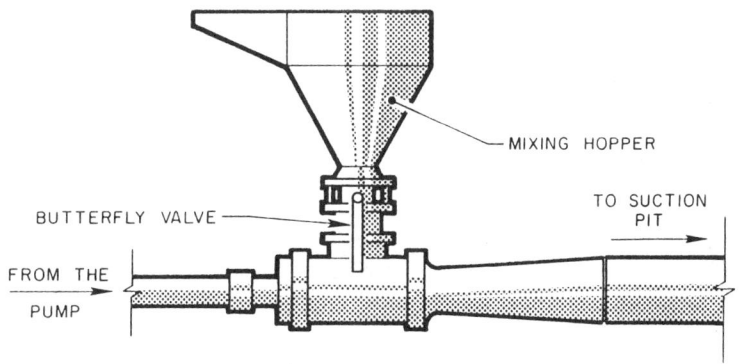

Figure 4-12. Jet Hopper for Mud Mixing

4.21 Suction Pit (Active Pit): Mud is stored and mixed here before returning to the borehole via the mud pumps and the kelly.

4.22 CASING AND CEMENTING

Drilling a hole to a gas or oil reservoir requires two operations. One is to drill the hole, and the second is to periodically line the hole with steel pipe (casing). Once installed, this casing is cemented in place to provide additional support and a pressure-tight seal.

4.23 Casing

Casing in a well has a number of functions:

- Prevents caving of the hole
- Provides a means of containing well (formation) pressure by preventing fracturing of upper, weaker zones
- Provides a means for attaching surface equipment (blowout preventers and production tree)
- Confines production to the wellbore
- Allows segregation of formations behind the pipe and thereby prevents interformational flow, and permits production from a specific zone
- Permits installation of artificial lift equipment for producing the well
- Provides a borehole of known diameter for further operations

One or more of the following strings of casing is required in every well:

- Conductor pipe
- Surface casing
- Intermediate casing
- Liner string
- Production casing (oil string)

These are illustrated in Figure 4-13. Casing size is defined by its outside diameter (o.d.) in inches, and its weight per foot (lb/ft).

4.24 Conductor Pipe: This is needed as a conduit to raise the circulating fluid high enough to return the fluid to the pit; it prevents washing out around the base of the rig; and it sometimes provides for attachment of a blowout preventer where gas sands may be encountered at shallow depth. Hole for the conductor can be drilled in the usual fashion, but often the pipe is driven with a pile driver (especially in swamps and offshore locations on a jack-up or fixed platform). The conductor pipe is usually from 30 to 42 inches in diameter offshore and 16 inches onshore.

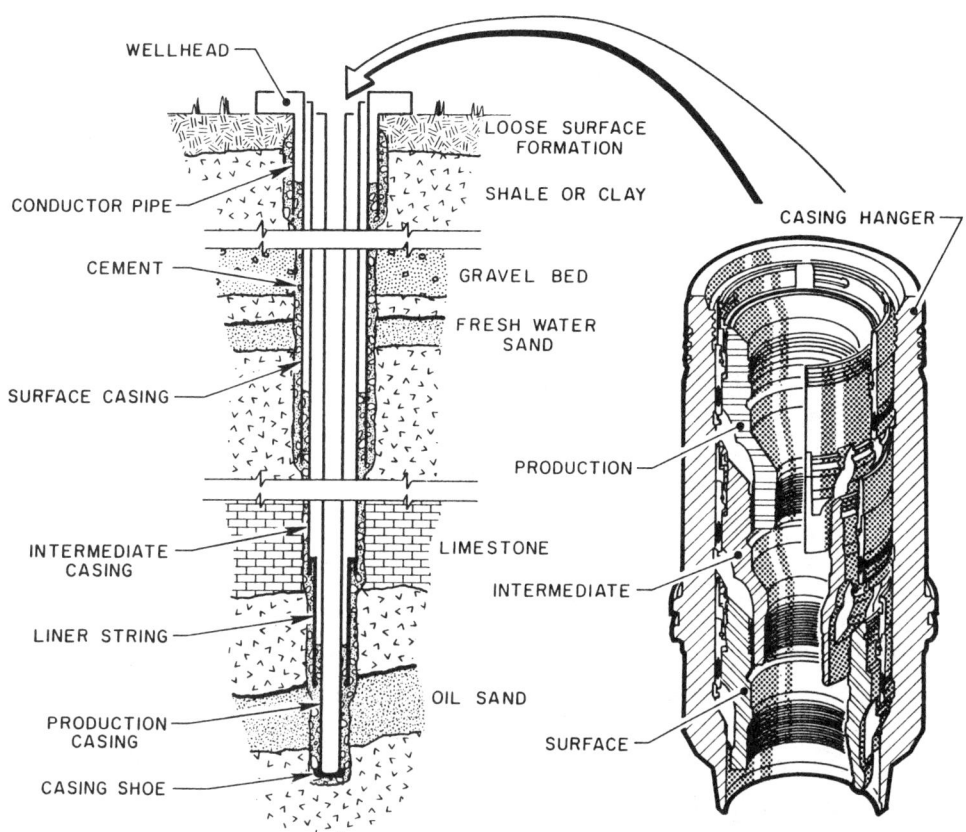

Figure 4-13. Casing Strings and Hanger

4.25 Surface Casing: This is set deep enough to protect the well from cave-in of loose formations that are often encountered near the surface. It is the starting point for the wellhead in most cases and serves as a base for the B.O.P. stack during drilling operations and for the "Christmas tree" if the well is completed. An important factor concerning the amount of surface casing needed (the setting depth) is that the string should be deep enough to reach formations that will not fracture or break down with the maximum expected mudweight at the depth where the next string is to be set. All later strings of casing are suspended and sealed at the top of the surface string by means of a casing hanger which is described further in paragraph 4.31.

4.26 Intermediate Casing: The primary purpose of an intermediate casing string is to protect the hole; such strings are sometimes termed "protection casing." The usual function of this string is to protect against loss of circulation in shallow formations when heavy mudweight is needed for drilling deeper because higher pressures are normally encountered. However, intermediate casing is sometimes set through high-pressure zones so that lighter drilling fluid can be employed for drilling deeper. In general, intermediate casing is set to seal off or protect some problem area and provide safety for further drilling.

4.27 Liner String: Unlike casing which is run from the surface to a given depth and overlaps the previous casing, liner is run only from near the bottom of the previous string to hole bottom. Liners are suspended from the previous string by means of a hanger device. They are often cemented in place, but production liners are sometime suspended in the well without cementing. Any type of casing may be employed for this purpose.

The main advantage of a liner is the lower cost, because only a short string of pipe is run instead of a complete string back to the surface. Liners are sometimes set in a deep hole as a protective string, serving the same purpose as intermediate casing. A tieback string may be run after the hole is drilled to total depth, thus connecting the liner to the surface.

4.28 Production Casing (Oil String): This serves to isolate the oil from undesirable fluids in the producing formation and from other zones penetrated by the wellbore. It is the protective housing for the tubing and other equipment used in a well.

4.29 Running the Casing: Prior to running casing, wireline logs are usually run in the open hole by a contract company specializing in that particular service. (This is explained in greater detail in Section 5, paragraph 5.57). During this operation the drill crew prepares the rig in order to run casing.

Once logging is over, the drillpipe elevator is removed and a heavy duty "slip" elevator is installed to fit the casing (Section 3, Figure 3-24). A "casing slip" is also installed over the rotary table. A board is rigged up in the derrick (a "stabbing board") so that the derrickman can stand and handle the individual joints of casing and guide the elevators into position on the pipe. A pick-up line attached to the drilling hook raises the individual casing lengths into the derrick preparatory to stabbing (connecting one joint of casing to the other). The casing elevator is not latched to the full casing string until the joint is made up. Following this, the casing string is lowered through the rotary table and wedged with casing slips ready to receive the next length of casing. Powered casing tongs are used to ensure proper make-up of each threaded joint. There is normally an arrangement for intermittently filling the casing with mud as the string is run into the hole. This is to prevent collapse due to insufficient hydrostatic pressure inside the casing.

4.30 Casing Accessories: The accessories described below are illustrated in Figure 4-14.

- Guide shoes. As the name suggests, a guide shoe is attached to the first length of casing to be lowered into the hole, and it has a rounded nose to guide the casing around obstructions.

- Float collars. These multipurpose devices permit the casing literally to float into the hole, by virture of being partially empty. The backpressure valve is closed by the pressure of the outside fluid column, thereby preventing entry of the fluid as the casing is lowered into the hole. It

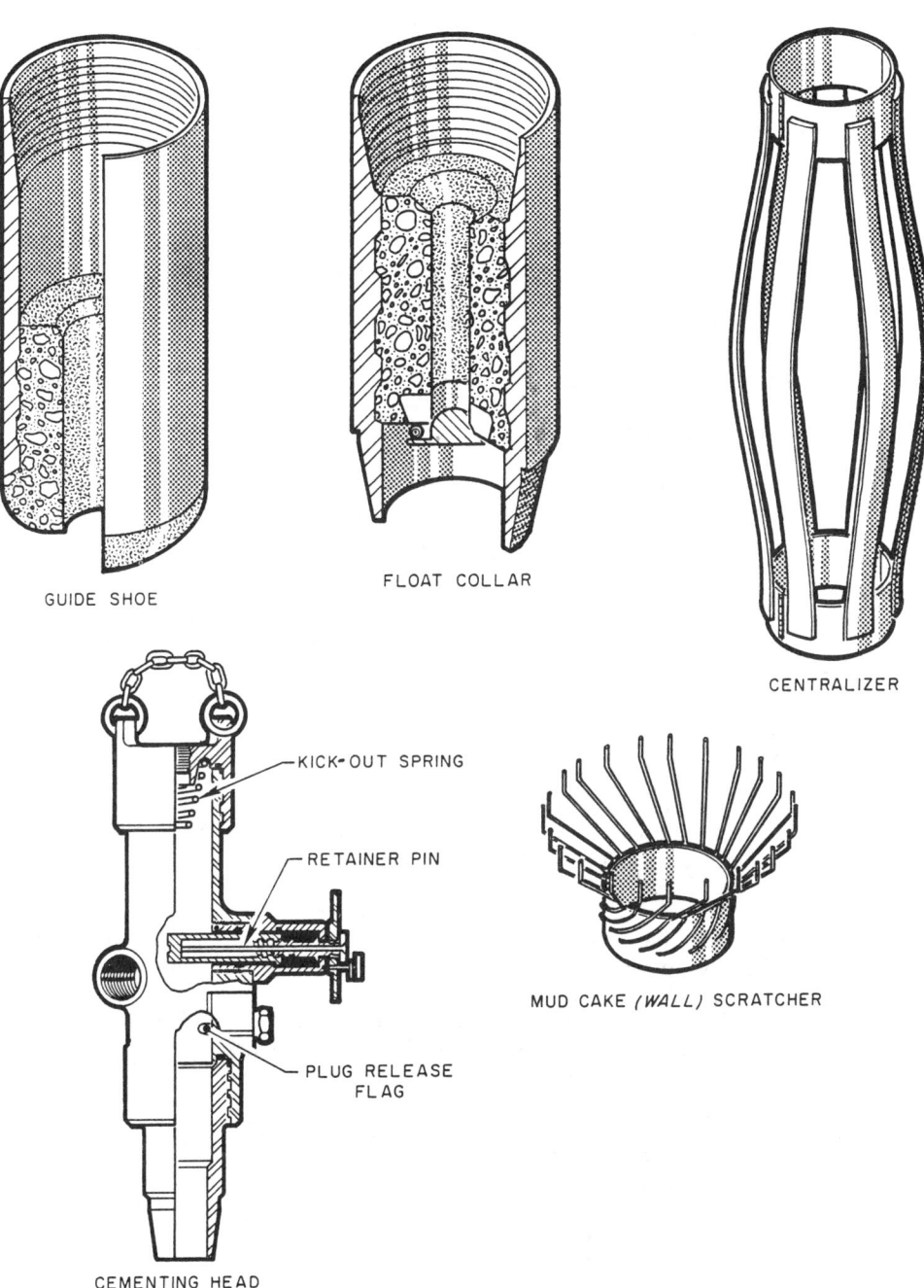

Figure 4-14. Casing Accessories

also serves as a check-valve in the string to prevent backflow of cement after being pumped outside the string. This is important because the density of the slurry is always greater than that of the mud. Also, the backpressure valve serves to prevent a blowout through the casing if a kick should occur during the operation — a most important function when high-pressure formations are exposed in the open hole.

A float collar also serves as a "stop" for the top plug when cement is displaced (this is discussed in more detail in paragraph 4.33). This action allows a quantity of slurry to stay inside the string at the casing shoe so that the operator has reasonable assurance of there being good quality cement outside the casing at that point.

Formerly, float collars and shoes were attached to the casing by welding, but this practice has been generally discontinued in favor of thread-locking compounds to prevent unscrewing while drilling out.

- Centralizers and wallcake scratchers. Centralizers in conjunction with scratchers are attached to casing for two main purposes:

    (1) To ensure a reasonably uniform distribution of cement around the pipe

    (2) To obtain a competent seal all the way around the casing and with the adjoining formation

    Centralizers hold the casing away from the wall of the hole and therefore also serve to prevent differential pressure sticking of the casing (paragraph 4.60). Wallcake scratchers are mechanical wall-cleaning devices, attached to casing, that abrade the hole when worked by reciprocating or rotating the casing string. This helps to provide a more suitable surface to which the cement can bond.

- Cementing heads. These provide the union for connecting the cementing lines from the service unit to the casing. This type of head makes it possible to circulate the mud in a normal manner, release the bottom plug, mix the cement and pump it down, release the top plug, and displace the cement without making or breaking any connections. This is explained further under paragraph 4.33 (Single-Stage cementing).

4.31 The Wellhead: This is the casing attachment to the B.O.P. or production ("Christmas") tree. It is a permanent fixture, bolted or welded to the conductor pipe or surface casing. Wellheads installed on the conductor pipe or surface casing are located in the cellar deck of land rigs, jack-ups and other fixed platforms. Those installed on the surface string are located on the seabed when used with barges, semi-submersibles and drillships.

The wellhead is the starting point for most blowout-preventer assemblies and is the vital link between the casing and the B.O.P. equipment. The conductor pipe is not usually provided with a pressure connection to the wellhead and preventer equipment. The surface casing is nearly always welded to the wellhead, and subsequent casing strings are inserted inside a wellhead housing and supported in a casing hanger (Figure 4-13). Hanger assemblies are designed to latch and seal inside the wellhead housing, so protection for the sealing surfaces is needed while drilling through the wellhead. To achieve this, removable sleeve-type protectors (called "wear bushings" or "bore protectors") are installed in the wellhead above the hanger subsequent to each casing string installation. These are removed prior to running the next string of casing in the well.

4.32 Cementing

Oilwell cementing is the process of mixing and displacing a cement slurry down the casing and up the annular space behind the pipe where it is allowed to set, thus bonding the pipe to the formation. Cementing procedures may be classified into primary and secondary phases. Primary cementing is performed immediately after the casing is run in the hole. Its objective is to obtain an effective zonal separation and help protect the pipe itself. Cementing also helps in the following ways:

- Bonds the pipe to the formation
- Protects the producing strata
- Helps in the control of blowouts from high-pressure zones
- Seals off "lost circulation zones" (LCZs) or other troublesome formations as a prelude to deeper drilling
- Provides support for the casing
- Prevents casing corrosion
- Forms a seal in the event of a kick during further drilling

4.33 Primary Cementing: Most primary cement jobs are performed by pumping the slurry down the casing and up the annulus; however, there are modified techniques for special situations. The various placement methods include:

- Single-stage cementing through casing (normal displacement technique)
- Multi-stage cementing (for wells having critical fracture gradients or requiring good cement jobs on long strings)
- Inner string cementing through drillpipe (for large-diameter pipe)
- Outside or annulus cementing through tubing (surface pipe or large casing)
- Multiple string cement (for small-diameter pipe)
- Reverse circulation (critical formations)
- Delayed setting (critical formations and to improve placement)

The single-stage and multi-stage cementing methods are discussed below.

1) <u>Single Stage</u> (normal displacement technique)

With reference to Figure 4-15, the following procedures are conducted when completing a primary cement job. The general practice is to pump 10 to 15 barrels of water or chemical "spacer" ahead of the bottom plug which is followed immediately by cement. The water

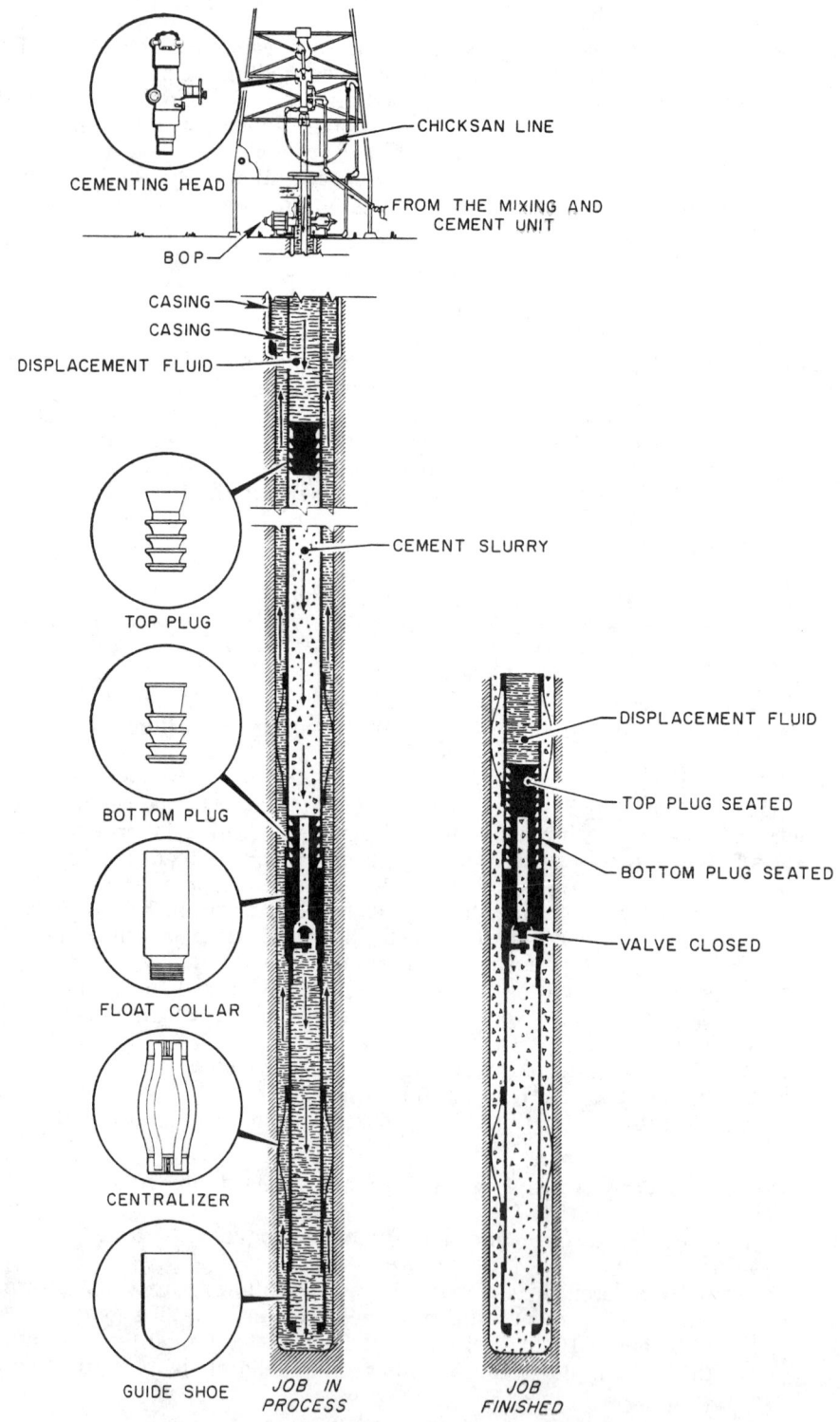

Figure 4-15. Primary (Single-Stage) Cementing

serves as a flushing agent and provides a space between the mud and cement slurry. It assists in the removal of wall cake and flushes the mud ahead of the cement, thereby lessening contamination.

Cement plugs usually consist of an aluminum body encased in molded rubber, cast in the desired shape. When the bottom plug reaches the float collar, the diaphragm in the plug ruptures to permit the cement slurry to proceed down the casing and up the annular space outside the pipe. The top plug, which is solidly constructed, is release when all the cement has been mixed, and it is followed by drilling mud or other fluid so as to displace the cement down the casing. This plug causes a complete shut-off when it reaches the float collar. A plug container cementing head is usually employed in order to facilitate releasing the plugs. Pumping is stopped as soon as there is positive indication (pressure increase) that the top plug has reached the float collar. Figure 4-16 shows a record of circulating pressures while mixing cement and displacing from casing to annulus. To ensure good cement circulation and drilling mud displacement, movement of the casing by either reciprocation or rotation may be continued throughout the time needed for circulation, cement mixing and displacement.

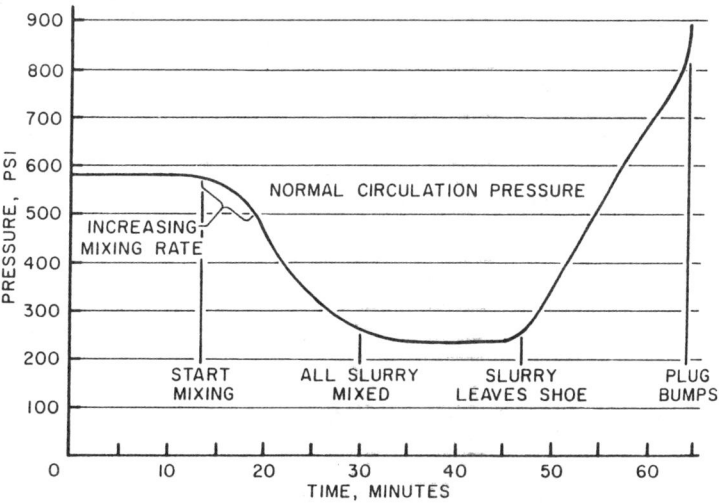

Figure 4-16. Record of Circulating Pressures While Cementing

2) Multi-Stage

Devices are used for cementing two or more separate sections behind a casing string, usually for a long column that might cause formation breakdown if the cement were displaced from the bottom of the string. The essential tool consists of a ported coupling placed at the proper point in the string. Figure 4-17 shows the steps involved on a multi-stage cementing job. Cementation of the lower section of casing is done first in the usual manner, using plugs that will pass through the stage collar without opening the ports. The multi-stage tool is then opened hydraulically by special plugs, and fluid is circulated through the tool to the surface. Placement of cement for the upper section occurs through the ports which are subsequently closed by the final plug pumped behind the cement.

4.34 Secondary Cementing: Secondary cement work is done after the primary job and includes

- Plugging to another producing zone
- Plugging a dry hole
- Formation "squeeze" cementing

The most important use of squeeze cementing is to segregate hydrocarbon producing zones from those formations producing other fluids. Squeeze cementing is also used to:

- Supplement or repair a faulty primary cementing job
- Repair defective casing or improperly placed perforations
- Minimize the danger of lost circulation in open hole while drilling deeper
- Abandon permanently a nonproductive or depleted zone
- Isolate a zone prior to perforating for production, or to fracture the formation

Injection of the slurry is done under pressure through perforations. The pumping rate is slow enough to allow for dehydration and initial setting, or both. Pumping is continued until the desired "squeeze" pressure is reached.

4.35 Cement Classifications: As a result of efforts to find hydraulic cements which could be used under water, it was discovered that limes produced from impure limestones yielded mortars which were superior to those produced from the more pure limestones. Such discoveries led to the burning of blends of calcareous and argillaceous materials. This process was patented and called "portland cement" because concrete produced from it resembled stone quarried on the Isle of Portland off the coast of England.

The portland cements for oilwell cementing carry the API classifications illustrated in Figure 4-18. The usable depth of API cements is a function of temperature and pressure. In areas of subnormal temperatures, API cements can be used at greater depths whereas, in areas of abnormally high temperatures, they may be limited to shallower depths. Normal API temperature gradient is considered to be $1.5°F/100$ ft of depth (Figure 4-19).

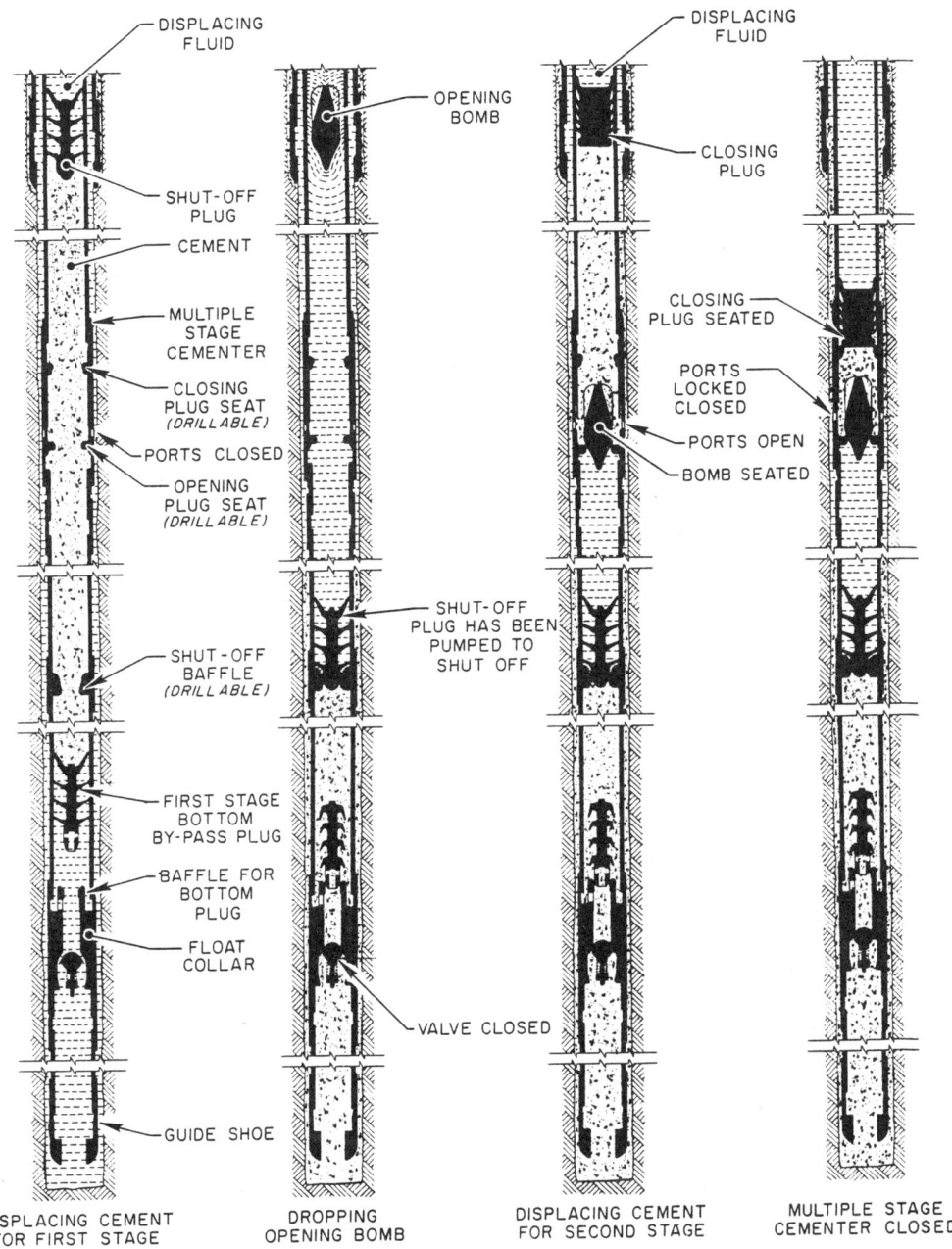

Figure 4-17. Successive Steps for Multi-Stage Cementing

API Class	Mixing Water Gal/Sack	Slurry Weight Lb/Gal	Well Depth Ft	Static Temperature °F	Conditions of Use
A	5.2	15.6	0 - 6,000	80 - 170	
B	5.2	15.6	0 - 6,000	80 - 170	
C	6.3	14.8	0 - 6,000	80 - 170	When high early strength is required.
D	4.3	16.4	6 - 10,000	170 - 230	At moderate high temperature and pressure.
E	4.3	16.4	6 - 14,000	170 - 290	At high temperature and pressure.
F	4.3	16.4	10 - 16,000	230 - 320	At extra high temperature and extra high pressure.
G	5.0	15.8	0 - 8,000	80 - 200	Can be used as basic cement or with accelerator and retarder for a wide range of uses.
H	4.3	16.4	0 - 8,000	80 - 200	

Figure 4-18. API Cement Classification

Well Depth Ft	Bottomhole Temperature Static °F	Bottomhole Circulating Temperature °F		
		Casing	Squeeze	Liner
2,000	110	91(9)*	98(4)*	91(4)*
6,000	170	113(20)	136(10)	113(10)
8,000	200	125(28)	159(15)	125(15)
12,000	260	172(44)	213(24)	172(24)
16,000	320	248(60)	271(34)	248(34)
20,000	380	340(75)	-----	-----

*Time to reach BHC temperature.

Figure 4-19. Basis for API Well Stimulation Test Schedules

4.36 Cement Additives and Their Effects: A large percentage of the world's cementing jobs utilize bulk systems rather than manual handling in sacks. Such systems enable the preparation and supply of compositions tailored to suite requirements of any well condition. This is accomplished by the use of additives with API Classes A, B, G or H cements. Some of the additives are accelerators, some are retarders. The additives are used to:

- Reduce slurry density and increase slurry volume
- Increase thickening time and retard setting
- Reduce waiting-on-cement time and increase early strength
- Reduce water loss, help sensitive formations, and help prevent premature dehydration
- Increase slurry density to restrain pressure

1) Accelerators

   Conductor and surface casing cements have lower temperatures and require an accelerator to promote the setting of cement and reduce excessive waiting time. The most common accelerators are:

   - H A-5
   - Calcium chloride
   - Sodium chloride
   - Diacel A
   - Cal seal

2) Retarders

   For deep wells, cement retarders help extend the pumpability of the cement. The primary factor that governs the use of additional retarders is the temperature of the well. As temperature increases, the chemical reaction between cement and water is accelerated which, in turn, reduces the thickening time or pumpability. Pressure has some effect, but an increase in temperature of $20°F$ may mean the difference between an unsuccessful and successful cementing job. Materials commonly used to retard or increase the setting times include

   - Calcium lignosulfonate
   - Sodiumcarboxymethylhydroxyethylcellulose derivatives
   - Blends of lignin materials with organic acids

3) Lightweight Additives

   Additives used to reduce slurry density include

   - Bentonite
   - Pozzolans
   - Diatomaceous earth
   - Expanded perlite and gibsonite

4) Heavyweight Additives

These additives are normally added when abnormally high pressures are expected. The most common materials are

- Hematite
- Barite
- Sand

5) Lost-Circulation Additives

During drilling, the problem of lost returns or circulation is fairly common. In most instances additives are used with the drilling mud; however, under certain circumstances it is necessary to use cement containing lost-circulation materials to maintain circulation. These materials are generally classified as fibrous (shredded wood, bark, sawdust), granular (gibsonite, nut shells, plastics, perlites) and laminates (mica, cellophane flakes and related products).

6) Fluid-Loss Additives

The application of low-water-loss additives in oilwell cements to reduce filtration rates is similar to that of drilling muds. While low-water-loss additives (like gel bentonite) were designed primarily for squeeze cementing, they are widely used today in high column cementing, particularly on deep liners.

7) Friction Reducers

Additives or thinners reduce the apparent viscosity of the slurry. The lower viscosity slurries will go into turbulence at lower pumping rates, reducing circulation rates and allowing cement to be pumped in turbulent flow at less than formation breakdown pressures. Contrary to the practice when pumping mud, cement is intended to be in turbulent flow to ensure better flushing of mud from the annulus. Additives currently used with cementing slurries for promoting turbulent flow at low displacement rates include

- Organic dispersants
- Salt
- Calcium lignosulfonate
- High-molecular-weight cellulose material in gel cement

8) Salt Cements

Salt-saturated cements were originally developed for cementing through salt zones because fresh-water slurries do not bond properly to salt formations — the water from the slurry dissolves or leaks away the salt at the interface, thus preventing an effective bond. Salt slurries also help protect shale sections that are sensitive to fresh water.

4.37 Mixing and Other Surface Equipment: The mixing system on any cementing operation proportions and blends the dry cementing composition with the carrier fluid. When this is achieved, a cementing slurry with predictable properties can be supplied to the wellhead. The most widely used mixing method is the jet-type mixer (Figure 4-20). A stream of water mixes with cement by passing through the mixer tub, creating a vacuum which pulls the dry cement into the tub (per the illustration) from the hopper immediately above. As the cement enters the jetstream of water, it is thoroughly mixed by the turbulent flow that occurs in the discharge pipe.

Control of mixing speed is regulated by the volume of water forced through the jet and by the amount of cement in the hopper while mixing. The mixed cement is pumped from the mixing tub to the cement head and down the casing.

The cementing unit is basically an assembly of special-purpose pumps. As the casing is sometimes reciprocated or rotated while cement is being pumped, the pipeline from the cement placement line on the rig floor to the cementing head must be flexible. This is achieved by use of steel pipe with swivel joints commonly called "chicksan line" (Figure 4-15, p. 4-28).

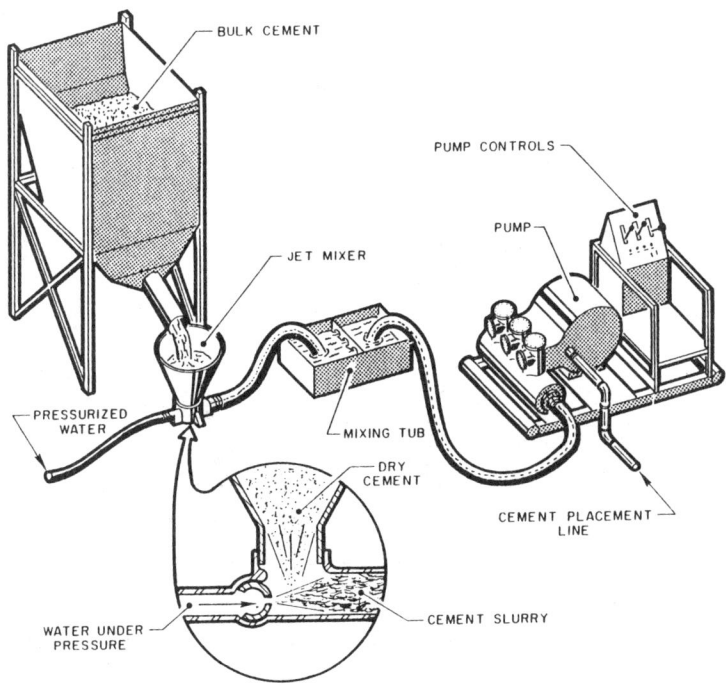

Figure 4-20. Typical Cement Mixing and Pumping Operations

4.38 Job Considerations: Cement systems cover a density range from 10.8 to 22 pounds per gallon. Slurry density is directly related to the amount of mixing water and additives in the cement and the amount of slurry contamination from drilling mud or other foreign material. In field operations, slurry control is usually maintained by measuring the density with a standard mud balance with cement taken from the mixing tub.

The volume of cement required for a specific fill-up on a casing job is based on field experience and regulatory requirements. In the absence of specific guides, a volume equal to 1.5 times that calculated from the wireline caliper survey is normally used.

To determine the height of the cement column in the annulus, a temperature log is run 12 to 24 hours after placement of the cement. Because cement generates heat when it hydrates, it is possible to locate the top of the cement by the anomaly in the temperature log, as seen in Figure 4-21. This log can also be used to determine the quality of the bond between the casing and the bore wall. A poor bond is shown as a variation in temperature which is out of line with the normal temperature gradient. A more sophisticated tool, the Cement Bond Log, measures the attenuation of acoustic signals and may determine cement bonding not only to the casing but to the formation as well. It requires skilled interpretation, but under favorable conditions even the compressive strength of the cement may be determined.

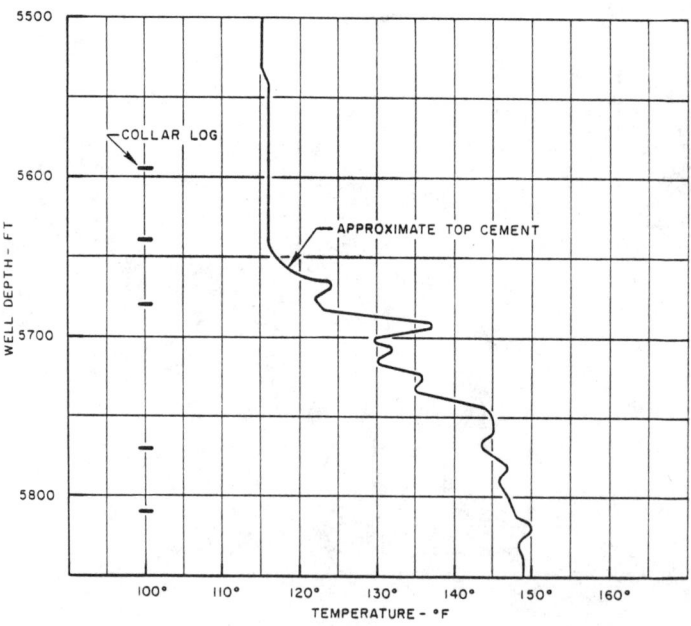

Figure 4-21. Temperature Survey Showing Top of Cement

4.39 CORING

At various points in a well, and particularly at potential producing horizons, it may be necessary to obtain more detailed information concerning the lithology than can be deduced from cuttings. Cores of 2 to 5 inches in diameter may be recovered in normal cases and they are a reliable source of data such as porosity and permeability, both horizontal and vertical. Although the relative saturations of gas/oil/water in the rock are altered during coring, information of this type may also be obtained.

Two basic types of coring are used:

- Conventional (at the time of drilling)
- Sidewall (after drilling while wireline logging)

4.40 Conventional Coring

Presently, there are four types of conventional coring tools:

- Diamond bit core barrel
- Wireline core barrel (with diamond or roller bit)
- Three-cone roller bit core barrel
- Pressure-sealed core barrel

4.41 Cutting the Core: The diamond tools are best for coring (both ordinary and wireline) and are used almost exclusively because of the expense of coring, their long downhole durability, and their reliable cutting and recovery capability.

Figure 4-22 shows a diamond-bit core barrel where the core is retrieved by pulling the barrel to the surface with the drillstring. This figure also shows a wireline core barrel where the core is retrieved while leaving the bit and drillstring in the hole, but this is no longer in common use in the oil industry.

Although every method of coring stresses the necessity for a clean hole, it is especially true in diamond coring operations because of the expense of the bit. For this reason most operators run a "boot basket," which is attached just above the bit, and circulate for a substantial period of time. Pieces of metal which would damage a diamond bit may then be jetted offbottom and hopefully collected in the basket.

Once the tool is onbottom and before coring commences, circulation is through the inner barrel. Just prior to coring, a metal ball dropped down the drillstring engages a valve in the tool and diverts the drilling fluid to the outside of the inner barrel. The fluid is then discharged through water courses in the face of the bit. The ordinary barrel (retrievable with the drillstring) is made up in 30-ft lengths, so 30, 60 or 90 feet can be cored at one time, with diameters up to 5 inches.

When wireline coring, only 15-ft lengths are retrieved at a time by use of an overshot, but when a new barrel is replaced, coring can continue. This method is often used on deep-sea projects where hole reentry is difficult. Using this method, however, cores of only 1-1/8 to 1-3/4-inch diameter are possible. In unconsolidated formations, a rubber sleeved inner barrel may be used to prevent collapse of the core during drilling and retrieval.

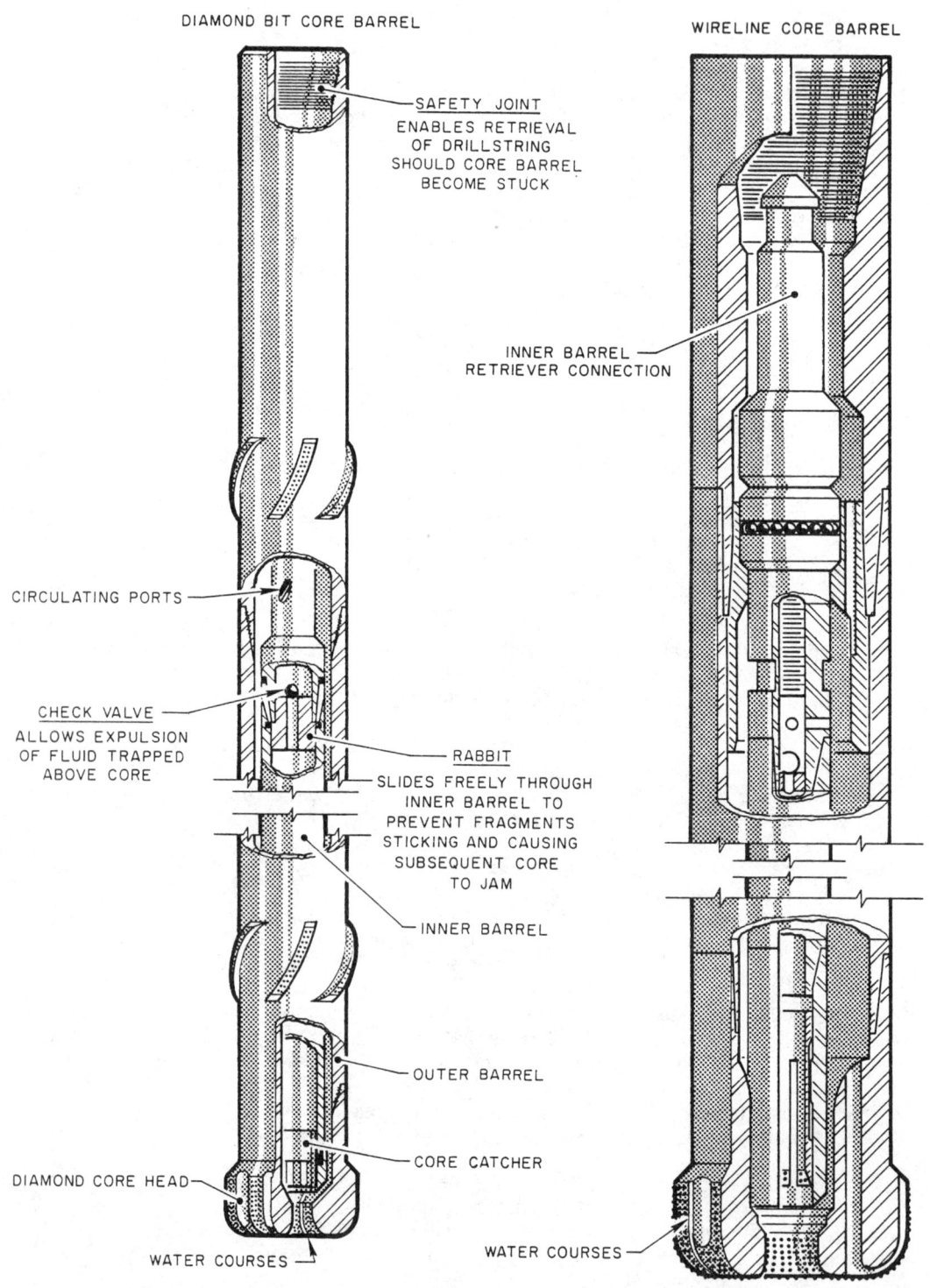

Figure 4-22. Diamond Bit Core Barrels

While the core is being cut, the logging geologist is primarily concerned with the drill rate (recorded over 1-ft intervals) and gas recorded, but perhaps more important, he must continue to collect cuttings samples at the shakers because of the possibility of incomplete or negligible core recovery.

The drilling rate during coring varies with the type of bit, the pump pressure, weight, rotary rpm, and formation drilled. The gas curves recorded by the mud-logging instruments are severely damped since there are fewer cuttings and drilling is slow.

4.42 Receiving the Core: When the core barrel arrives at the surface it is often the logging geologist's responsibility to supervise the collection of the core from the barrel. The barrel is usually hung in the derrick while special tongs, designed to grip the core, permit recovery of the core in sections. Core boxes are arranged on the drillfloor in the order in which they are to be filled.

To speed up the manipulation of the core and ensure its correct orientation, the boxes should be premarked. The system for marking the boxes varies from company to company and operation to operation, but they are generally marked on the end with the core number and box number written underneath, starting with number 1. On the side of the box an arrow is used to show the orientation (top to bottom) of the core in the box, and sometimes a "T" for top and a "B" for bottom on the other end. As the pieces of core are removed from the barrel they should be placed immediately into the boxes. (While collecting the core, never place your hands directly under the barrel, as a sudden drop of the core may cause injury.) Box 1 will contain the lower three feet of the core, as illustrated in Figure 4-23.

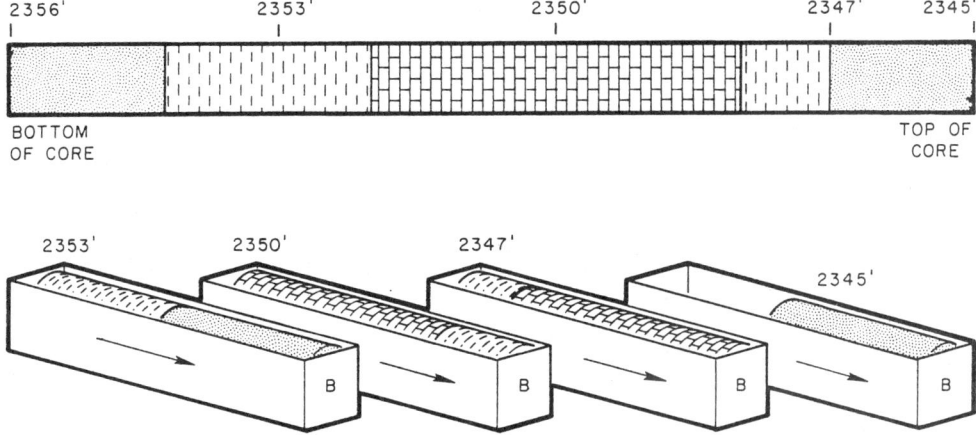

Figure 4-23. Suggested Method of Boxing Cores

When all the core has been removed from the barrel, the length recovered should be measured accurately. If the recovery does not equal the cored interval, it must be assumed that the missing portion was lost at the bottom unless there are circumstances indicating otherwise. At this same time, any adjustments necessary to fill the boxes can be made so that any partially filled box will be at the top of the core.

4.43 Sampling the Core for Special Purposes: Immediately upon removal from the core barrel, the core is wiped (not washed) free of drilling mud; the wellsite geologist then conducts an initial examination to decide whether core analysis is warranted. If the core appears to be of a potential reservoir rock, analysis will almost certainly be necessary. If the equipment for this is available onsite, the logging geologist may be required to perform this service. There is a manual that accompanies all core analysis equipment, and you should consult this manual for further details. However, for a brief outline of core analysis procedures, review the flowchart in Appendix B.

Even with core analysis equipment onsite, the client may wish to have the analysis performed in a laboratory where more detailed and precise determinations can be made. Generally, analysis includes fluid saturation, porosity and permeability. But no matter where the analysis is performed, samples for measurement of fluid saturation must be taken before the core is washed.

Normally, one sample per foot is selected, chosen on either side of changes in lithology (if detectable under the mud coating). If the analysis is not performed onsite, the samples are either placed in individual plastic bags and "canned" or they are wrapped in foil and dipped in wax for shipment.

When the saturation samples have been taken or when samples of the core are to be taken only for porosity and permeability measurements, the core may be washed clean of mud. As a rule, samples for porosity and permeability measurements should be taken at the top and bottom of several porous beds and at 1-ft intervals between. When core analysis equipment is available onsite, the entire core may be sampled by "plugging" which involves using a diamond-edge drill press; but when analysis of these measurements are to be performed elsewhere, the core is usually left intact.

4.44 Examining the Core: After the core has been washed or wiped clean, a geological description can be performed. This is mainly the responsibility of the wellsite geologist, but the logging geologist is advised to make a description also to be included on the mud log and/or attached to the bottom on a supplemental log (illustrated in Appendix A).

Gross characteristics such as dip, fractures, bedding irregularities, mottling, etc., should be noted first, along with the thickness of each bed measured to the nearest inch. A more detailed description is made from examination under the microscope for lithological alterations and under ultraviolet light (Section 5) for evidence of oil fluorescence. The description of cores of potential reservoir rocks should of course be more detailed than those of cores with no prospects.

4.45 Packing and Shipping: Cores are commonly very heavy and must be carefully prepared for shipment so that they cannot break or fall out of their boxes. Some form of packing inside the box, such as rags or paper, will prevent the core from moving around during shipment and help prevent breakage. Once the core is sealed inside, each box must be clearly marked with the following information, roughly in the order given here (also see Figure 4-23):

- Well name
- Core number (in the sequence of cores taken in the well)
- Box number
- Total number of boxes containing the whole core

A copy of the field description of the core should accompany the shipment. The outside of boxes should never display depth, unless specified by the client, as this information is proprietary and secret.

4.46 Sidewall Coring

Sidewall coring is a supplementary coring method used in zones where core recovery by conventional methods was small or where cores were not obtained as drilling progressed. Sidewall coring is useful in paleontological work, for it is possible to get shale samples for micropaleo analysis at definite depths.

The sidewall coring device, a CST (chronological sample taker) tool, is lowered into the hole on a "wireline cable" and a sample of the formation is taken at the desired depth. This is done by shooting a hollow "bullet" into and pulling it out of the wall of the hole. Usually there are as many as thirty bullets per gun. Since two guns can be used, up to sixty cores can be obtained during one run. If an electric log has been obtained previously, a spontaneous potential (SP) or gamma-ray (GR) curve run in conjunction with the samples can position the samples by direct log correlation. (See Section 5 on Wireline Logs.)

Sidewall cores taken with CSTs are small (1 x 2-1/2 inches), and in some cases the recovered material consists largely of mud cake. Sidewall coring is usually unsuccessful in very hard rocks. Nevertheless, cores of this type provide a means of examining the rock in portions of the section on which information may be extremely scanty. Sidewall cores are sometimes taken with the intention of evaluating the porosity, permeability and saturation characteristics of the rock. However, because compaction occurs as the bullet enters, the results are inevitably less reliable than those from a conventional core.

In some cases the logging geologist may be requested to perform gas analysis on sidewall cores. The CST gas analysis procedure is identical to that performed on gas samples from DSTs and wireline formation test tools except for the method of collecting the sample.

The escape of gases when CSTs are transported to the surface and when removed from the core chambers influences the gas readings somewhat, but with proper allowances, CST gas analysis can provide useful information. The test is conducted after the cores have been removed from the core chambers and placed into small glass containers. The procedure is as follows:

1. For each sample, make a hole large enough for free movement of a syringe needle through the lid of the glass container, and seal the hole immediately with tape.

2. With a syringe, take 10 cc of the gas sample from the container and seal the hole again.

3. Follow the Formation Test Gas Analysis procedure in Appendix E as of Step 2.

4.47 VERTICAL CONTROL AND DIRECTIONAL DRILLING

Controlled directional drilling is the science of directing a wellbore along a predetermined course to a target located a given distance from the vertical. Regardless of whether it is used to to hold a wellbore as close as possible to the vertical, or to deliberately deviate from the vertical, the principles of application are basically the same.

4.48 Vertical Control

No hole is drilled exactly vertically from top to bottom. It is desirable on most wells, however, to drill as near to vertical as possible to ensure that vertical depth is close to measured depth and restrict the hole problems that can occur with deviated holes.

It is generally accepted that a straight or vertical hole is one that

- Stays within the boundary of a "cone," as specified by the client (usually about $3°$)

- Does not change direction rapidly (no more than 3 degrees per 100 ft of hole) and hence form a "dogleg"

In order for the driller to be sure he is maintaining a vertical hole within the limits set out in the drilling contract, periodic measurements must be taken by the drilling crew. If any deviation has occurred, it must be recorded and compared with the amount of deviation permissible in that part of the hole. In straight-hole drilling, the measuring device is used only to determine inclination or drift, as compass headings of hole direction are not necessary. Deviation surveys are usually included in the Remarks Column of Exlog's Mud Log.

4.49 Measuring Inclination: The drift survey instrument (Figure 4-24) can be (1) run into and pulled out of the drillpipe on a "sand line," (2) dropped into the pipe and retrieved with an overshot assembly, or (3) dropped into the pipe and recovered by removing the drillstring (tripping) from the hole. The record of the inclination angle is made when a paper disc is punched by a pendulum-balanced stylus inside the instrument. Concentric circles printed on the discs are marked to show the angle of inclination from the vertical. In this example the hole inclination is $4°$.

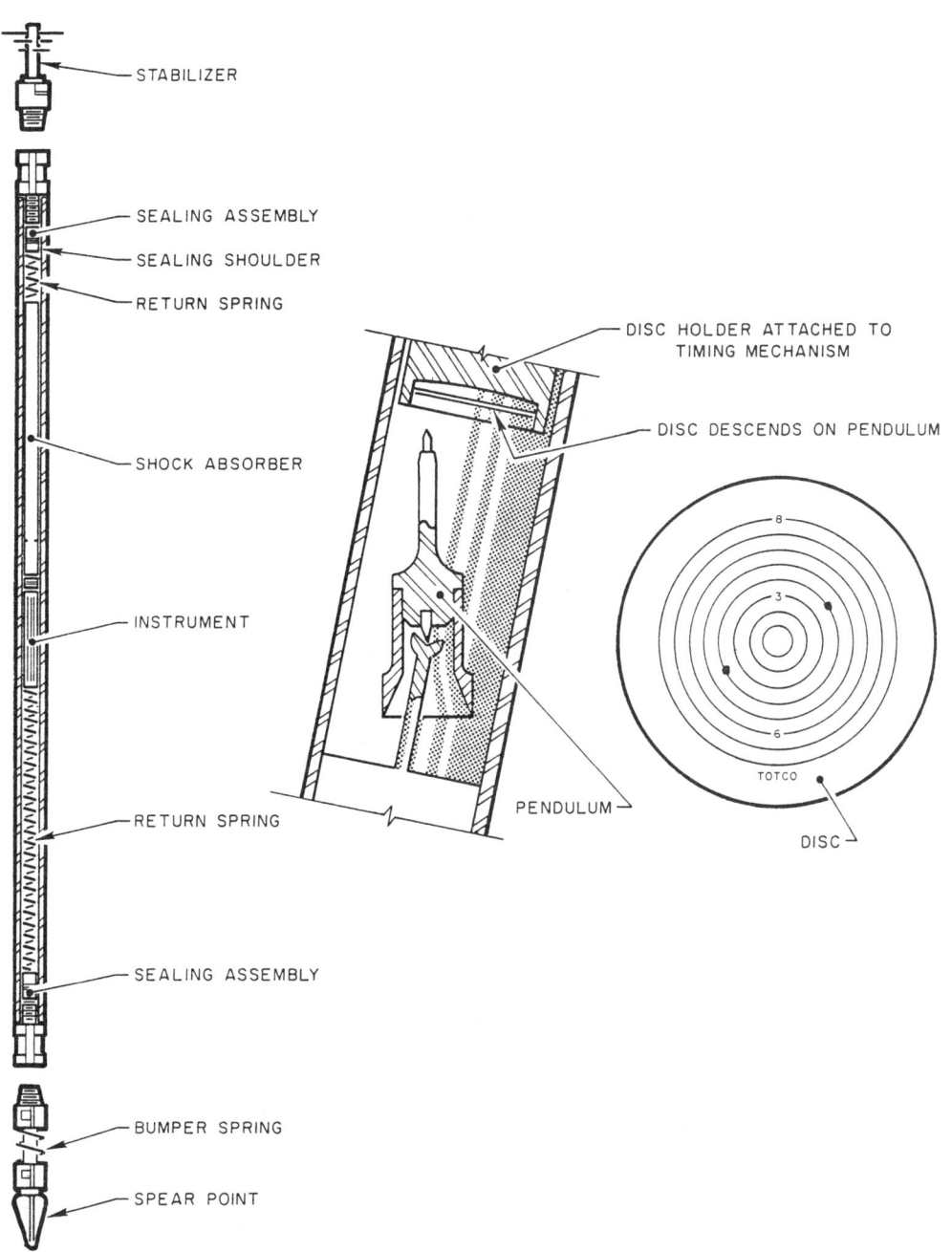

Figure 4-24. Drift Survey Instrument

4.50 Preventing and Correcting Deviation: Where deviation is expected, it can be inhibited by use of a stiff bottomhole assembly (a bottomhole assembly with many stabilizers). This will align the bit with the hole already drilled and resist any change in direction.

To straighten a crooked hole, the pendulum principle is often utilized. By removing any stabilizers placed just above the bit and retaining an upper stabilizer, and by using light weight on bit (as shown in Figure 4-25), the drill collars below exert a pendulum effect to straighten the hole.

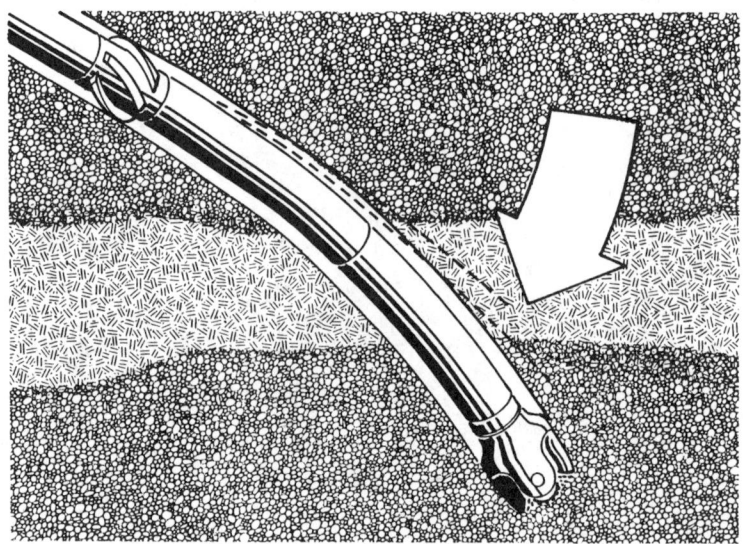

Figure 4-25. The Pendulum Effect

4.51 Directional Drilling

The most common applications of directional drilling are illustrated in Figure 4-26 and discussed briefly below.

- Multiple wells from artificial structures. Today's most common application of directional techniques is in offshore drilling where an optimum number of wells can be drilled from a single platform. This operation greatly simplifies production techniques and gathering systems, a governing factor in the economic feasibility of the offshore industry.

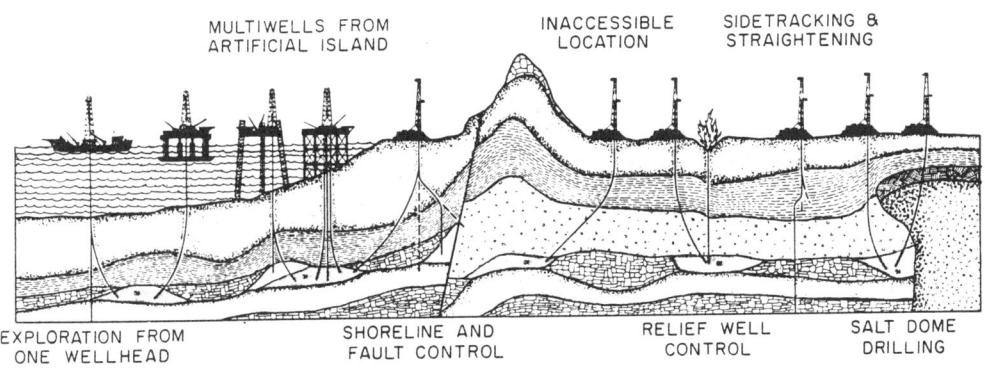

Figure 4-26. Applications of Directional Drilling

- Fault drilling. Another application is in fault control where the wellbore is deflected across or parallel to the fault for better production. This eliminates the hazard of drilling a vertical well through a steeply inclined fault plane which could slip and shear the casing.

- Inaccessible locations. The same basic techniques are applied when an inaccessible location in a producing zone dictates remote rig location, as in production located under river beds, mountains, cities, etc.

- Sidetracking and straightening. This is used as a remedial operation, either to sidetrack an obstruction by deviating the wellbore around and away from the obstruction, or to bring the wellbore back to vertical by straightening out crooked holes.

- Salt dome drilling. Directional drilling programs are also used to overcome the problems of salt dome drilling, to reach the producing formations which often lie underneath the overhanging cap of the dome.

- Relief wells. Directional drilling was first applied to this type of well so that mud and water could be pumped in to kill a wild and cratered well.

4.52 Basic Hole Patterns: A carefully conceived directional drilling program based on geological information, knowledge of mud and casing program, target area, etc., is used to select a hole pattern suitable for the operation. However, experience has shown that most deflected holes fit one of the types illustrated in Figure 4-27.

- Type I is planned so that the initial deflection is obtained at a shallow depth (approximately 1000 ft), and the angle is maintained as a "locked in," straight angle approach to the target. This pattern is mainly used

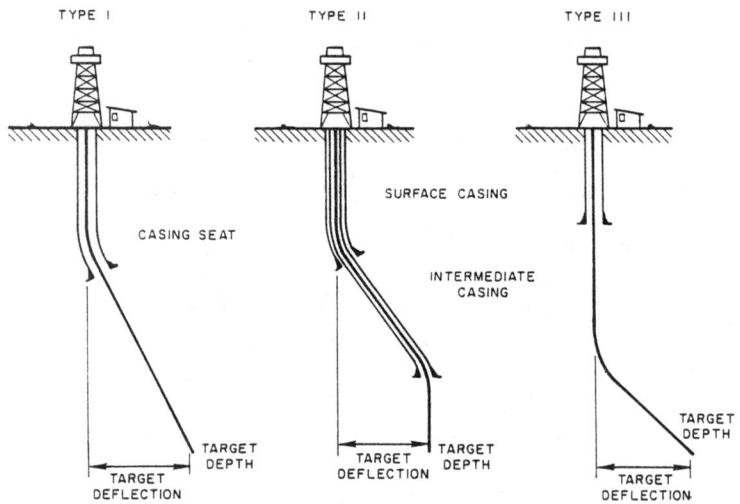

Figure 4-27. Basic Hole Patterns

for moderate depth drilling in areas where the producing formation is located in a single zone location and where no intermediate casing is required. It is also used to drill deeper wells requiring a larger lateral displacement.

- Type II, called the "S" curve pattern, is also deflected near the surface. The drift is maintained, as with Type I, until most of the desired lateral displacement is obtained. The hole angle is then reduced and/or returned to vertical in order to reach the target.

- Type III is planned such that the initial deflection is started well below the surface and the hole angle is maintained to bottomhole target. This pattern is particularly suited to special situations such as fault or salt dome drilling, or to any situation requiring redrilling or repositioning of the bottom part of the hole.

4.53 Deflection Tools: A prime requirement for directional drilling is suitable deflection tools, along with special bits and other auxiliary tools. A deflection tool is a mechanical device that is placed in the hole to cause a drilling bit to be deviated from the present course of the hole. There are numerous deflection tools available for deflecting a hole or correcting direction. The selection of a deflection tool depends upon several factors, but principally upon the type of formation at the point where the hole deviation is to start. The most common tools used for deflection are

- Downhole hydraulic motors (with a "bent sub")
- Jet bits
- Whipstocks

1) Downhole Hydraulic Motors

The downhole motor with a bent sub is the most widely used deflection tool. It is driven by drilling mud flowing down the drillstring to produce rotary power downhole, thus eliminating the need for rotating the drillpipe.

The first variation of the downhole motor (illustrated in Figure 4-28) is the "turbine" type motor, or "turbodrill." It consists of a multistage vane-type rotor and stator, a bearing section, a drive shaft and a bit rotating sub. The first stage is comprised of a rotor and a stator of identical profile. The stator is stationary and deflects the flow of drilling fluid to the rotor which is locked to the drive shaft and thus transmits the rotary motion to turn the bit.

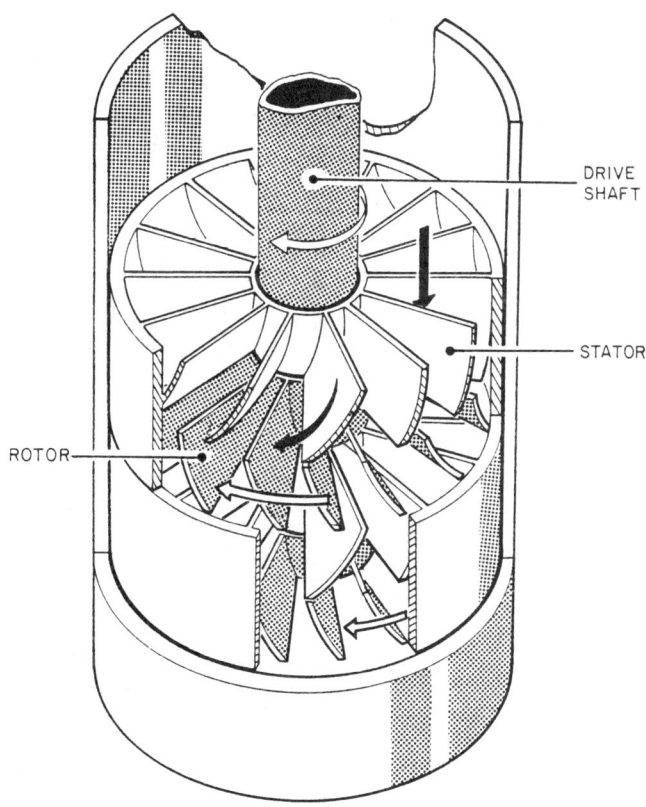

Figure 4-28. Schematic Cross-Section of a Turbine Motor

The second variation of downhole motor is the "positive displacement" or "helicoid hydraulic" motor. It consists of a two-stage helicoid motor, a dump valve, a connecting rod assembly, and bearing and shaft assembly. The helicoid motor has a rubber-lined spiral cavity with an elliptical cross-section which houses a sinusoidal steel rotor. As the mud is pumped under pressure from above, it is forced downward between the rotor and the spiral cavity. The rotor is thus displaced and turned by the pressure of the fluid column, powering the drive shaft and resulting in a rotational force that is used to turn the bit.

The bent sub (as seen in Figure 4-29) is used to impart a constant deflection to the tool. Its upper thread is cut concentric to the axis of the sub body, and its lower thread is cut with an axis inclined 1 to 3 degrees in relation to the axis of the upper thread. In addition, the "hydraulic bent sub" can be locked into position for straight drilling, or unlocked and reset for directional drilling.

Both downhole motors can be used with the following assembly which consists of a full-gauge bit, the downhole motor, a bent sub or hydraulic bent sub, a nonmagnetic drill collar, and the normal drillstring.

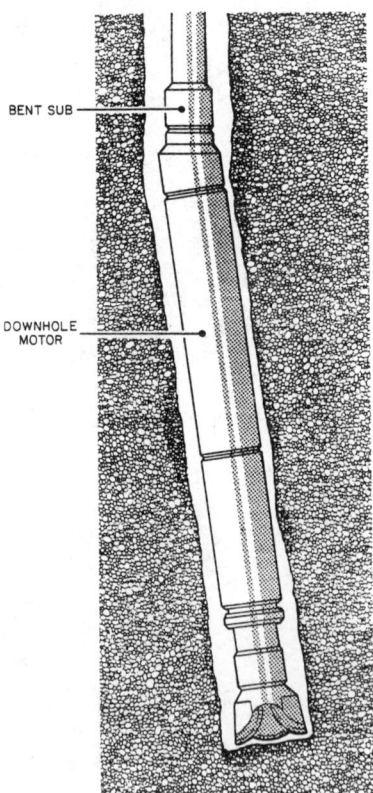

Figure 4-29. Downhole Motor Deflected with a Bent Sub

2) Jet Bits

Where subsurface formations are relatively soft, the hole can be deviated by using a jet bit. In this method, all but one of the jet openings in a conventional bit are closed off or substantially reduced in size. The jet left open has a very large nozzle. This nozzle is oriented in the proper direction onbottom and the pumps are started, but the drillstring is not rotated. Instead, it is usually worked up and down slowly, approximately 10 feet offbottom. Then weight is applied on the drillstring and bit while jetting, and the jetting action literally washes the formation out from under the jet (Figure 4-30). After the jetting has set a proper course, the drillstring is rotated. Since the washed-out section is the path of least resistance, the bit will follow it. Extra weight is then applied to bow the collar, and the drilling continues until the correct hole angle is attained.

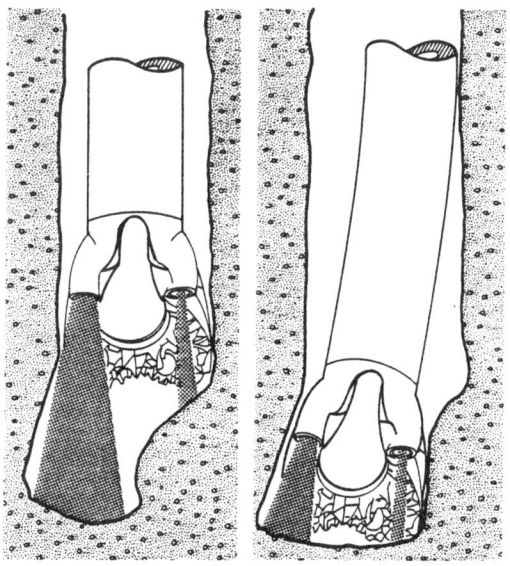

Figure 4-30. Deflection by Jetting

3) Whipstocks

The standard "removable" whipstock (Figure 4-31) is used to initiate the deflection and direction of the well, sidetrack cement plugs, or straighten crooked holes. It consists of a long inverted steel wedge that is concave on one side to hold and guide a whipstock drilling assembly. It also has a chisel point at the bottom to prevent the tool from turning, and a heavy collar at the top to help withdraw the tool from the hole.

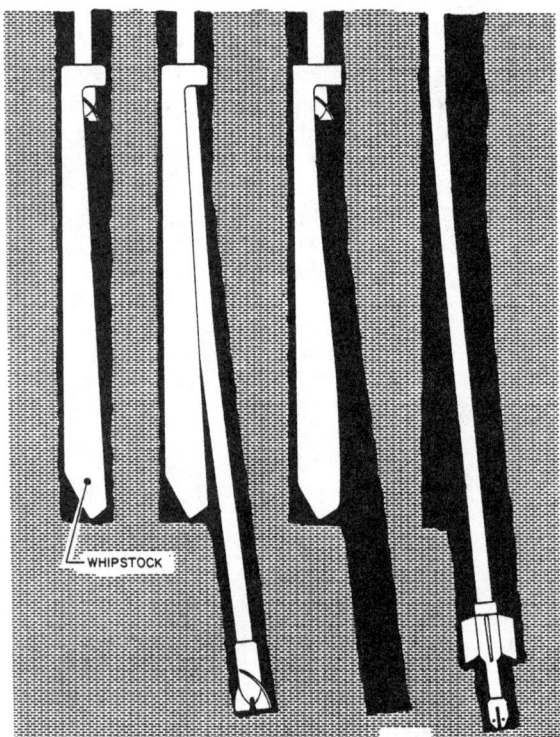

Figure 4-31. Deviating with a Whipstock

The "circulating" whipstock is run, set and drilled like the standard whipstock. But in this case, the drilling fluids flow through a passage to the bottom of the whipstock and circulate the cuttings out of the hole, ensuring a clean seat for the tool. It is most efficient for washing out bridges and bottomhole fills.

The "permanent casing" whipstock is designed to remain permanently in the well. It is mainly used to bypass collapsed casing or junk in the hole, or to reenter and drill out old wells. After the bit has been drilled below the whipstock, increased weight is applied until approximately 20 feet of pilot hole has been drilled. The whipstock is then retrieved and the pilot hole opened to full gauge with a pilot bit and hole opener.

4) Deflection Tools in General

Downhole motors present many advantages over the whipstock. When jetting becomes impractical, they permit drilling a full-gauge hole at the kick-off point, thus eliminating costly follow-up trips to widen the hole. Orientation is also more accurate since the motors penetrate along a smooth gradual curve in build-up and drop-off portions. Corrections, if needed, can be made downhole without making a trip. Finally, downhole motors eliminate the need for clean-up trips due to bridges, doglegs (abrupt hole angle changes), etc., since the tool can be circulated, rotated and drilled to bottom.

4.54 Orientation of Deflection Tools: The most widely used and practical way of determining orientation is with the "Single Shot Direct Orientation Instrument" which is run down the drillpipe on a wireline (usually the "sand line"). A nonmagnetic (monel metal, an alloy of copper and nickel) collar is used directly above the deflection tool so that the mechanisms in the tool indicate the true inclination and direction of inclination. When the tool is retrieved at the surface, the recording disc is developed and the relationships between hole direction, inclination and direction of the tool face (as they appear at bottomhole) can be readily seen. Only minor calculations, including an allowance for declination, are needed to determine how much the drillstring must be rotated to position the tool face in the desired direction.

After the drillstring has been rotated the calculated number of degrees, a check survey is run to determine whether the deflection tool also turned the same number of degrees and is facing correctly. Several attempts at rotating the tool may be needed before it is facing properly, and each attempt is checked with a survey. Once the deflection tool is facing properly, the kickoff can begin.

When downhole motors are used to "kick off" a directional hole, the drillstring does not rotate during drilling and therefore the continuous surface readout method can be employed. This method utilizes an instrument that transmits continuous data on hole direction, inclination, tool-face position and bottomhole temperature to a surface readout device on the rig floor via wireline. By observing a surface readout, the operator can keep a constant check on the direction in which the downhole motor is facing. As drilling proceeds, hole direction can be maintained on course and direction changes made when necessary. For this reason, a deflection tool is often referred to as a "steering tool."

4.55 Drilling the Deviated Section of Hole: Once the initial deflection and direction of the well are established, directional control is, for the most part, accomplished with conventional drilling techniques. The directional engineer determines the inclination and course direction of the wellbore at specific depths by the "single shot" survey (i.e., using the same instrument used for the orientation of the initial deflection). Following the directional engineer's advice, it is up to the driller to apply certain drilling techniques to drill the hole so that it is maintained on course to the target. Once the hole is kicked off from the vertical, these techniques are used to increase or maintain the inclination angle through the proper use of bottomhole tool assemblies, the application of weight-on-bit, and adjustment of rotary speed and rate of circulation. If the hole changes direction, the directional engineer must use deflection tools or other methods to bring it back on course.

When a section of hole has been drilled, and before it is cased, a "multiple shot" survey is usually run to obtain a complete directional survey. Records are taken at regular time intervals as the tool is pulled out of the hole. The tool operator takes note of the time and depth of each survey station and uses this information when reading the film upon which the surveys are recorded.

4.56 Engineering and Formation Evaluation Considerations: There is a tendency in a deviated hole for the drillstring to rest on the side of the hole. This tendency increases with inceased inclination, and it has been estimated that, in a $70°$ deviated hole, over 90 percent of the string will be resting on the side of the hole. Rotation of the string causes erosion and therefore contamination of the cuttings samples, a fact which has to be taken into account when making a geological evaluation. Engineering factors to be considered include

- Increased vertical drag
- Increased rotary torque
- Increased possibility of differential sticking
- Excessive wall friction that creates rolling action and affects directional control

4.57 FISHING

When great stress is placed on downhole subsurface equipment, the probability exists that sooner or later there will be a mechanical failure and some part of the equipment will be left in the borehole. Another common source of trouble is the pipe and associated equipment becoming "stuck" in the hole. The technique of removing this is called "fishing."

4.58 Situations Requiring a "Fishing Job"

These fall into three categories:

- Fatigue failures
- Stuck pipe
- Foreign objects in the hole

4.59 Fatigue Failures: The following are some of the possible causes.

- Fatigue failures caused by excessive stress in the drillstring, as when the rotary table continues to turn when the lower portion of the drillstring becomes stuck -- commonly called a "twist-off."

- Parting of the drillstring because of excessive pull when attempting to free equipment which has become stuck.

- Mechanical failure of parts of the drilling bit, causing some part of the bit to become lost.

4.60 Stuck Pipe: The following are some of the possible causes.

- Mechanical sticking because of stuck packers or other downhole assemblies, crooked pipe when a string of pipe has been dropped, and junk in the hole. (See paragraph 4.63.)

- "Key seating" caused when the drillpipe under tension wears a slot into the wall of the hole, as seen in Figure 4-32.

- Bridging, due to caving or swelling formations

- Differential sticking, caused when the pipe comes in contact with a permeable formation and the string is held in place by the differential pressure existing between the mud column and the formation. A example of this is shown in Figure 4-33.

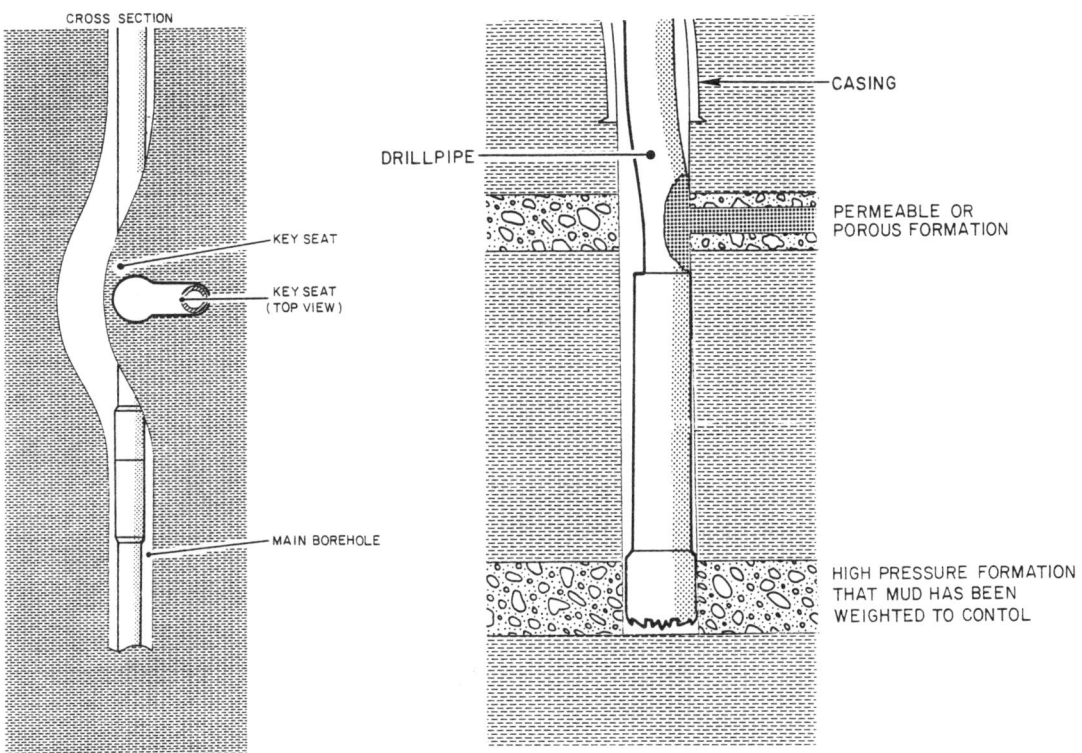

Figure 4-32. Key Seating   Figure 4-33. Differential Sticking

- Sloughing hole sticking, a problem usually associated with shale, results when shale absorbs water from the drilling fluid. This reduces stability of the shale section, causing it to expand perpendicular to the bedding plane and slough off into the borehole.

4.61 Foreign Objects in the Hole: These include tools and other undrillable objects which have been dropped into the hole, and parts of the drilling bit which, because of mechanical failure, have become lost.

4.62 Fishing Tools

Many of the tools used to recover equipment are specially designed for the particular job. However, due to the similarity of equipment used in most drilling operations, certain more-or-less standard fishing tools have been developed. A broad classification of fishing tools divides them into two groups:

    1) Those used to recover miscellaneous equipment (junk)
    2) Those used to recover pipe (fish)

4.63 Fishing for Junk: When a relatively small piece of equipment (junk) is lost in the hole, it may be retrieved using one of the following tools (Figure 4-34).

- "Junk" or "boot" sub. This is run immediately above the bit to catch small junk thrown up by turbulence. It is normally run before running a diamond bit so that no fragments can damage the bit.

- "Finger-type" or "poor boy" junk basket. This cuts a small core, after which weight is applied to the tool and bends the beveled fingers inward to trap the junk inside. This can be made "on the spot" from casing.

- Core-type junk basket. This is essentially a finger-type junk basket but has a mill shoe. Instead of applying weight to contain the core, this tool has "catchers" which grip the core and junk on the trip out.

- Fishing magnet. This is used for picking up steel fragments.

- Jet bottomhole cutter. This is used when junk is so large or oddly shaped that it cannot be readily retrieved with regular junk baskets. It breaks up the junk into small pieces by use of a shaped explosive charge. The junk may then be retrieved using one of the above tools.

- Grapple or rope spear. This is used to retrieve wireline in the hole.

4.64 Fishing for Pipe: When the drillstring has actually parted or is stuck in the hole, the operation for correcting the situation is called "fishing." (If the fish cannot be recovered, it must be cemented off and the hole sidetracked.) Some of the tools used for fishing are described below and illustrated in Figure 4-35.

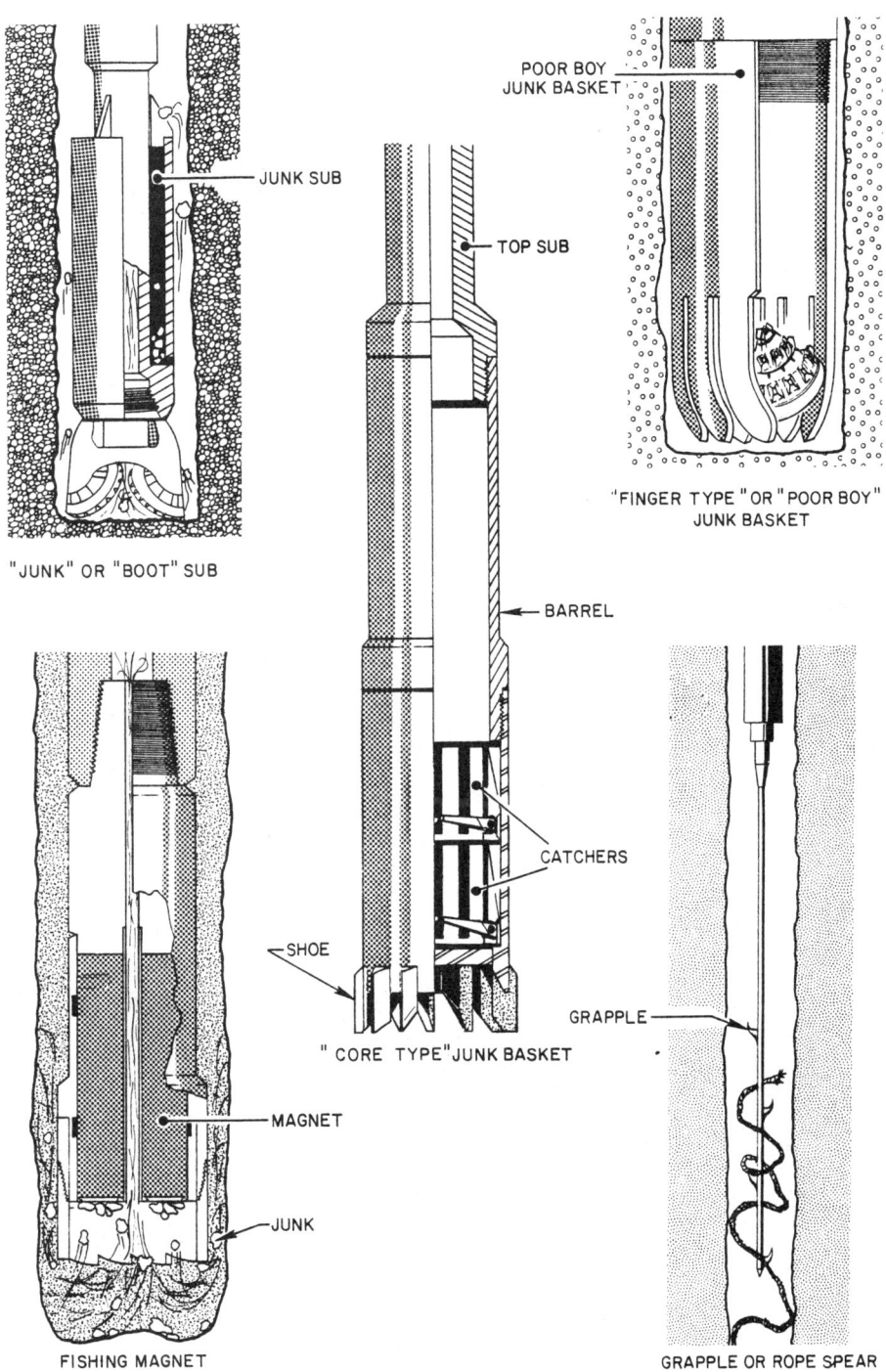

Figure 4-34. "Junk" Fishing Tools

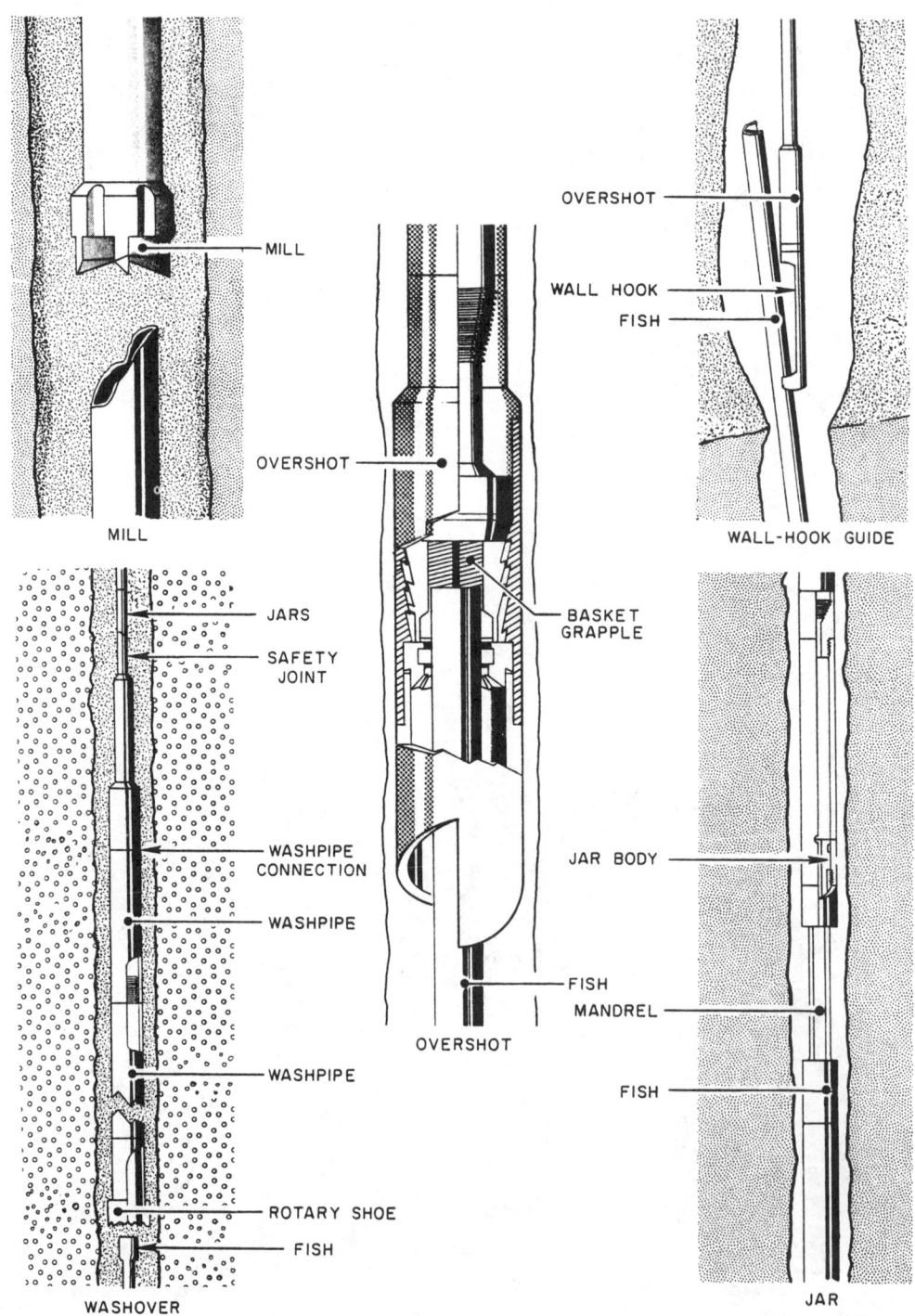

Figure 4-35. "Pipe" Fishing Tools

- Mill. Milling is sometimes necessary in order to dress the top of a fish so that the selected fishing tool is able to make a firm positive catch.

- Overshot. This is probably the first tool to be used when it is established that the top of the fish is fairly smooth. It can be a rotary taper tap or die or a more modern type which works like a set of "slips" in a core barrel to engage the top of the fish.

- Wall-hook guide. This is used if the top of the fish is in a washed-out section of hole, and it takes the place of the regular guide on the bottom of an overshot. It engages the fish and guides it into the overshot.

- Jar. This is used when a drillstring is stuck or when a "fish," caught in an overshot, cannot be pulled from the hole. In a normal drillstring a jar may be included in the middle of the collars, whereas in a fishing string it is placed immediately above the fishing tool. Jarring provides a method for giving an upward jerk to free the pipe. It works similar to a trip-hammer.

- Free-point indicator and string shot. When jarring has not been successful, this is used to determine at what point in the hole the fish is stuck. It is an electronic instrument that can sense torque or movement; it is lowered by wireline as far as possible into the hole and raised slowly while the string is stressed. Below where the pipe is stuck no torque will be sensed, but the instrument gives a positive indication as soon as the free point is reached.

  The free point indicator is raised until the string shot is positioned opposite a tool joint, one or two joints above the stuck point. Left-hand torque is applied to the drillpipe, and the primacord string shot is exploded. Loss of torque in the drillpipe is a definite indication that the tool joint has been loosened. The "backoff" is completed by further left-hand rotation and by picking the pipe up a few feet.

- Washover. This is a large-diameter pipe with a rotary cutting shoe on the bottom. It is run over stuck pipe in order to free it before fishing.

- Spotting. This is used when jarring alone will not free the fish. Oil or special chemicals are spotted around the fish in an attempt to penetrate the wall cake, causing it to deteriorate and make the pipe slick. Spotting with water when differentially stuck, and acid spotting when stuck in limestone, are often used in an attempt to free the pipe.

  If spotting and jarring do not free the fish, the "free point" is located and the portion of the drillpipe above is "backed off." Washover operations can then be carried out to retrieve the stuck portion of fish.

- Safety joint. This is a coarse-threaded joint which may be easily released and run above a fishing tool in case it should happen that the fish cannot be freed and the fishing tool cannot be released.

5
FORMATION EVALUATION PROCEDURES

5.1   **WELL LOGGING**

5.2 INTRODUCTION

There is no instrument available which indicates the presence of oil underground. The methods of geology and geophysics can suggest the most probable geographical location and geological time periods in which oil is likely to be found in significant accumulations, but it is the exploratory well that determines whether the estimates made from surface measurements will be borne out in fact. At the time of drilling a well, the most important thing to those who are directly concerned with the drilling operation is to drill a straight, vertical, true-gauge hole as quickly and safely as possible at the specified location and to the specified depth. However, the drilling practices that are necessary for the accomplishment of these ends often act as a barrier to the discovery of hydrocarbons. For example, it is essential that the hydrostatic pressure created by the density of the drilling fluid in the hole be sufficient to overcome the pressure exerted by fluids in the formations; the alternative can be a costly and extremely dangerous kick or blowout. Yet this same overbalance causes filtration of the drilling fluids into the formations and pushes the formation fluids (where permeability exists) away from the wellbore. Thus the composition and concentration of formation fluids can be determined only with difficulty. In exploratory drilling it is necessary to have a group of methods and tools capable of locating and evaluating the commercial significance of the sedimentary rocks penetrated by the drill bit. We call the use and interpretation of these methods "formation evaluation."

Formation evaluation methods can be classified broadly according to whether they are used (1) as drilling is in progress or (2) after the hole (or at least a portion of it) has been drilled.

In the first classification are

- Drilling fluid and cuttings analysis logging
- Coring and core analysis

In the second classification are

- Wireline logging
- Sidewall coring
- Wireline formation testing
- Drillstem testing

Of the many methods available for formation evaluation, no one is of any great value on its own; each must be used to complement the others.

Exploration Logging's role in the oil industry is primarily in the field of formation evaluation, and we are specifically involved in the method of drilling fluid and cuttings analysis (mud) logging. With the need for complementary interchange of evaluation methods, Exploration Logging's activities are closely associated with most of the others.

Any tabular or graphical portrayal of drilling conditions constitutes a well log. Below is a list of well logs most widely used for formation evaluation.

- Formation Evaluation Log (Mud Log)
- Pressure Log
- Core Log
- Wireline Log

Mud Logging is the continual inspection of the drilling mud and cuttings for traces of oil and gas and, in part, serves as a primary lead to coring and testing. It has an added usefulness as a safety measure for the early detection of hazardous drilling conditions which could result in a blowout.

Also, the Pressure Log and computer analysis of certain drilling parameters have recently been recognized as useful methods and are now established as two of Exploration Logging's main areas of interest.

Probably the earliest of the formation evaluation methods was coring. It is expensive, but provides samples of subsurface formations for detailed lithologic examination at the wellsite. Exploration Logging is usually called upon to handle the core and prepare it for examination by the wellsite geologist or petroleum engineer. However, we can frequently be expected to conduct a detailed examination ourselves which could involve the use of sophisticated core analysis equipment at the wellsite.

Exploration Logging, although not directly involved in the collection of wireline data, is involved with the analysis of this information to support some of the services we do provide, e.g., interpretation of lithologies and stratigraphic horizons for wellsite geologists, and pore pressure evaluation which is performed by our Pressure Evaluation Geologists (PEGs). Wireline data is of great importance because it reflects the post-drilled behavior of a formation. These particular characteristics and those determined from evaluation methods used while drilling usually provide enough information for the petroleum geologist and engineers to make worthwhile conclusions about the prospects of a well.

If a formation looks promising, a Drillstem Test (DST) may be run (paragraph 5.66). This method of evaluation simulates the conditions that will prevail when the well is completed, and so it is the deciding factor in determining whether the production will be adequate to justify the additional cost of completion. Another service company is responsible for the DST, but the Exlog logging geologist records the test intervals and results on the Mud Log. (See Appendix A for an example.)

5.3 DRILL RETURNS LOGGING

5.4 Theory

Drill returns (mud) logging provides continuous onsite inspection, detection, and evaluation of the rock units as they are being drilled with regard to potential oil and gas production. Correct methods of obtaining this data and its subsequent evaluation are very important factors in all exploratory drilling programs, and their effectiveness depends primarily upon the logging geologist.

The crushed "cylinder" of formation which is drilled to make the hole is released into the mudstream. Once released, the formation and any contained fluids, gas or oil are carried to the surface by the mud. Mud logging largely becomes a matter of extracting this information in terms of restoring (recording on the Mud Log) the original in-place characteristics of the formation as much as possible.

The first disturbance of the subsurface strata as a result of being drilled is that of varying amounts of flushing by the mud filtrate. Ordinarily, the drilling mud exerts a hydrostatic pressure on the formation in excess of the formation pressure. The formation serves as a filter medium upon which wall cake is deposited and through which the filtrate water permeates, flushing interstitial fluids away from the wellbore. On the bottom of the hole where new formation is being continuously exposed and wall cake is not permitted to accumulate, the rate of filtration of mud fluids is always at its maximum.

Parameters that affect the amount of oil and gas remaining in the formation after being flushed and which, in turn, affect the amount of oil and gas entrained in the drilling mud are listed below. Also, see paragraph 5.15 for other effects of these parameters.

- Depth
- Rate of penetration (ROP)
- Size of hole
- Volume of drilling fluid circulated
- Physical properties of the formation
- Properties of the drilling mud

These parameters do not vary so much that these changes would account for the wide variations of gas shows obtained from wells that may be productive of gas and saltwater or gas and distillate. It is reasonable to assume that in many cases the formations have been flushed to the extent of being completely depleted of producible hydrocarbons before they are drilled. More often, many formations are flushed to varying degrees.

After undergoing flushing, the formation is subjected to the bit action, then released into the mudstream in the form of cuttings chips under hydrostatic pressure of the mud column. During their travel time from the bottom of the hole to the surface, the cuttings undergo a normal production cycle in that the pressure on them, caused by the hydrostatic pressure of mud, is reduced to atmospheric. Gases, if present, and liquids (to a small degree) expand due to the pressure reduction and cause the cuttings to release into the drilling mud any fluids which they contained. Thus, upon reaching the surface the cuttings will have been depleted, either by flushing or production. For this reason a great deal of importance is placed on the hydrocarbon content of the mud as the source of information for evaluating the productive possibilites of the formation being drilled.

The fluids released from the cuttings and conveyed to the surface by the mud are the bases for several parameters recorded by well logging instruments and methods. The parameters are important considerations in the continuous evaluation of the productive possibilities of the formation as it is being penetrated. This is not to discount the importance of the cuttings in formation evaluation. The cuttings are samples of potential reservoir rock -- the containing media for oil or gas, or both. Aside from their importance as a basis for correlation and stratigraphic purposes, they afford the means of the first study of the reservoir characteristics of the formation. However, they must be studied and evaluated, bearing in mind that they have been extensively flushed and produced.

5.5   Application

Mud logging is applied generally on wildcat or exploratory wells and on field development wells where specific problems are involved such as described below.

- Wildcat wells -- specifically in areas where detailed subsurface information is lacking.

- Field development or delineation wells -- in areas where lensing sands and folding and faulting make subsurface correlation difficult.

- Wells which are expected to encounter high-pressure formations -- in addition to supplying complete downhole information, a warning of pending blowouts may be possible thus aiding in controlling and drilling high-pressure zones.

- Areas where electric-log interpretation is difficult -- information is provided for muds having high saline content, or sands containing fresh or brackish waters.

- Mud logging is used as a correlative tool to enable operators to change or modify their daily well programs so that possible productive formations may be immediately evaluated by one of the following methods:

    -- Conventional coring. Drilling breaks may be evaluated by the logging geologist and the need for coring determined or drilling resumed with a minimum loss of time. Coring of nonproductive formations is reduced to a minimum.

    -- Wireline logging. In expensive wildcat wells, suites of wireline logs may be run if the mud log depicts shows prior to TD (total depth).

- Sidewall coring. In some areas, wells may be drilled to the planned total depth without taking conventional cores. After an electric-log run, the mud log may be used for correlating interesting zones for sidewall samples.

- Drillstem tests. It is desirable in many areas (mainly in the U.S.A.) to test possible productive formations as the well is being drilled; therefore, drilling may be halted when shows occur. Testing is restricted to those zones showing some promise of production.

Using mud logging to the greatest advantage depends on several factors. The most important determinant is the client's knowledge of the subsurface in the area in question. Consideration should be given to downhole mechanical factors -- how safely and conclusively such operations as testing may be carried out in open hole, etc. The dependability of correlation with other holes and the oil company's philosophy toward the well in question are important considerations.

Essentially, there are two general methods of utilizing mud logging. In one, it is used primarily for evaluating the formation as the hole is being drilled and for controlling points for casing, coring, testing, or further evaluation. If this is to be the general purpose of the logging operation, drilling proceeds and the hole is mud-logged without interruption until a significant increase in drilling rate, or other indicators such as increase in total gas and/or the appearance of heavy gases, indicates that a possible reservoir has been encountered. (A carbonate reservoir, however, may give a decrease in drilling rate.) After approximately 10 to 15 feet of the porous section have been penetrated, drilling is halted and the cuttings from the break are pumped to the surface. The usual interval is 10 to 15 feet, but the interval depends on a number of considerations including anticipated payzone thickness and the thickness of the overlying stringers which may give a false alarm. If no show is indicated, drilling can be resumed with a minimum loss of time, but if analysis of the mud and cuttings indicates possibilities of oil or gas, a core or drillstem test may be justified. Thus, coring or testing is kept to a minimum consistent with not overlooking any productive possibilities. Most wildcats are handled in this manner.

In other applications for mud logging, the client may wish to drill a well to the planned total depth without intermediate coring or testing. In this type of program, the mud log is often a valuable aid in interpreting the electric log. The well is drilled and mud-logged to total depth without interruption. After an electric log is run it is correlated with the mud log, and sidewall samples may be taken in all interesting sections of the hole.

5.6   Mud Logging Techniques and Equipment

Mud logging is not complex in principle and does not interfere with the drilling process, and the results are available almost immediately. The log is recorded simultaneously with the drilling of the hole. Detailed data on the physical characteristics of the subsurface strata is collected and analyzed as it becomes accessible at the surface. This information is continuously evaluated, and control of certain phases of the drilling operation is exercised by the Operator based on interpretation of the results. Besides almost immediately indicating the presence of any potentially productive zone, the mud log serves as a basis for tailoring and altering the drilling program efficiently and is an important corroborative and correlative tool.

The Exploration Logging mud log furnishes the following information:

- Direct measurement of hydrocarbon gases from the drilling mud
- Chromatographic analysis of the drilling mud for individual hydrocarbon gas content (C1-NC4) in ppm
- Total combustible gas from drill cuttings
- Oil from drilling mud and cuttings
- Detailed rate-of-penetration curve
- Lithology log and description (including estimated visible porosity)
- Drilling mud characteristics
- Data pertinent to the well's operation such as coring points, trips for a new bit, drillstem tests, etc.
- Bit data, carbide information, deviations, and other pertinent engineering information.

5.7 Lag Determinations: A definite time interval is always required for pumping the samples from a particular depth to the surface where they become accessible. This critical time interval, when definitely determined, is called the "lag" and is measured in terms of cycles of the mud pump. Applied in this manner, the mud pump serves the purpose of a displacement meter. The lag applies to all downhole information, the formation cuttings and the fluids (gas, oil and water) which they contain. The lag always exists and changes continuously as the hole becomes deeper. It is necessary always to know the lag and apply it continuously to returning samples. Because of the factors present which cause it to change, the lag must be frequently checked and corrected.

The lag can be determined by placing a tracer in the drillpipe at the surface when the kelly is "broken off," allowing the tracer to be pumped through the hole and back to the surface, and counting the number of strokes required of the circulating pump to make this circulation. From this total pump-cycle count, the number of cycles required to pump the tracer down through the pipe to the bottom of the hole is subtracted. This figure is calculated on the basis of the capacity of the drillpipe (see Figure 5-2) and the displacement of the circulating pump. The result is the lag time or strokes.

Various materials (such as whole oats or barley or strips of colored cellophane) may be used as tracers and picked up on the shaker screen for approximating the lag. Under ordinary circumstances, however, calcium carbide placed in the drillpipe will react with the mud to form acetylene. This gas will be picked up by the mud gas detector and is the most convenient and reliable method for determining the lag. Acetylene gas appears as wet gas (PV) on the gas detector and is easily distinguished from methane produced from the formation.

Determining and using lag in terms of pump strokes has distinct advantages over lag determined on a time basis. The counters tracking the cuttings up the hole stop automatically when the pump is stopped. Clocks would continue to run, and some subtractive factor would have to be introduced. The most important advantage, however, lies in accuracy. A lag determined in terms of an interval of time is correct for only one speed of the circulating pump (that speed at which the lag determination was run), whereas the lag in pump cycles is accurate for any pump rate.

If it is not possible to run a carbide, then a theoretical calculation of lag time must be made. This is rarely accurate except after setting casing since it must be assumed that the hole is exactly in gauge throughout, and this is not usually so. The greater the amount of open hole, the less accurate the calculation will be. The lag time calculation is merely a calculation of the time required to displace mud in an annulus of known inner and outer diameters:

$$\text{lag time} = \frac{\text{depth (ft)}}{\text{annular velocity (ft/min)}}$$

$$\text{annular velocity} = \frac{\text{pump output (bbl/min)}}{\text{annular volume (bbl/ft)}}$$

$$\text{annular volume} = \text{hole capacity} - \text{pipe (capacity} + \text{displacement)}$$

The hydraulic slide rule is designed to perform these three operations and will give a theoretical annular velocity if pump size and rate, annulus, and drillpipe sizes are known.

With variations in hole size or casing size, and when a tapered string is being used, separate calculations must be made for each individual annulus size. For example,

        depth = 10,780 ft
        bit size = 8-1/2 inches
        drillpipe = 3-1/2 inches
        drill collars = 6 inches/900 ft

pump 1:   stroke = 18 inches
           liner = 6 inches
           rate   = 45 spm

pump 2:   stroke = 18 inches
           liner = 5-3/4 inches
           rate   = 50 spm
           last casing = 9-5/8 inches (40 lb/ft) @ 8000 ft

Therefore,

1. Always use the casing I.D. (inside diameter), i.e., 9-5/8 inches (40 lb/ft) = 8.835 inches I.D. (N.B. In the field, casing is usually referred to by its <u>outside</u> diameter.)

2. Construct a diagram of the borehole and drillstring to assist you in deciding how many separate calculations must be undertaken. This is illustrated in Figure 5-1.

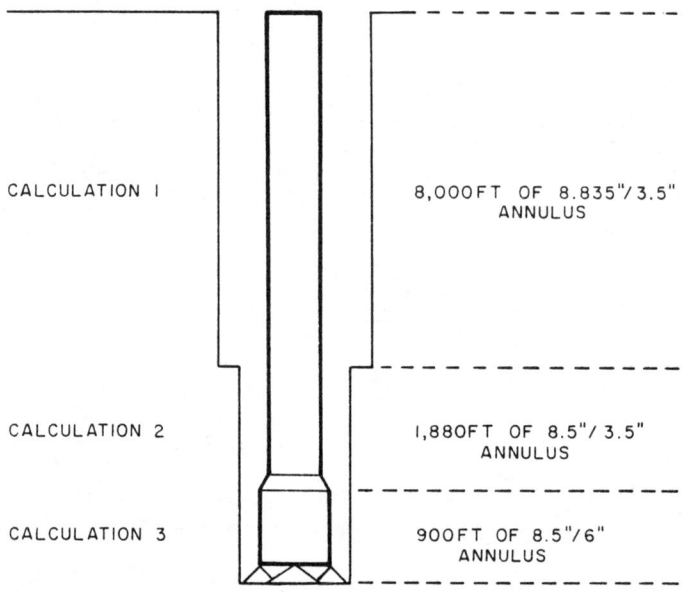

Figure 5-1. Lag Calculation

3. Use a hydraulic calculator to calculate the total pump output.

   pump 1   = 316 gal/min
   pump 2   = 325 gal/min
   total output = 641 gal/min

4. Calculate: hole I.D. plus pipe or collar O.D., and hole I.D. minus pipe or collar O.D., for each section of hole.

   calculate (1) 8.835 + 3.5 = 12.335 inches
                  8.835 - 3.5 = 5.335 inches

   calculate (2) 8.5 + 3.5 = 12.00 inches
                  8.5 - 3.5 = 5.00 inches

   calculate (3) 8.5 + 6.0 = 14.5 inches
                  8.5 - 6.0 = 2.5 inches

5. Use the hydraulic calculator to calculate individual annular velocities.

   calculate (1) annular velocity = 240 ft/min
   calculate (2) annular velocity = 261 ft/min
   calculate (3) annular velocity = 434 ft/min

6. Calculate individual lag times (round off to nearest whole minute).

   calculate (1) lag time = 8,000 ft @ 240 ft/min = 33 min
   calculate (2) lag time = 1,880 ft @ 261 ft/min = 7 min
   calculate (3) lag time =   900 ft @ 434 ft/min = 2 min

   Total lag time = 42 min

Using the reverse of this calculation from a carbide lag time, it is possible to calculate an average true hole size and hence the degree of hole enlargement (only for open hole section).

On most rigs, both pumps will be of equal pump output and efficiency. In this case, pump output can be expressed in gallons per stroke (gal/stroke); therefore, if you know the number of gallons in the annulus, the number of strokes needed to pump it to the surface (lag strokes) can be calculated.

Using the same example,

$$\text{output in gal/stroke} = \frac{\text{total pump output (gal/min)}}{\text{pump rate (spm)}}$$

$$= 641/95 = 6.75$$

annular volume = hole capacity − pipe (capacity + displacement)

Using this formula and referring to Figure 5-2, with pipe capacity and displacements the total annular volume can be calculated. Consider each section separately, from top to bottom:

   Section (1): (8000 x 3.185) − (8000 x 0.500) = 21,480 gal
   Section (2): (1880 x 2.948) − (1880 x 0.500) = 4,602 gal
   Section (3): (900 x 2.948) − (900 x 1.469) = 1,331 gal

   total annular volume = 27,413 gal

$$\text{lag pump strokes} = \frac{\text{annular volume}}{\text{pump output}}$$

$$= 27{,}413/6.75 = 4061$$

$$\text{at 95 spm, lag time} = \frac{\text{lag strokes}}{\text{pump rate}} = \frac{4061}{95} = 43 \text{ min}$$

Note that by using data directly from Figure 5-2, the annular volumes of sections (1) and (2) have been slightly overestimated since no account has been taken of the larger diameter tool joints on each joint of drillpipe. Section (3) is correct since drill collars do not have external tool joints. If tool joint size and length are taken into account, this example yields a lag of 3910 strokes (41 minutes).

	A.P.I. STANDARD SIZE		BARRELS PER FOOT			GALLONS PER FOOT		
	O.D.	I.D.	CAPACITY	DISPLACEMENT (METAL ONLY)	TOTAL DISPLACEMENT (PIPE AS SOLID)	CAPACITY	DISPLACEMENT (METAL ONLY)	TOTAL DISPLACEMENT (PIPE AS SOLID)
DRILLPIPE	2.375	1.815	0.00320	0.00261	0.00548	0.1344	0.1036	0.2301
	3.5	2.602	0.00658	0.00532	0.01190	0.2762	0.2236	0.4998
	4.0	3.340	0.01084	0.00471	0.01544	0.4551	0.1977	0.6528
	4.5	3.640	0.01287	0.00680	0.01967	0.5406	0.2855	0.8261
	5.0	4.276	0.01776	0.00652	0.02428	0.7460	0.2739	1.0199
	5.5	4.670	0.02119	0.00820	0.02938	0.8898	0.3443	1.2341
COLLARS	3.5	1.5	0.00219	0.00971	0.01190	0.0918	0.40800	0.4998
	5.0	2.25	0.00492	0.01937	0.02428	0.2065	0.8134	1.0199
	6.0	2.25	0.00492	0.03005	0.03497	0.2065	1.2622	1.4687
	6.5	2.813	0.00769	0.03335	0.04104	0.3228	1.4009	1.7237
	8.0	2.813	0.00769	0.05448	0.06217	0.3228	2.2882	2.6110
	8.25	2.813	0.00769	0.05843	0.06611	0.3228	2.4539	2.7768
	9.0	2.813	0.00769	0.07099	0.07868	0.3228	2.9817	3.3046
	9.5	3.0	0.00874	0.07892	0.08767	0.3672	3.3148	3.6819
CASING	5.0	4.408	0.01887	0.00541	0.02428	0.7927	0.2272	1.0199
	5.5	4.892	0.02324	0.00614	0.02938	0.9764	0.2577	1.2341
	6.625	5.675	0.03128	0.01135	0.04263	1.3139	0.4768	1.7906
	7.0	5.92	0.03404	0.01356	0.04760	1.4298	0.5694	1.9991
	9.625	8.835	0.07582	0.01467	0.08999	3.1847	0.5950	3.7795
	13.375	12.615	0.15458	0.01919	0.17377	6.4928	0.8059	7.2982
	16.0	15.125	0.22222	0.02645	0.24867	9.3336	1.1108	10.4441
	20.0	19.124	0.35526	0.03328	0.38854	14.9216	1.3979	16.3188
HOLE	3.875	—	0.01458			0.6125		
	6.0	—	0.03497			1.4687		
	7.625	—	0.05648			2.3720		
	8.5	—	0.07018			2.9476		
	8.625	—	0.07226	———	———	3.0349	———	———
	8.835	—	0.07582			3.1845		
	12.25	—	0.14576			6.1221		
	17.5	—	0.29748			12.4941		
	26.0	—	0.65664			27.5788		

Figure 5-2. General Capacities and Displacements of Drillpipe, Collars, Casing, and Hole

The slight difference in lag time between the two methods can be expected as the hydraulic calculator cannot be read very accurately. The data worksheet illustrated in Figure 5-6 shows where pump-stroke count and lag values should be recorded.

For a rough estimate of hole volume a simple rule to remember is

$$d^2 \simeq bbl/1000 \text{ ft}$$

where

d = diameter (inches)

Derivation:

$$\pi r^2 h = \text{volume}$$

$$\pi \frac{d^2}{4} h = \text{volume} \quad (h = \text{height})$$

where

$$r^2 = \frac{d^2}{4}$$

Convert to feet:

$$\pi h \frac{d^2}{4} \times \frac{1}{144 \times \phi}$$

where

$\phi = 5.6146$ (cu ft $\rightarrow$ bbl)

Therefore,

$$\frac{\pi d^2 h}{144 \times 4 \times \phi} = \text{vol (cu ft)} = Kd^2$$

$$K = \frac{\pi h}{144 \times 4 \times}$$

If $h = 1000$ ft,

$$K = 0.9714$$

Therefore, if vol = Kd^2,

$$\text{vol} = 0.9714 \, d^2 \text{ per 1000 ft}$$

5.8 Pump Stroke Counters: An important instrument in the well-logging unit is the Pump Stroke panel. The pickup unit (sensor) consists of a microswitch attached to the mud pump by means of a C-clamp and is tripped by the action of the pump rod. The circuit runs from the pump stroke panel instrument to the pickup unit and returns to the instrument panel. This panel includes a flashing light synchronized with the count, a digital counter for each pump, and a stroke-per-minute (spm) meter. In normal operation, these counters are connected so that both simultaneously advance one number with each cycle made by the mud pump. The spm meter is calibrated to read the pump rate (or number of complete cycles the mud pump is making per minute) directly.

The pump stroke counters and spm meter serve a very important function in the logging operation. Used together they are the means of timing the cuttings and mud samples up the hole, thus affording the logger an easy, accurate means of correlating the data from these samples back to their correct depths. The pump-stroke count, depth, drilling rate, and the correlation of the samples with depth are each function-dependent upon the other. The pump-stroke counters serve to integrate this data into its proper perspective for the logging geologist.

5.9 Depth and Drill Rate Recorder: The depth and drill rate are available simultaneously with drilling an interval of hole, whereas the rest of the subsurface data pertaining to the interval is not available until returns from that interval can be pumped to the surface. The drilling rate is determined for individual regular intervals of hole called the "logging interval." In most areas the logging interval is 5 or 10 feet, or 1, 2 or 3 metres, meaning that data regarding the hole is collected or determined for each 5 or 10 feet (or 1, 2 or 3 metres) of hole that is made. (During coring, however, the interval is usually reduced to 1 foot or 1/2 metre.) Whether the drilling rate is calculated in minutes-per-foot, feet-per-hour, minutes-per-metre or metres-per-hour depends upon the particular client.

To determine depth and rate of penetration, Exploration Logging uses a system which senses the height of the kelly. This system responds to changes in hydrostatic pressure of a column of water maintained between a sensing device and a kelly chamber (Bristol bottle) located near the kelly swivel.

In logging units where the Bristol recorder is used the hydrostatic head acts against a diaphragm, and its pressure changes are mechanically transferred by connecting levers to a pen which records on a circular chart. In other logging units the hydrostatic pressure variation is detected by a differential transducer, and an appropriate electrical signal is produced and sent to the logging unit and recorded on a linear chart. The Bristol depth recorder system utilizes a circular clock-chart, while the latter uses a continuous-strip chart. The plots from both systems are the same in that they show the kelly position at any given time; from this the depth and drill rate can be determined.

By first securing the depth from the driller at a "kelly down" position, and by adding the measured length of successive joints of drillpipe, the logging geologist can determine the exact depth of the hole at any given time (Figure 5-3). On a floating rig corrections must occasionally be made for tidal changes. The depth at the kelly-down position is the <u>sum</u> of the length of the kelly, all drillpipe, drill collars, subs and bit that are in the hole below the kelly bushing. To obtain the 10-ft increments as shown on the charts in Figure 5-3, start from the kelly-down point where the exact depth is known and count back up the curve. In order to keep a current and immediate track of the kelly-down position of the kelly being drilled, it must be projected and marked back up the chart. By using this method, the kelly-down depth will always be identical with the driller's depth at that point. To

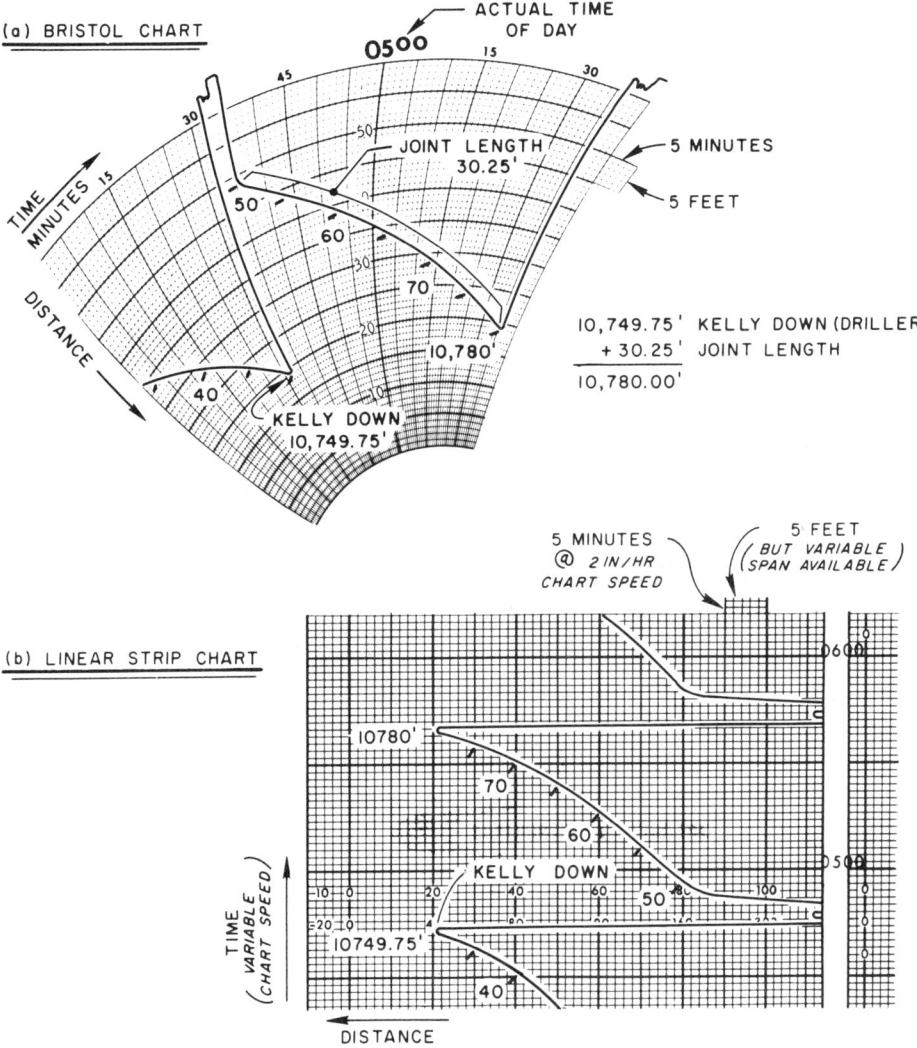

Figure 5-3. Kelly Height, Recorder Chart, Example No. 1

determine the drilling rate for a given interval simply count the minutes taken to drill the interval; then if required, with the aid of charts, convert this to feet per hour (see Figures 5-4 and 5-5) or use the following equation:

$$\frac{60}{\text{minutes}} \times \text{feet} = \text{ft/hr}$$

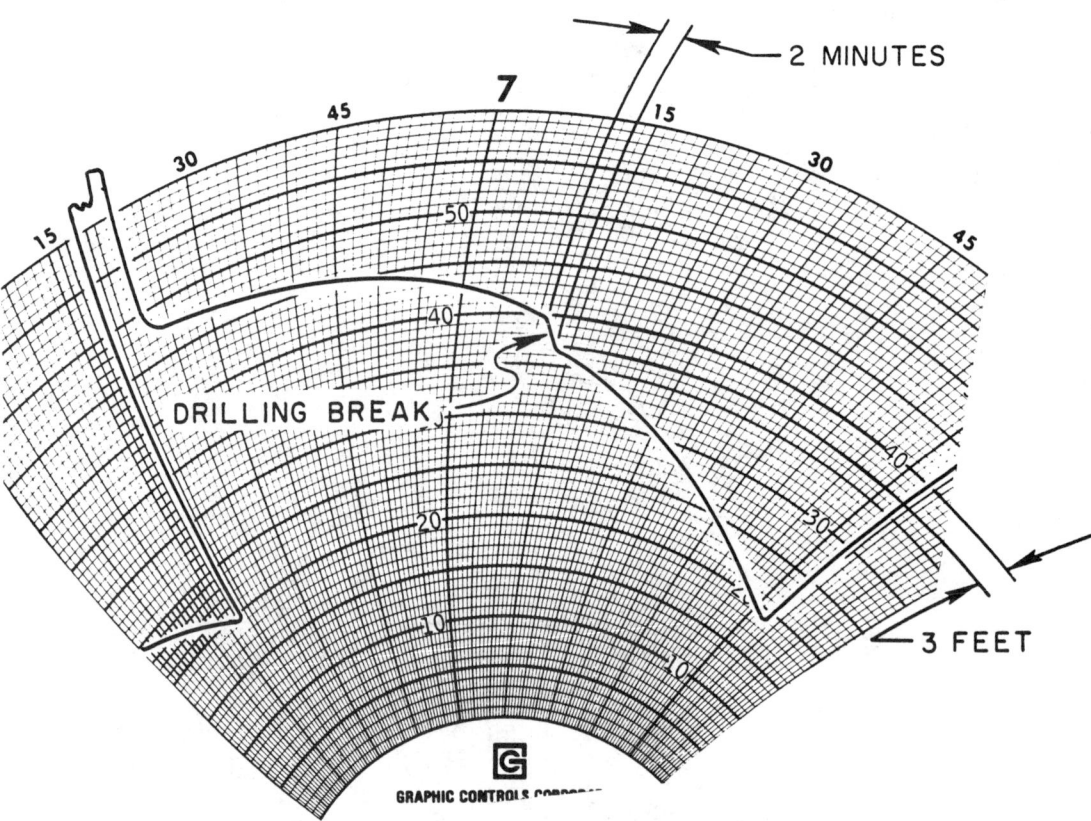

Figure 5-4. Kelly Height, Recorder Chart, Example No. 2

FACTOR	MINUTES*	FOOTAGE INTERVAL									
		1	2	③	4	5	6	7	8	9	10
(x60)	1	60	120	180	240	300	360	420	480	540	600
(x30)	②	30	60	90	120	150	180	210	240	270	300
(x20)	3	20	40	60	80	100	120	140	160	180	200
(x15)	4	15	30	45	60	75	90	105	120	135	150
(x12)	5	12	24	36	48	60	72	84	96	108	120
(x10)	6	10	20	30	40	50	60	70	80	90	100
(x8.6)	7	9	17	26	34	43	51	60	69	77	86
(x7.5)	8	8	15	22	30	38	45	52	60	68	75
(x6.7)	9	7	14	20	27	33	40	47	53	60	67
(x6.0)	10	6	12	18	24	30	36	42	48	54	60

(Feet Per Hour)

To use, locate time spent in drilling the interval in "Minutes" column, then read horizontally across to vertical column under Footage Interval and read drilling rate in ft/hr. The column of Factors at the left is used for calculating drilling rates of intervals larger than 10 feet.

* Time spent in drilling interval.

Figure 5-5. Drill-Rate Chart for Random Intervals

As drilling progresses, the depth and the pump-stroke counters advance. When the end of a logging interval (sample interval) is reached, the readings on the pump-stroke counters are totaled and entered on the Data Worksheet (Figure 5-6) in the Sample Finish column opposite the depth at the time. The lag strokes (the strokes needed to displace the annular volume) are then added to this and entered in the Sample Up column. When the pump strokes have advanced to this Sample Up figure, the sample for that particular interval will be at the surface and ready to be collected.

The kelly height curve (Figure 5-3) is marked according to the sample interval and frequency at which drill rate is to be recorded. In this particular example, the sample interval is 10 feet and the drill rate interval is 5 feet.

Major factors that control the drilling rate include lithology or rock type, porosity, bit type, weight on the bit (WOB), rotary table speed (rpm), pump pressure, and drilling fluid properties. Figure 5-7 shows typical drilling rates that should be computed and plotted on the log.

Sharp increases in drilling rate are known as "drilling breaks," and sharp decreases are "reverse drilling breaks." Extra cuttings should be collected and analyzed from these changes in drilling rate. However, lithology changes may occur with no change in drilling rate. Therefore, although the drilling rate normally reflects changes in formation, it cannot be relied on entirely. After the cuttings have been examined, the type of rock can be noted directly on the chart (Figure 5-7). Refer back to Figure 5-3 for an example of constant drilling rate and lithology.

160

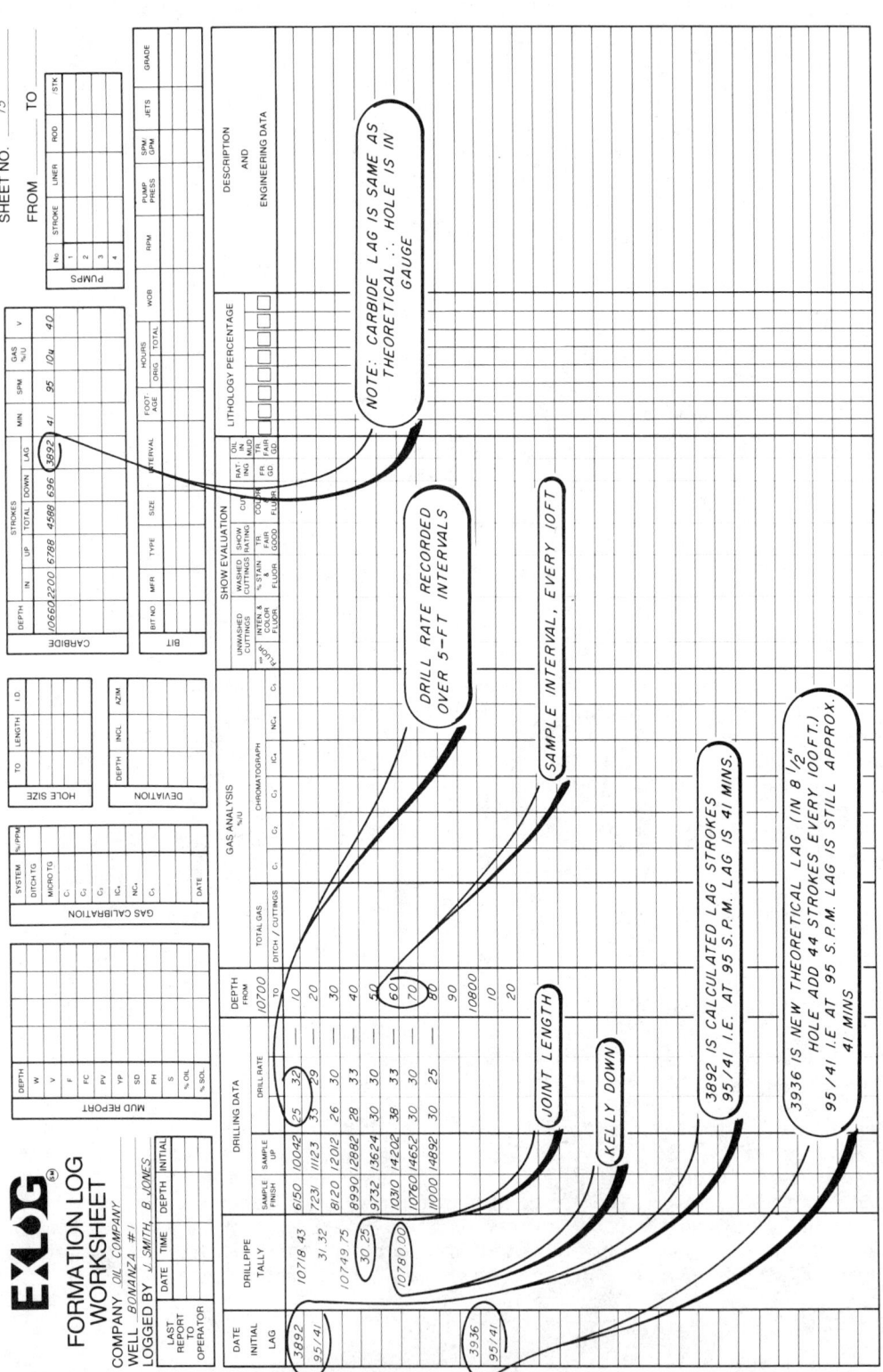

Figure 5-6. Example Data Worksheet

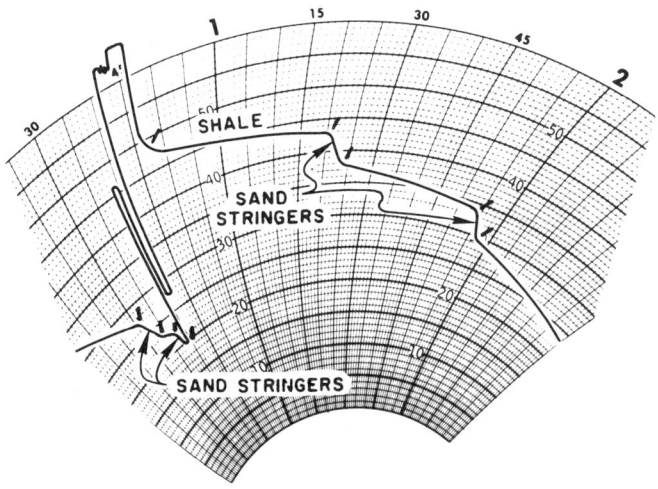

Figure 5-7. Kelly Height, Recorder Chart, Example No. 3

General rig operations can also be interpreted from the kelly height chart using the kelly position (height or other character of the curve, or both). Operations other than routine drilling should be noted directly on the chart for later reference (Figure 5-8). Also, any other items that may be helpful can be noted (lag time, pump off, etc.).

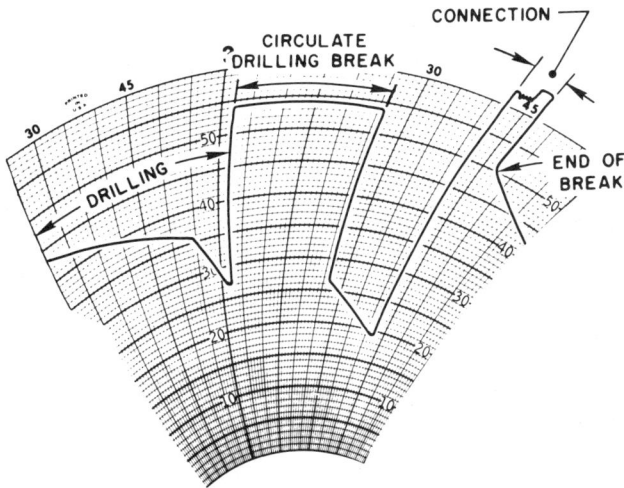

Figure 5-8. Kelly Height, Recorder Chart, Example No. 4

Figure 5-9 represents a well on which drilling is underway and a Data Worksheet and Log are in the process of being compiled. It presents diagrammatically the flow pattern whereby information as to the physical in-place characteristics of the formation is transformed into representative parameters on the log. The wellbore is a pipeline from the bottom of the hole through which the lithology and related information flows to the surface. The cuttings and fluid dispersed from them in the mud represent a continuous column of subsurface information moving up and out of the hole. This continuous flow of data is being fed at the bottom of the hole by the bit and is transferred onto the Data Worksheet and Log by the logging geologist.

Referring to the example data Worksheet (Figures 5-6 and 5-9), when bottomhole depth was 10,710 ft, the Total Pump Strokes read 6150, the lag was 3892 strokes, and cuttings from newly drilled formation of that depth started up the hole. By the time a depth of 10,720 ft was reached, the counters registered 7231. As this 10 ft of hole (10,710 to 10,720) was being drilled, the cuttings from this interval were released into the mud and became dispersed in the annulus. The bit continued to drill ahead while the mud and cuttings from this interval were coming up the hole. By the time the pumps make 2811 more cycles, the counters will have advanced to a reading of 10,042 and cuttings from the top of the interval (10,710 to 10,720) will begin to pass over the shaker screen. Cuttings and mud from this interval will continue to arrive at the surface until the counters read 11,123, after which returns will be from the next logging interval (10,720 to 10,730). The mud gas readings and the data from a sample collected when the counters read 11,123 may properly be recorded for the interval 10,710 to 10,720.

A check in the Sample Finish and Sample Up columns on the Data Worksheet reveals that, while sample returns from the 10,710-to-10,720 interval were coming to the surface, the bit had progressed to a depth below 10,780 ft. If the same procedure is followed for successive samples, all the subsurface information will have been restored to its correct in-place depth.

5.10 Gas Determination from the Drilling Mud: Mud logging is performed by using the returning mudstream as a medium of communication with the bottom of the hole (bottomhole). There is a relationship between the kind and amount of gas or oil (or both) in the drilling mud arriving at the surface, and the gas and oil (or both) that was in place in the formation being drilled at the time that portion of mud was passing bottomhole.

The gases, if present, are released by the cuttings into the mudstream and entrained, probably in solution, in the drilling mud. It remains, then, to remove and detect this parameter from the mudstream. To do this, we used the Total Gas Detector and chromatograph and their associated pieces of equipment consisting primarily of three components:

- The gas trap which continuously samples the drilling mud and simultaneously removes the gases from it

- The equipment which transports and regulates the air-gas mixture from the trap to the detector in the logging unit

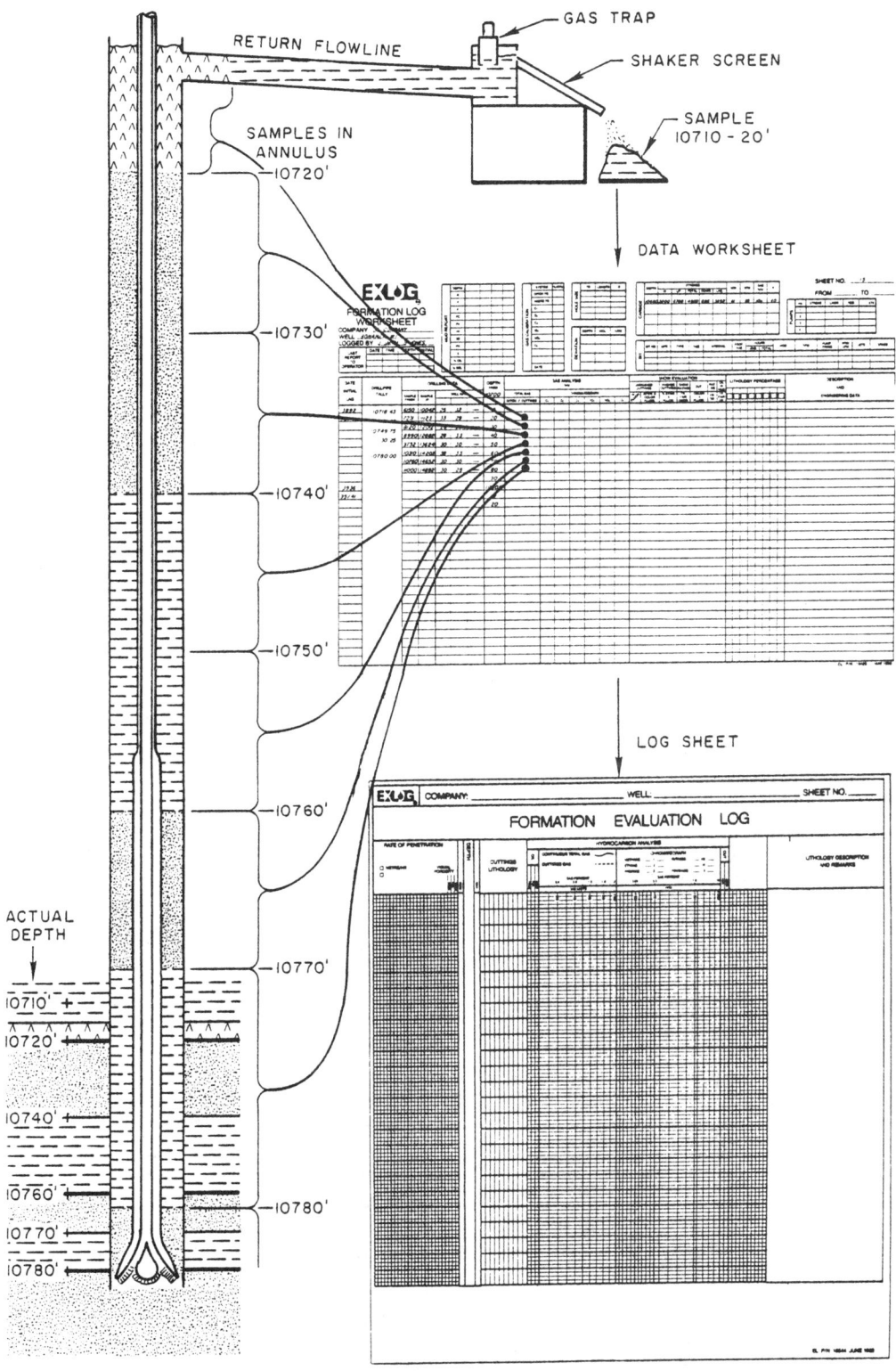

Figure 5-9. Depth Correction of Mud and Cuttings

- The gas detector and the chromatograph proper which process the air-gas mixture into quantitative and qualitative gas readings, respectively

The entire process, which is continuous and automatic, is illustrated in Figure 5-10. In Logging Units 3 through 59, however, the vacuum pump is situated after the gas detector and chromatograph.

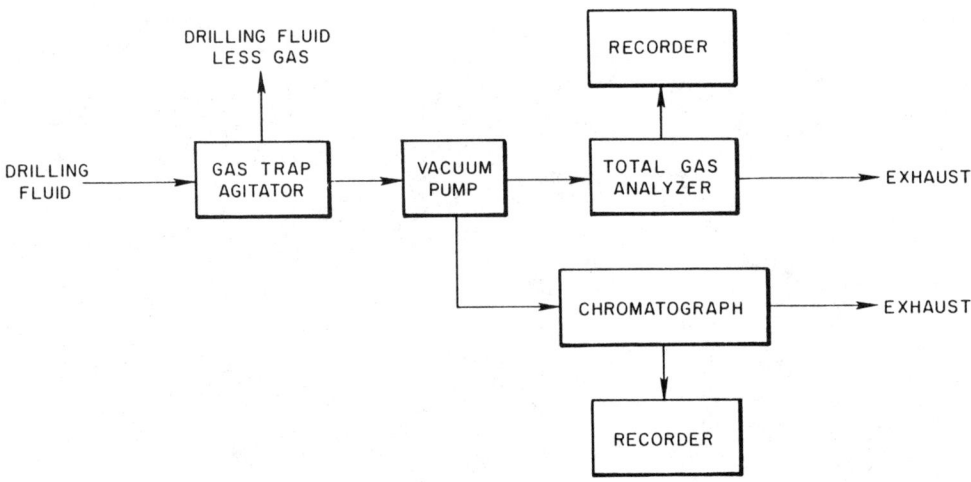

Figure 5-10. Gas Detection Flowchart

5.11 The Gas Trap: To meet the unique requirements of mud logging, this device must perform important functions:

- Extract the gas contained in the drilling mud, independent of such variables as density, viscosity, and gel strength of the mud

- Sample consistently, regardless of the flowrate of mud through the circulating system

Exploration Logging's gas trap (Figure 5-11) is a rectangular steel box that sits in the mud ditch (as near the flowline exit as possible, but before the shakers) and allows the drilling fluid to continuously pass through it by means of slots in the base. The mud level should always be 1 to 2 inches above these slots. An agitator motor sits on top of the gas trap and has a propeller shaft extending into the trap. The propeller continually agitates the drilling fluid as it passes through the trap. On the instrument panel a circuit breaker switch controls the motor, and an ammeter and warning light indicate whether the motor is on and running properly.

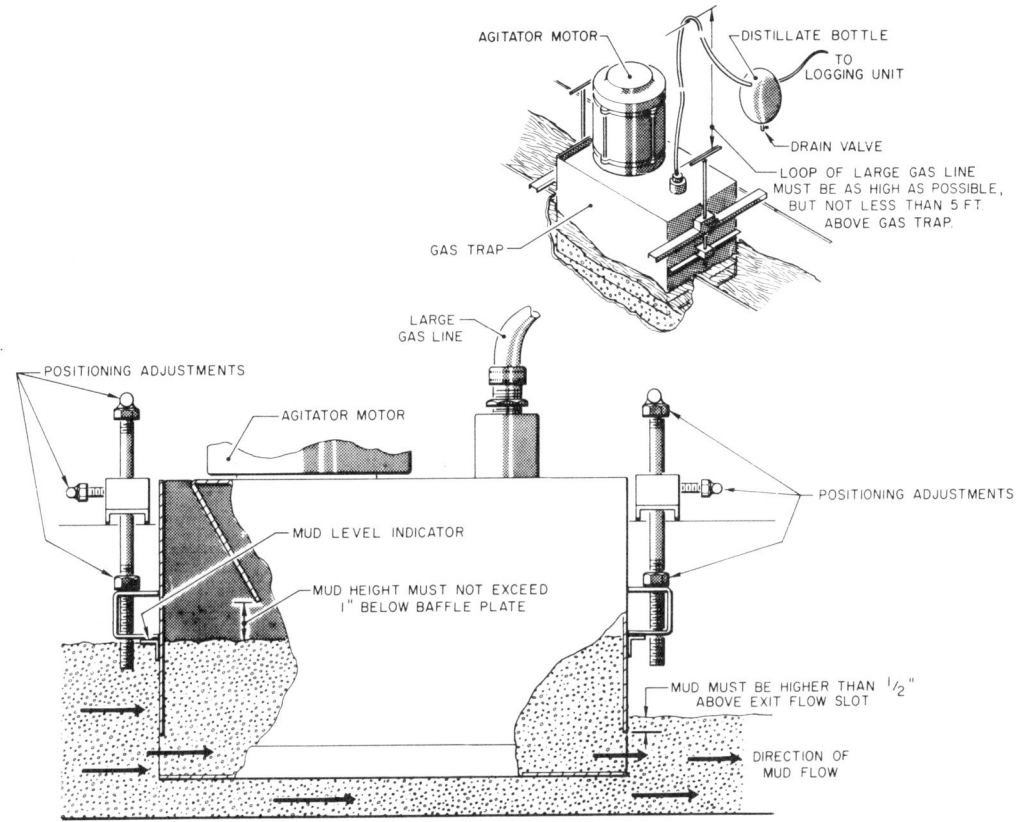

Figure 5-11. Gas Trap

The efficiency of this trap is excellent, particularly with the low-molecular-weight hydrocarbon gases. A continuous flow of air enters through a vent in the top of the trap and is whipped through the mud where the maximum mud surface is exposed. The efficiency for methane separation from the drilling mud is not seriously affected by normal changes in mud properties. As the drilling fluid occupies approximately one-third of the gas trap's volume, any gas that is agitated from the mud will merge with the air that is occupying the remaining two thirds. It is this air-gas mixture that is subsequently drawn into the Gas Detector.

5.12 The Vacuum System: After the gases are removed from the mud, they are transported to the Gas Detector in the logging unit. This is accomplished by a vacuum pump which is connected to the trap by a length of hose. Through this hose the pump pulls a continuous measured stream of fresh air in through the vent of the trap. Because the gases (if present) are being continuously extracted from the mud in the trap, they are mixed with this stream of air and carried into the logging unit via a condensate bottle, where water vapor is extracted. There, the flow of air or air-gas mixture passes through additional flow-regulation equipment, plumbing and instruments and arrives at the filament where a continuous gas reading is obtained. The vacuum system in Logging Units 59-149 is presented in Figure 5-12. Subsequent units are slightly different, and units prior to these utilize the vacuum pump after the Gas Detector. The various components of the vacuum system are described in detail in the Care and Operations manual appropriate for the Logging Unit number.

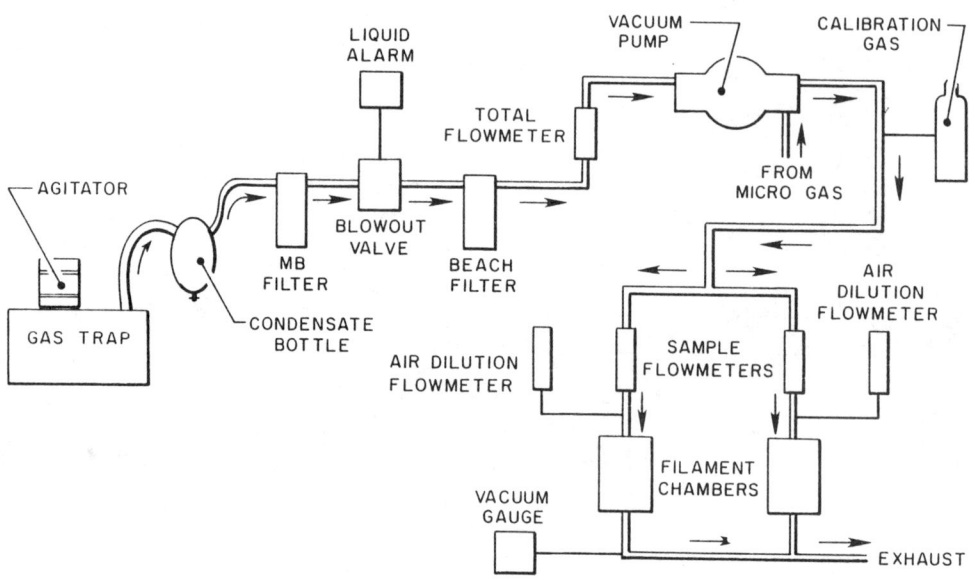

Figure 5-12. Vacuum System

5.13 The Gas Detector: Logging units prior to number 157, the first ALFA-GEMDAS X, have catalytic gas detection systems, and unit numbers 157 and up have flame ionization systems.

- Catalytic Gas Detector. This instrument functions on the principle of catalysis, i.e., the catalytic oxidation of gases on a filament in the presence of air. Figure 5-13 is a schematic diagram of such a detector. It is an application of the Wheatstone bridge measuring circuit in which a resistance (the detector filament), which varies according to the concentration of gas, is balanced against a fixed standard (the reference filament). The reference filament is coated with an inert compound to seal the catalytic surface from the atmosphere, and the imbalance is measured. With the normal voltage applied across the entire bridge, both filaments are heated sufficiently to oxidize all gaseous hydrocarbons.

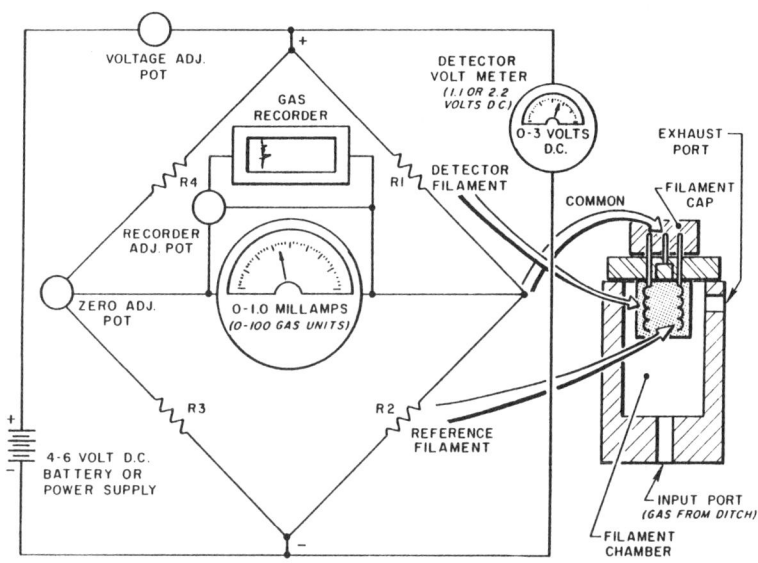

Figure 5-13. Schematic Diagram of a Simple Catalytic Gas Detector

Before the gas detector is placed in operation, it is calibrated using air as a standard. A valve (Zero Adjust) is opened to admit fresh air to the system, which places both filaments in a like atmosphere in which the gas concentration is zero. By adjusting the "Zero" potentiometer (pot), the gas meter, which is a sensitive milliammeter, is adjusted to read zero. Electrically, the gas detector is then said to be in balance. The Zero Adjust valve is then closed, and the filaments will be in whatever atmosphere is being created in the trap. As long as no gas is being liberated from the mud, the filaments remain in an atmosphere of air and the detector reads "0" gas. However, as soon as any gas from the mud becomes mixed with the air being drawn through the trap, the filaments are surrounded by this atmosphere. Having free access to the detector filament, this mixture oxidizes. The oxidation creates heat; the detector filament temperature is increased (increasing resistance); the electrical balance is upset; and current flows through the milliammeter. The electrical imbalance is greater and the gas reading is higher in proportion to the greater amount of gas present. The resultant gas reading is in "units" relative to the concentration of gas-air mixture from the trap.

With the normal or "high" voltage applied, the filament is heated sufficiently so that all the hydrocarbons present in the air-gas mixture react, and the resultant reading is that of all the gases, or the "total gas" (TG) reading. When the filament voltage is reduced to the "low" value (1.4 volts), the temperature of the filament is accordingly reduced to the point at which only the "wet gases," i.e., petroleum vapors (PVs), react; and the resultant reading is PV. Subtracting the wet gas reading from the total gas reading gives the methane value. For this purpose two meters are provided, one reading TG and the other reading PV.

A concentration of 2 percent methane in air gives a reading of 100 milliamps or 100 units on the milliammeter. Therefore, 1 gas "unit" is equal to 0.02 percent methane in air. If the gas concentration becomes greater than 2 percent, the mixture must be diluted so that the readings will be "on-scale." This is accomplished by introducing air from the atmosphere into the air-gas mixture and is controlled volumetrically by the air flowmeters. However, the flowrate through the filament chamber must be kept constant at 2 standard cu ft/hr (2 scfh). Thus, when a volume of air is introduced, a corresponding reduced volume of the air-gas mixture is being drawn through the system. When the volume of air-gas mixture is reduced by one-half, the scale of the milliammeter (and the recorder) is effectively doubled and the number of units shown must be multiplied by two.

- Flame Ionization Gas Detector (F.I.D.). With this system a continuous sample is fed into a regulated, constant-temperature hydrogen flame. The flame is situated in a high-potential (300 volts) atmosphere between two electrodes. As combustion occurs, the gas ionizes into charged hydrocarbon residues and free electrons. A predictably constant ratio of these charged particles moves immediately to the positive electrode, inducing a current at that probe. The amount of current induced is proportional to the total ion charge produced in the flame and increases as the percentage of hydrocarbons in the sample increases. Current flow is amplified in the instrument electronics, graphically recorded on a chart recorder, displayed on the front panel meter, and digitized by the meter electronics. This meter displays the percentage of methane-equivalent (C1) hydrocarbons present in the ditch sample. It is factory-calibrated to read 1.00 when a 1% methane calibration gas burns in the FID. When burning a ditch sample containing heavier petroleum vapors (those with a greater number of carbon atoms in the molecular structure than in methane), the meter displays a reading reflecting the proportionately greater number of carbon atoms. For example, when burning a 1% concentration of pentane (C5), the meter reads 5.00; when burning a 2% pentane or a 10% methane, the meter reads 10.00: 2% pentane = 2x5 carbons = 10; 10% methane = 10 x 1 carbon = 10. Each of these readings indicates that the relative concentration of combustible hydrocarbons is 10 times greater than that in the calibration gas. Sample gas can be furnished in two concentrations; undiluted (X1), called normal, and diluted (X10), which contains nine parts air to one part sample gas.

- Recording Ditch Gas. Since the gas being analyzed and recorded is coming from an interval in the hole that has been penetrated previously, it is necessary that the time shown on the recorder chart be synchronized with the kelly height recorder. The intervals (or depths), correctly lagged, are noted directly onto the recorder chart. This can be done either by marking the gas curve at certain drilled intervals or by marking gas increases and decreases, or both. Other rig operations may also be noted on the chart (Figure 5-14). In this example, "bottoms-up" (circulating for the period of one lag time when not drilling) was not circulated before making a trip.

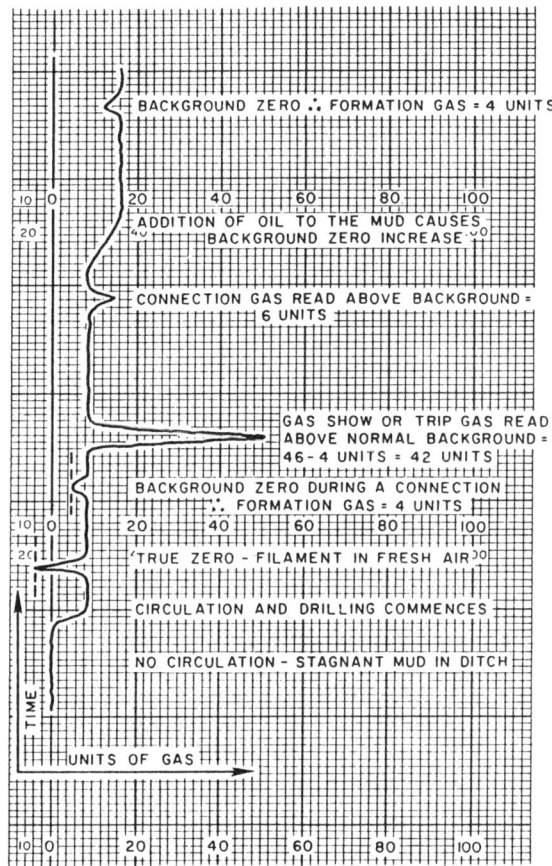

Figure 5-14. Specimen Ditch Gas Recording Chart

Be sure you take readings from a correct baseline. At all times there is background presence of hydrocarbons in the mudstream due to recirculation of trip gas, addition of diesel oil contamination from previous oil-bearing formations, etc. For meaningful formation evaluation it is necessary to plot only the gas released from the formation drilled; therefore, you <u>must</u> establish a true baseline, or zero.

5.14 The Microgas Analyzer: The microgas analyzer, sometimes referred to as the blender gas analyzer, is used to check the combustible hydrocarbon content of the drilling mud and cuttings. It differs from the ditch gas analyzer in that it is a <u>batch</u> system. Samples of the drilling fluid and cuttings are collected and checked periodically — and <u>always</u> during any ditch gas shows. These samples (approximately 100 cc, but <u>always</u> a consistent amount) are placed in the blender jar and agitated for a standard length of time, and the resultant air-gas mixture is drawn through the microgas analyzer. The vacuum is provided by the vaccuum pump, and the filament power is provided by the 6-volt power supply.

The gas combustion, air dilution, milliammeters, voltmeters and flowmeters are all identical to (and are employed in the same manner as) those in the ditch gas analyzer. But as this is a batch system, no recorder is used and the gas readings are read directly from the milliammeters as gas units.

On prospective gas wells the microgas results are used mainly as a check on the ditch gas analyzing system. On prospective oil wells and wildcat wells, the microgas from the cuttings is extremely important as it may form the basis for further evaluation as an indicator of porosity and permeability.

5.15 Recording Gas Information: The following factors affect gas recordings:

- Diameter of the hole — i.e., a greater volume of cuttings produces a greater volume of gas for a given mud volume

- Rate of penetration (ROP) — same as above

- Differential pressure — gas may be bleeding into the borehole from the formation

- Mud properties — density, related to formation pressure; viscosity and gel strength, which affect trap efficiency; and oil content, which also affects viscosity and gel strength

- Geometry of the ditch and position of the gas trap in the ditch

- Distance between the gas trap and the borehole, especially if an open flowline is used

- Nature of the formation

5.16 Trip and Connection Gases: After a trip has been made and drilling is resumed, a period of time equivalent to the lag must transpire before any cuttings or gas shows from formation drilled after the trip may appear at the surface. It is quite common for an increase in the mud gas reading to occur sometime between the time drilling is resumed and the time the first sample from newly drilled formation is due at the surface. This occurrence is commonly referred to as "trip gas." Usually, trip gas makes its appearance toward the end of this period, just before the first newly drilled sample is due.

Trip gas is gas from the formation. It may be from some previously drilled gas-bearing zone. Frequently, however, it may appear after every trip in holes where no significant show has been previously encountered and may be from some section of very low permeability containing gas under fairly high pressure. The presence of trip gas is not fully understood, yet there are a number of theories as to why it occurs. The following theory is one of the more likely.

Visualize what happens as the drillstring is pulled out of the hole, for it is during this operation that the gas which is subsequently labeled "trip gas" gains entry to the mud system. In the process of tripping out, the drillstring is pulled through a mud-filled cylinder of a diameter only slightly greater than itself. A swabbing action of the formation takes place because of a momentary reduction in hydrostatic pressure immediately adjacent to the drillstring. This permits gas to enter the borehole and become enveloped by the mud. The mud column then remains static until the trip is completed. When circulation is resumed, the gaseous interval is pumped to the surface where the gas is detected as trip gas.

The fact that trip gas most often makes its appearance near the expiration of the lag period, indicating it to be from near bottomhole, may be accounted for in at least two ways. First, at the time a trip is started, the bottom section of hole will have been only recently opened up and exposed to the mud column; wall-building and invasion forces of the mud will have been at work only a short time. The hole wall near the bottom will not be nearly as well sealed against entry of the gas as are sections less deep. Because the bottom section has not been as thoroughly invaded by the mud filtrate, the gases will have only a short distance to travel to reenter the hole, compared to shallower zones. Secondly, the mechanical forces which result in the accumulation of trip gas may be expected to be greatest where the hole is nearest to the gauge. The most recently drilled section of hole near the bottom is more likely to be in-gauge than shallower hole which has had the chance to wash out and cave off. Therefore, conditions are much more conducive to creating trip gas near bottomhole than at shallower depths.

It is important to remember that this trip gas will usually accompany the returns for formation that were drilled <u>prior</u> to the trip. There is always the possibility that the gas is not trip gas but <u>rather</u> is a legitimate show which was encountered just prior to making the trip and is coincidental with the appearance of trip gas. The gas reading should be watched closely to see whether it persists as a legitimate show might. Trip gas will usually build up rather rapidly to a peak and begin decreasing almost immediately. The cuttings and other available data should be carefully scrutinized to ascertain definitely whether the gas reading is due to trip gas. The mud and mud-pit level should be watched for indications of a possible kick or for saltwater accompanying the trip gas.

Trip gas will sometimes be observed to recirculate once and possibly several times. This recirculation may be recognized on the recorder by the occurrence of regularly spaced peaks in the gas curve. These peaks tend to broaden when recycled. The time interval between peaks is equivalent to the time required to make one complete circulation of the entire system.

Conditions resulting in the occurrence of trip gas may be aggravated by the air introduced into the mud column if the drillpipe is "floated" in when run back into the hole. This huge air bubble will assist in collecting gas in the mud as it is circulated out. For the same reason, but to a lesser extent, this same accumulation of gas and air may prevail as a result of making a connection.

5.17 Chromatography: The chromatograph separates and analyzes hydrocarbons in the ditch gas sample to determine how much of each hydrocarbon is contained in the sample. Exploration Logging employs two types of chromatograph: the catalytic (Standard) detector, and the flame-ionization detector (FID). Each separates and records the gases in a similar manner, but the difference between the two is the way in which the various gases are detected once separation has occurred.

- Catalytic Standard Chromatograph. The standard chromatograph separates the hydrocarbons by passing the sample through a compound of hexadecane and firebrick. The compound is housed in coiled aluminum columns (upper diagram, Figure 5-15), and a predetermined quantity of the sample is cycled through the columns at 5-minute intervals. The principle of chromatography is that, when forced through a certain medium, different compounds move at different rates depending on their molecular weight. Lighter hydrocarbons pass through the columns first: methane (C1), ethane (C2), propane (C3), isobutane (IC4), and normal butane (NC4). During the first half of the 5-minute cycle, the instrument is in the forward-flow mode and hydrocarbons C1 through C4 pass to column B. During the second half of the cycle the instrument is in the backflush mode. The sample in column B is passed to the filament block for testing while the heavier hydrocarbons, still in column A, are purged from that column by carrier air. The columns are held at a constant temperature ($100°$ to $125°F$) inside the oven to ensure a constant flowrate through the columns. The hydrocarbons flowing to the filament block catalyze on the active filament.

  When the hydrocarbon to be tested enters the chamber, the carrier air and the hydrocarbon combine on the filament. The filament remains unchanged, but the catalyzation causes the filament to heat in proportion to the hydrocarbon concentration in the sample. The active filament is an integral part of a balanced resistance bridge which has a normal output of 0 volts. When catalyzation occurs, both the current through and resistance of the filament change, and the output of the bridge varies. The output of the bridge then goes to the recorder. If higher than C4 analyses are required, the chromatograph can be set to HOLD before backflushing occurs and a single analysis continued for the required length of time.

  The theoretical upper limit of sensitivity of the hot-wire filament for methane is 9.5 percent. At higher concentrations, reversals occur due to insufficient oxygen being available for complete combustion, and the excess methane cools the filament.

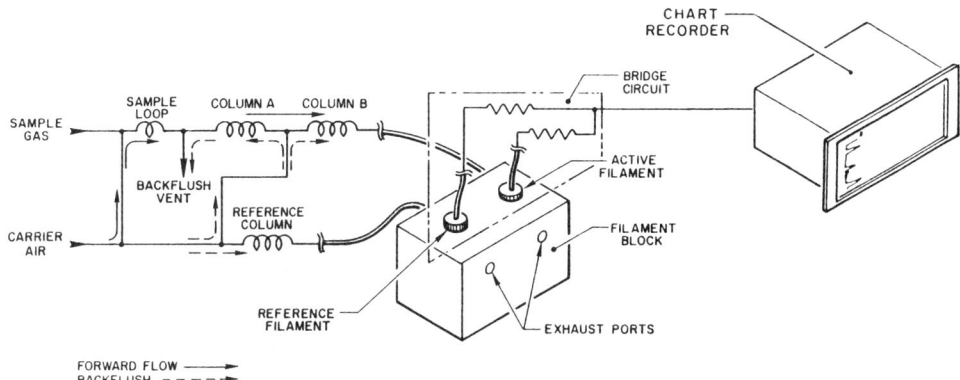

(a) CHROMATOGRAPH COLUMNS AND CATALYTIC FILAMENT BLOCK

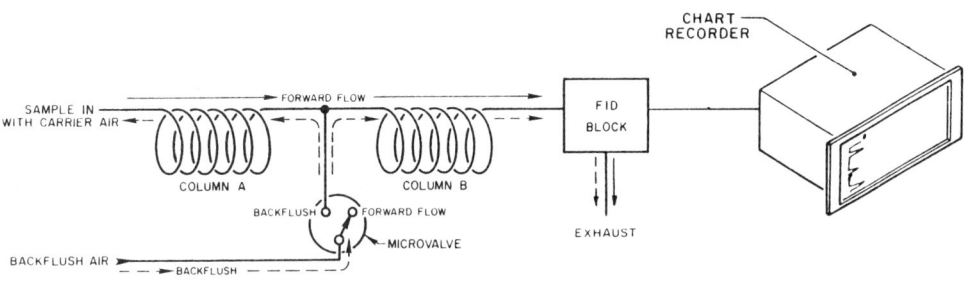

(b) CHROMATOGRAPH COLUMNS AND FID BLOCK

Figure 5-15. Standard (Catalytic) and F.I.D. Chromatographs

There are a few disadvantages to the catalytic chromatograph:

-- It has a limited dynamic range with large sample preparation.

-- It has a negative response to carbon dioxide.

-- It is affected by large amounts of nitrogen and suffers thermal drift due to temperature changes.

- FID Chromatograph. Once separation has occurred, the individual hydrocarbons go to a circular chamber inside an aluminum block for detection, as illustrated in the lower diagram of Figure 5-15. This chamber (the FID chamber) completely encloses a hydrogen flame which is not affected by logging unit pressure or by normal amounts of carbon dioxide and nitrogen. The hydrocarbons are mixed with the hydrogen flow and heated in the chamber. The detector response is essentially proportional to the carbon content of a molecule and depends upon the quantity of gas entering the flame per unit of time. Mixing hydrocarbons with the hydrogen flame produces ions which are attracted to a probe in the FID chamber. The ions then flow to a high-gain amplifier, then to a chart recorder and digital meter.

> The FID has a greater dynamic range and has a wider linear range than the catalytic chromatograph. It is also less likely to be affected by temperature change.

- Recording. The recording (or "signature") of the original mixture is termed a chromatogram (Figure 5-16). The sensitivity of the detector to each gas is established on a regular basis by passing a calibrated sample through the column. Variations in the ratios C1/C2, C1/C3 and C1/C4 are thus readily available. For further information about both chromatographs, refer to the Mud Logging: Principles and Interpretation, Standard Chromatograph and FID Chromatograph manuals.

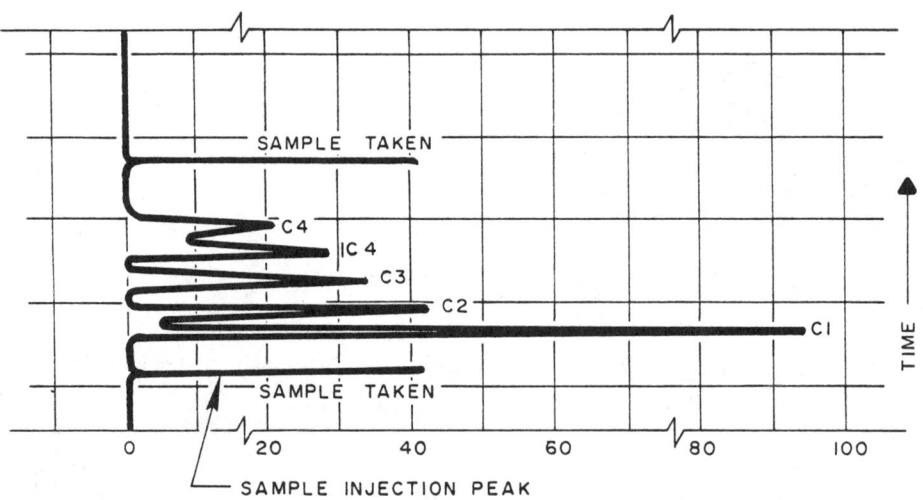

Figure 5-16. Sample Chromatogram

Although more accurate and informative than the continuous ditch gas detectors, chromatography is at a disadvantage when gas shows of short duration occur. If the duration of the gas show is shorter than the cycle time of the chromatograph, the show may be passed without a sample being taken. If a gas show does occur when the chromatography is in mid-cycle, it is possible to manually abort and restart the cycle. To do this, set the chromatograph to backflush for 30 seconds. This allows the sample loop to be filled with fresh sample. The chromatograph cycle can then be restarted. Since the previous cycle was not completed, some heavy hydrocarbons, normally removed by the backflush, may be retained in the column and be carried to the detector during subsequent cycles, causing anomalous peaks. This procedure is not recommended for regular use but may be necessary when a gas show occurs, peaks, and begins to decline before the chromatograph has had time to sample it.

5.18 Ultraviolet-Light Box: This is a self-contained unit employing four 8-watt ultraviolet (UV) tubes and a white light mounted within the cabinet. It is used for determining percentage, physical character, color, and intensity of hydrocarbon fluorescence in drilling fluid and cuttings. It is also used for determining minerals that fluoresce. The upper half of the unit can be removed from the base and used as an ultraviolet scanning device for cores. Further information can be found in paragraphs 5.29, 5.30 and 5.31.

5.19 Mud Press: Each logging unit is equipped with a filter press to which air pressure is supplied by the logging unit pneumatic system. Filtrate and filter cake are significant properties of the drilling fluid, and the filtrate itself is used for various chemical analyses. A good drilling fluid should deposit a good filter cake on the wall of the hole to consolidate the formation and to retard the passage of fluid into the formation. It must also have a minimum "water loss" to avoid clay swelling problems or excessive invasion. To test these properties, it is necessary to determine the rate at which fluid is forced from a filter press and the thickness of the residual solid film deposited on the filter paper by the loss of fluid. A number of tests can be conducted on the filtrate, including determination of resistivity, chloride content and nitrate ion content as indicated below:

- Resistivity from Filtrate. The resistivity of the filtrate may be measured periodically during the course of drilling and later used for interpretation of electric log resistivity curves. More commonly, however, the resistivity test is conducted on a sample of the mud in the hole prior to running a suite of E-logs. A portable resistivity meter is usually available to conduct this test.

- Chloride Content from Filtrate. The salt or chloride test is very significant in areas where salt can contaminate the drilling fluid. (This would include the majority of the world's oilfields.) The salt may come from make-up water, salt stringers or beds, or from saltwater flows. The procedure for this test is outlined in Appendix B.

- Nitrate Ion Test. Water recovered from a formation or drillstem test cannot always be identified as invasion filtrate or natural connate. The usual method is to determine the salinity (or resistivity) of the recovered water and compare it to the salinity (or resistivity) of the mud filtrate from a filter press. However, in some areas and in some circumstances, the salinity of the naturally occurring formation (connate) water may be indistinguishable from that of the mud filtrate. Tests on the compositon of formation waters have shown that the natural occurrence of the nitrate ion (NO3) is very rare -- almost unknown. Therefore, if nitrate ions are added to the mud, the filtrate will contain nitrate ions and carry them into the formation during invasion. On a subsequent formation test, if the water recovered contains the same concentration of nitrate ion, the fluid recovered is solely filtrate and not formation water. Conversely, if the nitrate ion concentration is reduced, it can be concluded that the ion concentration has been diluted by the entry of formation water.

From this type of information the evaluation of a drillstem or wireline test can be made quickly and confidently at the wellsite. The method may also be used upon well completion for determining the oil-water contact and thus be of help in setting up production zones. The procedure for this test is included in Appendix B.

5.20 Pit Level Indicators: Two types of pit-level indicators are used by Exploration Logging, and they both work on the same principle — a float rises or falls with the mud level. This movement is detected by a change in resistance in the electrical current from the sensor to the logging unit. The basic pit-level indicator consists of a float on the end of an "arm" in the mud pit, attached to a variable resistor enclosed in an oil-filled housing. The pit level is recorded on a chart with as great a span as possible, assuring maximum sensitivity so that any changes in level are immediately recognized. With this system, usually only one (the active) pit is monitored.

Today it is common to monitor more than one pit. Using Exploration Logging's Pit Volume Totalizer (PVT), up to six pits with the basic system, and twelve pits with the microprocessor system (used standardly in GEMDAS X and higher configurations), can be continually measured in any drilling environment. Figure 5-17 shows the basic system. As with the basic Pit Level Indicator, the PVT is used primarily to detect any fluctuation in mud volume caused by hole fluid loss or gain. Transfer of mud between pits does not affect the total pit volume reading, but any significant loss or gain to the pit system causes a change in the reading. The PVT also provides information as to available mud, which can assist in the maintenance of the mud system during normal drilling. The total-pit-volume reading, displayed on a panel meter in digital form, is continuously recorded on a chart recorder and can be multiplexed to the computer if used with a GEMDAS unit. The volume of a pit is determined by using vertical float-actuated sensors to measure the mud level in the pit. This level is then compared to the known maximum capacity of the pit.

The PVT can be operated in one of two modes: (1) an uncompensated mode for use on fixed drilling platforms whether onshore or offshore, and (2) a wave-compensated mode for use on offshore floating rigs. Wave compensation is required when motion causes apparent changes in the level of the mud pits. Using a single sensor, the pit capacity readout fluctuates with the rise and fall of the float, but its effect is overcome by electrical damping.

Reporting changes in pit level is one of the logging geologist's greatest responsiblities. The driller and the derrickman will also observe the level, but their duties often prevent them from keeping a continuous watch. Normally, the logging geologist will be informed of any transfer of mud within the system. However, any change in level with no apparent reason should immediately be reported to the driller. Corrective action (if required) can then be initiated, whether it is from a rise in level due to an influx of formation fluid or gas (signaling the possiblity of a "kick"), or a loss in level (signaling loss of circulation). There are some simple rules

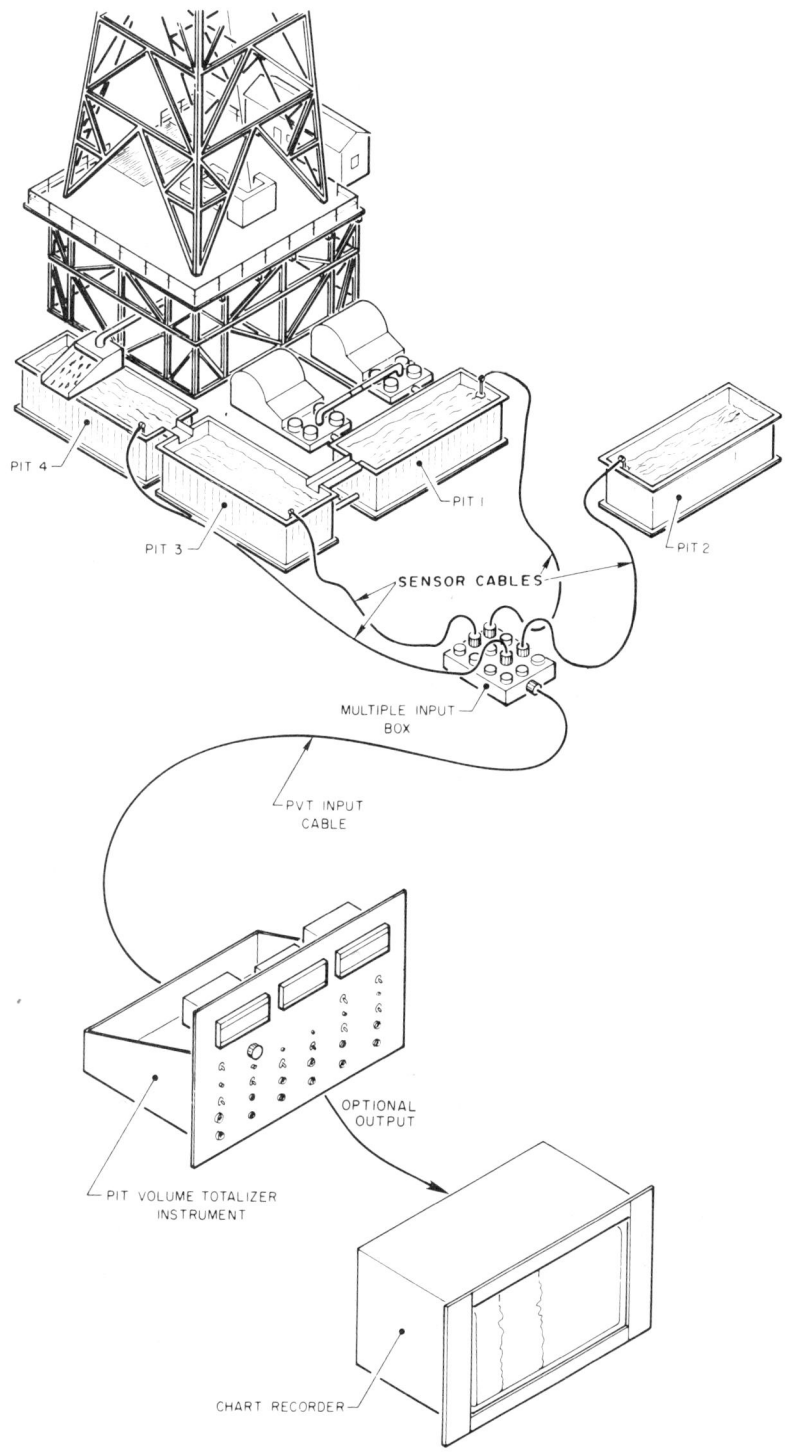

Figure 5-17. Basic PVT System

to follow and questions to ask when a change in pit level occurs. They are listed below, and should give the logging geologist an idea of how important it is to have good communications rapport with all wellsite personnel from the first day. Action should be taken after a 10-bbl gain or loss.

- Rule 1: Inform the driller. You must assume there isn't time to check out the source of the pit level change.

- Rule 2: Check the flowline for normal flow and have the driller stop the pumps if necessary.

- Rule 3: Check the pit level probe for normal operation.

  -- Pit-level losses (slow, gradual): What is the volume per foot or metre of hole being drilled? Can the loss be accounted for by this? Is the desilter, desander running? (You may not have been told that the centrifuge was kicked in; you may be in a sandy formation -- check it!) Slow, steady mud losses usually occur in loose, porous formations.

  -- Pit-level losses (rapid, dramatic): Are they about to trip? Is a pill/slug being prepared? Have the sand traps just been dumped? Are they displacing seawater after a cement job? Is it an interpit transfer? Rapid, dramatic losses usually occur in jointed, vuggy formations or just after casing if the hydrostatic pressure exceeds the formation fracture pressure.

  -- Pit-level gains (slow, gradual): Adding water? Adding diesel? What are the gas readings? Is the mud being cut?

  -- Pit-level gains (rapid, dramatic): Stopped pumping (mud draining out of the kelly hose)? You don't need to inform the driller he has stopped pumping -- he knows this. Degasser off? The degasser tank holds roughly 20 bbl or more; this may drain back into the system when it is switched off. Interpit transfer?

When tripping-in, mud displaced by pipe fills the pit. When tripping-out, mud may be swabbed into the pits by ascending pipe. Mud is transferred back to the annulus periodically from the trip tank or possum belly to keep the hole full. Check the total capacity and displacement volumes of the pipe to see whether it corresponds to the gain in mud when tripping in, or to the loss of mud when tripping out.

5.21 Mud Logging Equipment Flowchart: The pieces of equipment described thus far constitute the minimum requirements for a basic mud logging job. Their relationships are displayed on the flowchart shown in Figure 5-18.

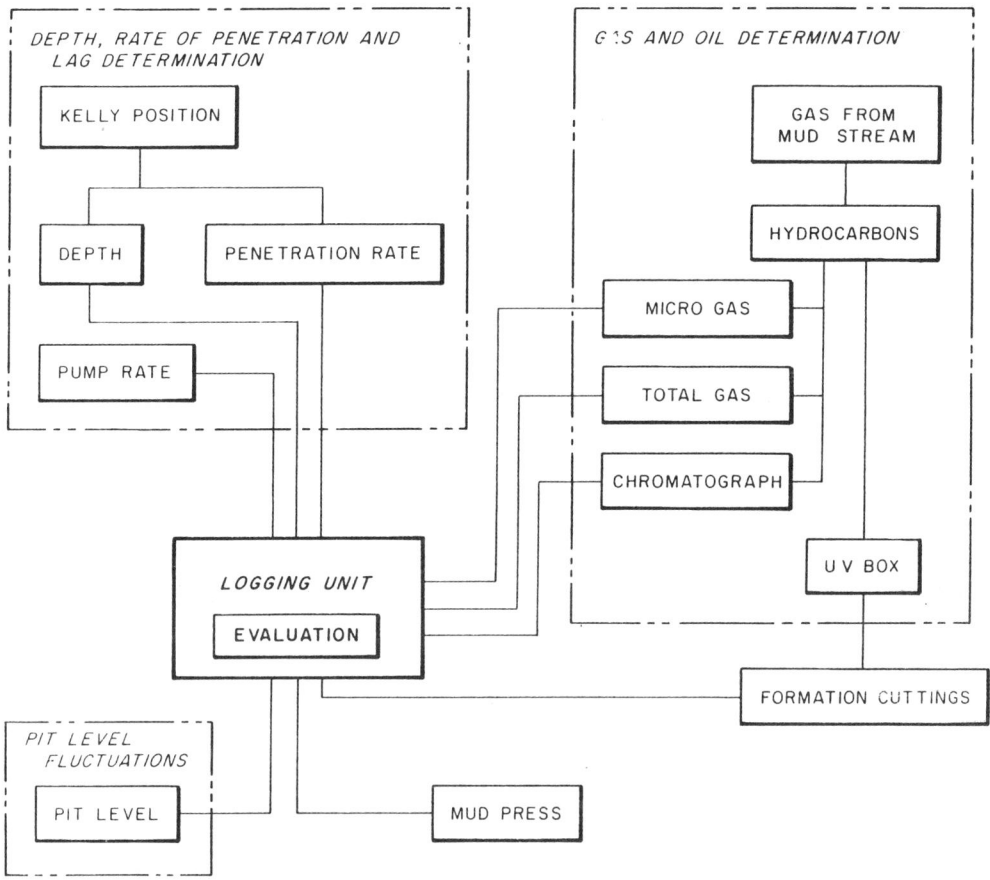

Figure 5-18. Logging Unit Basic Data Inputs

5.22 Logging Procedures: When first reporting to your shift each day, check that the following are operative and correct (don't assume someone else has taken care of this).

- Blow out the gas lines, drain the condensate bottles and check the filters.

- Clean the air slot on the gas trap and check that the level of the trap in the mud is correct; repeat cleaning the slot every two hours.

- Fill the kelly-height line and check that it has not shown any characteristics of leaking.

- Check the kelly-down figures with the driller, and be sure the marked depth is correct.

- Observe the sample and its description logged by the last-shift geologist, and discuss how you are each describing it.

- Check that the mud log is up-to-date.

- Be sure the gas alarm is turned ON and set about 0.2 to 0.4 percent above the background.

- Check to see when the last carbide was run (one must be run at least every 400 feet or 12 hours, whichever is shorter; see paragraph 5.7).

Commence logging: ALL DATA MUST BE RECORDED ON THE WORKSHEET!

- Every sample interval,
    1. Mark the kelly-height chart every 5 ft or less.
    2. Record the pump-stroke-counter reading.
    3. Calculate when the sample is due up.

- Keep the drill rate up-to-date on the Worksheet.

- When catching samples,
    1. Mark depth on the Total Gas chart, the chromatograph chart and other relevant charts. Record values immediately on the Worksheet (Total Gas, C1, C2, C3, C4, etc.).
    2. Check washed sample for fluorescence. If fluorescing, check for cut with chlorethene; then describe the sample and relative percent of fluorescence.

- Do not make the mistake of falling behind in drafting the log. Logs should be up-to-date at least half an hour before the end of the tour. The following guidelines will be helpful:
    1. Never let the drill-rate plot fall behind more than two singles.
    2. Never let the lithology column or gas curves get more than 100 feet behind unless drilling is extremely fast.
    3. Keep descriptions up-to-date. Whenever new lithology appears, immediately update the preceding ones.
    4. When typing log descriptions, follow the established format. Consistency is very important! Start descriptions as close to the left margin as possible and do not run over the margin. Describe lithologies as in paragraph 5.25. When the typed material does not concern lithology (casing, wireline logs, mud data, etc.), start a couple of spaces in from the margin so that a box can be drawn around the typing. Extend the box to near-margin ends.

- The client and wellsite geologist may have their own special requirements regarding procedures to follow upon encountering drilling breaks. Generally, the rig operators require notification to the driller upon making 2 to 3 feet of hole so that a flow check can be made. The wellsite geologist and company man probably will require notification and may circulate returns after the break has extended for 5 to 10 feet. Check with the individuals concerned.

- While logging, notify the drillfloor personnel of (1) rises and falls in the pit level, (2) excessive gas readings, and (3) trip gas. Again, you must follow the client's procedure.

- When there is a gas show follow these general procedures, but refer to paragraphs 5.29-5.31 and Figure 5-22 for more detailed information about what to do when encountering a show.

    1. Notify the driller of rise in gas.

    2. Attenuate all gas instruments and chart recorders appropriately; mark on charts the value of attenuation.

    3. Catch samples from the shakers at regular intervals and especially during gas peaks.

    4. Check for fluorescence, cut, etc.

    5. Mark all gas peaks with the depth interval from where they came.

    6. When the "show" is over, bring everything up-to-date as soon as possible.

5.23 Samples

The importance of the cuttings samples cannot be overstressed. There is no substitute for representative cuttings samples accurately correlated to the depth from which they came.

5.24 Collection and Preparation: Every rig has shaker screens for separating the cuttings from the mud as they reach the surface. If the screen mesh is small enough to remove small cuttings and the job is in an area where there is reason to believe that no unconsolidated sands will be encountered, the shaker screen will provide a collection point for composite sampling (i.e., interval sampling). However, when unconsolidated sands pass through the screen, they can be extracted from the mud by desanders and desilters and a sample collected from them for examination. This sample should be considered along with the shaker screen and composite samples when making an overall evaluation.

Cuttings samples should be taken at regular intervals as often as possible, and never at intervals greater than 15 minutes. Fill the sample bags progressively to give a representative sample of the whole interval. Mark the bag with the well name and depth interval. Also, take samples when changes in penetration rate or background gas are noticed as these often indicate a change in formation lithology or porosity.

Take care at the shale shaker to ensure that a representative sample is taken with minimum cavings. Check the desander and desilter outlets regularly for fine sand which passes through the shale shaker screen.

Washing and preparing the cuttings are probably as important as the examination itself. In hard rock areas, the cuttings are usually quite easily cleaned, in which case it is a matter of washing the sample in a sieve to remove the mud film. In many areas, however, particularly areas and zones of loose sands and shales, it is more difficult and requires several precautions. Primarily, the clays and shales are often soft and of a consistency which goes into suspension and "makes mud." Take care to wash away as little of the shale as possible; and, in determining the sample composition, take into account that which is washed away.

After the cuttings have been washed to remove the mud, wash them through a 5-mm sieve unless doing so will cause excessive loss of shale or clay. It is generally considered that newly drilled cuttings will go through the 5-mm sieve and that the material which does not is cavings and may be discarded.

Cuttings from wells drilled with oil-based or oil-emulsion muds are usually more representative of the drilled formation than cuttings drilled with water-based mud because the oil emulsion prevents sloughing and dispersion of clays and shales into the mud. At the same time, washing and handling cuttings drilled with this type mud poses somewhat of a problem; they cannot be cleaned by washing in water alone. It is usually necessary to wash the cuttings first in a detergent solution to remove the oil mud. (Refer to Appendix C for techniques used with oil-based muds; washing instructions are included.) Some of the liquid commerical detergents available may be used. In extreme cases, you may need to wash the cuttings first with a nonfluorescent solvent such as naphtha, then wash in a detergent solution to remove the solvent. Use a solvent only if absolutely necessary because you don't want to risk removing any oil staining present. Always follow the client's instructions.

An oven mounted on the wall of the logging unit is used to dry a portion of the cuttings sample after it has been washed, but some of the washed cuttings are examined wet under the microscope.

5.25 Examination of the Cuttings: Check them primarily for lithology, staining and porosity; the objective is to depict changes of formation and the appearance of new formational materials. The microscope and ultraviolet light are used as complementary tools in reconstructing the characteristics of the originating strata. Estimate the percentages of lithology, staining and porosity with great care since factors such as grain shape and size, color, distribution, etc., may affect the apparent relative percentages (Figure 5-19).

There are many potential sources of contamination to consider when undertaking estimates of lithology percentages, examples of which are:

- Cavings (cuttings from previously drilled intervals rather than from the current interval). Although ditch cuttings are first washed through a coarse sieve to remove cavings, some may remain in the sample. Cavings may be recognized as generally large, splintery rock fragments that are often concave or convex in cross-section. They are lithologically identical with formations from higher sections of the open hole. If found in large quantities, this may indicate a serious underbalanced mud condition or a situation where rotation is too fast and the stabilizers are catching on the side of the hole.

- Recycled cuttings. If cuttings are not efficiently removed from the drilling fluid at the shale shakers, desanders and desilters, they may be

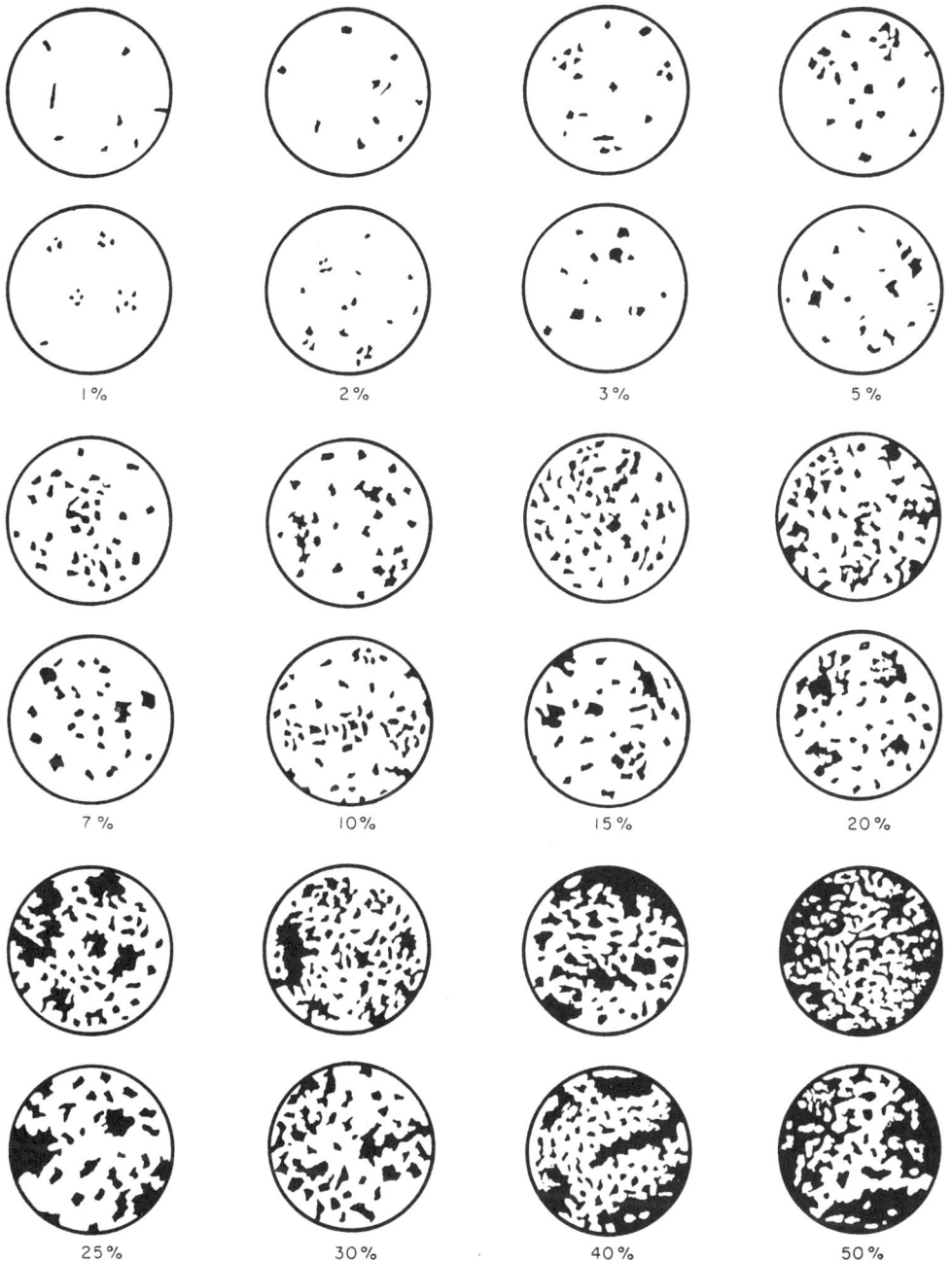

Figure 5-19. Comparison Charts for Visual Estimation of Percentage Composition.

recycled through the mud system. Recycled cuttings may be recognized as small, abraded, rounded rock fragments in the sample.

- Mud chemicals. Some mud chemicals may be confused with rock types. Lignosulphonate, for example, may resemble lignite, and bentonite gel may erroneously be identified as montmorillonite clay in a poorly mixed mud system. Moreover, lost circulation material (LCM) such as nut shells, fibers and mica flakes is a common source of contamination in lost circulation zones (LCZs), and these can possibly be removed from samples by flotation before study.

- Cement. Cement contamination is usually encountered when drilling after casing or while sidetracking. Cement may be mistaken for siltstone but can readily be identified by testing with phenolpthalein solution in which cement stains purple due to its high pH.

- Metal. Metal is occasionally found in samples and frequently originates from wear of the inside of casing by the drillstring. This is often remedied by the use of rubber drillpipe protectors. Exceptionally high amounts of metal should be reported to the drilling superintendent.

In some cases, samples may be totally unrepresentative of the formation at bottomhole. For example, in evaporite sections drilled with a water-base mud, salts dissolve and there is no lithological indication of their presence in lagged samples. However, evaporites can still be recognized by good logging practice:

- Evaporites generally drill at rates of 40 to 60 ft/hr
- Gas values through evaporites will be very low if not zero
- There will be poor or no returns at the shale shakers
- Limestones and dolomites are frequently found in association with evaporite deposits
- Anhydrite sections can usually be identified by $BaCl_2$ solution which produces $BaSO_4$ precipitate
- The chlorides content of the drilling fluids should increase very significantly. Refer to Appendix B for chloride test procedures.

Use only a single layer of cuttings for percentage estimation, and take care to select a representative sample from the sieve because a large degree of shape and density sorting occurs during washing. Once the percentages of the various constituents have been estimated, proceed with the sample description in logical order similar to that below:

    rock type   luster
    color   cementation or matrix
    hardness (induration)   structure
    grain size   porosity
    grain shape   accessories
    sorting   inclusions

5.26 Noncarbonate Clastics: These may be identified as follows.

- Color. The color of a sandstone may be caused by the colors of the constituent grains, color of the cement or by staining of the entire aggregate. The latter is most common in surface sections, but staining also may be prevalent at or near unconformities in the subsurface. Quartz sands may acquire a surface film of coloration during a period of exposure as free sand before final deposition either as a marine or nonmarine aggregate.

  In the subsurface, colors of shales and siltstones are often very significant, either in the correlation of stratigraphic units or in the determination of environments of sedimentation. It has been amply demonstrated in rocks of different ages and in widely separated regions that the colors of shales indicate relative positions in a sedimentation basin. The normal lateral sequence from the shore toward the basin deep is bright red to red and green, to green and gray, to light and dark gray, to dark gray and black. This does not imply that all black shales originate in basin deeps or than all red shales are near-shore deposits. It is, however, a normal arrangement.

- Induration. Induration is its resistance to physical breakdown or disaggregation. Induration does not necessarily refer to the hardness of the constituent's grains, though they may have considerable influence on the degree of induration. Common adjectives describing induration are dense, hard, medium-hard, soft, spongy, friable, compact, brittle, slatey.

- Grain size. Grain sizes of both carbonates and noncarbonate clastics are charted in Figure 5-20.

- Grain shape. There is an almost infinite number of variations in the shapes of clastic grains, but for the sake of simplicity and consistent recognition, only five general classes are considered and are defined as follows:

  — Sharp: conchoidal surfaces terminating in sharp edges and corners

  — Angular: flat, plane surfaces, generally terminating in acute or right angles; edges usually thin to sharp

  — Subangular: flat, plane surfaces terminating in well-rounded edges

  — Rounded: generally rounded surfaces, broadly rounded edges and corners

  — Well rounded: all surface convex; nearly equidimensional; spheroidal

- Sorting (texture). The degree of sorting is one of the most important features of a clastic rock. In petroleum geology the sorting of sands in a potential oil reservoir has particular significance, for it is one of the features which determines the effectiveness of porosity and permeability of the rock. A poorly sorted sand generally has low porosity and permeability. The texture of the rock is determined by its coarseness and by the sorting and arrangement of the grains. The relative porosity and permeability of a sandstone can be estimated by carefully placing a drop of water on a dry chip and observing under the microscope how rapidly the water is absorbed into the rock.

MM	NON-CARBONATE CLASTICS		CARBONATES			MM
			GRAINS	CRYSTALS		
256.0	BOULDER	CONGLOMERATE	CALCIRUDITE	COARSELY MEGACRYSTALLINE	MEGACRYSTALLINE	256.0
64.0	COBBLE					64.0
4.00	PEBBLE					4.00
2.00	GRANULE			FINELY MEGACRYSTALLINE		2.00
1.00	VERY COARSE SAND	SAND	CALCARENITE	VERY COARSELY CRYSTALLINE	CRYSTALLINE	1.00
0.50	COARSE SAND			COARSELY CRYSTALLINE		0.50
0.25	MEDIUM SAND			MEDIUM CRYSTALLINE		0.25
0.125	FINE SAND			FINELY CRYSTALLINE		0.125
0.062	VERY FINE SAND		CALCARENITE (very fine grained)	VERY FINELY CRYSTALLINE		0.062
0.031	COARSE SILT	SILT	CALCILUTITE	COARSELY MICROCRYSTALLINE	MICROCRYSTALLINE	0.031
0.016	MEDIUM SILT			MEDIUM MICROCRYSTALLINE		0.016
0.008	FINE SILT			FINELY MICROCRYSTALLINE		0.008
0.004	VERY FINE SILT			VERY FINELY MICROCRYSTALLINE		0.004
	CLAY	CLAY		CRYPTOCRYSTALLINE	CRYPTO-MICROCRYSTALLINE	

Figure 5-20. Grain Size Terminology

- Luster. The luster or surface texture of sand grains sometimes reveals the history of the grain before and possibly after, deposition. Below are definitions of commonly used terms.

    — Coated: precipitated or accretionary material on the surface of the grain. Iron oxide, calcium carbonate, sulfates, clays, pyrite, etc.

    — Pitted: solution or impact pits, often of pinpoint size

    — Frosted: deeply etched, frosty, translucent, usually white

    — Silky: lightly etched or scoured

    — Oily: greasy or oily sheen; common in hematite and magnetite grains

    — Vitreous: glassy, shiny

    The lusters of shales are earthy, resinous (usually dolomitic), waxy, soapy, oily, silky, velvety and sooty.

- Cementation. Identify the character and composition of cementing material whenever possible. The cementing of a sandstone has a significant bearing on its performance as an oil reservoir. The character of the cementing material may reveal much as to the rock depositional history and postdepositional alteration. Note the relationship of the cement to the constituent grains. Common cements are calcite, dolomite, sulfates, iron oxides, silica, pyrite, clays, silts and siderite. Some sandstones are compacted into firm aggregates, yet have no discernible cement.

- Structure. The structure of a sandstone as used in the description of well cuttings refers to such characters as laminations, fractures and fracture patterns, banding and nodular or concretionary characteristics. Common structures of shales and siltstones are massive or lumpy, platy, laminated, foliated, fissile, splintery, flakey, jointed and fractured. In some areas fractured shales serve as oil reservoir rocks; therefore, it is important to record the presence of fractures.

- Accessories. In addition to the minerals constituting the bulk of the rock, other minerals in very small amounts may also be present. Although they constitute a negligible portion of the aggregate, they are still very important. These minerals are often diagnostic of the environment of the sedimentation, source areas of the sands or the mode of transportation. Some of the minerals may be diagenetic and thus indicate the postdepositional history of the sandstone. Some of the more common but minor constituents of sandstones are biotite, muscovite, glauconite, pyrite, barite, siderite, cherts, coals, solid hydrocarbons, and a variety of hard, nonmetallic minerals.

  Fossils are generally more abundant in the fine clastics; in many sections they are limited entirely to the shale portions. They are often abundant in tophole sections. Whatever the minor constituents may be, they should be noted in the description.

- Inclusions. Shales and siltstones frequently contain fragments of reworked and redeposited shales, limestones and other types of rocks. They may also contain masses of gypsum, anhydrite, chert, iron oxides, nodules and pellets of barite, pyrite and mud, oolites and concretionary materials and grains of solid hydrocarbons and coals. Such extraneous grains or masses are termed, collectively, "inclusions."

5.27 Carbonates: The identification of rock type does not present any problems except in the case of carbonates. The Exploration Logging Stain Kit has been produced to help with these carbonate determinations. It is capable of differentiating thirteen carbonates and two of the common sulfates. Refer to Appendix B for detailed staining procedures.

Another useful tool when working with carbonates is "calcimetry." It is used to determine the percentage of calcium carbonate and dolomite in a sample. The procedure for this determination is also included in Appendix B.

There is almost an infinite number of gradations between the nearly pure end members of the basic carbonate group of rocks — limestone and dolomite being the end members. Four subdivisions can be determined by characteristic reactions in cold, diluted hydrochloric acid. The following reactions can be observed.

- Limestone. Violent effervescence; frothy audible reactions; specimen bobs about and tends to float to the surface.

- Dolomitic limestone. Brisk, quiet effervescence; specimen skids about the bottom of the container, rises slightly off bottom; continuous flow of CO_2 beads through the acid.

- Calcitic dolomite. Mild emission of CO_2 beads; specimen may rock up and down but tends to remain in one place.

- Dolomite. No effervescence; no immediate reaction; slow formation of CO_2 beads on the surface of the rock; reaction slowly accelerates until a thin stream of fine beads rises to the surface.

These reactions are somewhat modified by the presence of noncarbonate constituents or the physical characteristics of the specimen.

Acid reactions are unreliable when the test chips are wet with water before they are introduced into the acid. The film of water on the chip and the water filling the pores prevent immediate contact of the acid with the rock.

- Color. The normal colors of limestones and dolomites are gray, white, buff and brown. Less frequently they are red, orange, various hues of green, purple and black. These colors may occur in combination in a variety of patterns such as mottling, banding, speckling and grading. The manner of coloring should be stated in the sample descriptions. Red and orange speckling and mottling often occur above and adjacent to large structural uplifts. Red, green and orange are often associated with surface weathering, unconformities and subsurface oxidation through the action of circulating waters.

- Grain size. The grains of carbonate rocks vary greatly in size and general appearance in short lateral and vertical distances. It is necessary to describe the grains completely in order to determine the origin of the rock. The grains may be fragments of limestone, shells, microfossils, oolites, algal remains, crystals, or precipitated grains or any combination of these types (refer back to Figure 5-20).

- Character of grains. The origins of carbonate grains are widely different, and a description of such a rock is not complete unless the character of the constituents is included. It is usually not sufficient to state "fragmented limestone" or "oolitic limestone." The fragments and oolites should be described. Oolites may be classed as spheroidal, spherical, elliptical, irregular, flattened, etc. Fragments are described in much the same way as sand grains: sharp, angular, subangular, rounded, and globular (or spheroidal). When discernible, give the origin of the fragments also; i.e., coral, shell or limestone fragments.

- Texture. The textures of many types of limestones will be indicated if grain shapes and sizes are stated as suggested above. However, certain textural terms widely used in the descriptions of crystalline varieties are given in the following table. It has been observed that certain types of porosity are frequently associated with specific textures. While the textural and porosity types are not always associated, the frequency is sufficient to warrant attention. The porosity in a limestone is in most cases secondary from solution or fracturing.

Textural Description	Typical Porosity
Rhombic — perfectly formed rhombs of nearly equal size, medium to coarse (usually pure dolomite)	Vuggy, drusy crystals in vugs
Sucrosic — sugary, similar to rhombic, but finer, lacking the perfection of crystal form; friable (usually calcitic dolomite)	Interstitial, tubular to cavernous
Microsucrosic — very finely sugary, often quite friable (usually calcitic dolomite)	Tubular to cavernous
Grainy — not visibly crystalline but with definite grains, often chalky in part (limestone, or dolomitic limestone)	Pinpoint to tubular or vermicular
Subcrystalline — glassy or resinous mass (usually pure dolomite)	Sparse pinpoint, tendency to fracture
Slabby — very coarsely crystalline, uneven grain size (rarely dolomitic)	Usually nonporous
Oolitic — spheroidal or smooth-surfaced grains with concentric internal structure	Intergranular or isolated pinpoint
Pseudo-Oolitic — rounded clastic grains simulating oolites	Interstitial to isolated pinpoint

- Structure. Structural features of carbonate rocks include stylolites, fractures, microfractures, laminae, banding, crinkling, concretions, whorls, and brecciation. Many of the structural characters can only be inferred from the small chips in the cuttings sample.

- Inclusions. The term "inclusion" is used here with certain reservations. Masses of noncarbonate material commonly called "inclusions" are in many instances chemical replacements of the original rock. The distinction can usually be made when the rock is viewed in thin or polished section, but not in the rough chip. Anhydrite and gypsum frequently occur as small isolated masses in the carbonate rock. These masses may be either replacements or inclusions that are contemporaneous with the host rock. Cherts occur in a very similar form and are likely to have a similar origin. Unless these masses can be identified as replacements, it is better to describe them simply as inclusions -- a less specific term.

5.28 Evaporites: This group includes anhydrite, gypsum, halite, sylvite, and other chlorides and sulfates as can be identified.

Anhydrite is often amorphous appearing though it does possess perfect cleavage. It is considerably harder than gypsum; it cannot be scratched with the fingernail. It tends to be translucent with a pearly luster. Gypsum occurs as a fibrous to lacy mass of selenite crystals, as a glassy solid mass with a subvitreous luster or as a snowy, earthy to massive rock (alabaster). Any form of gypsum can be scratched easily with the fingernail. Anhydrite is brittle and gypsum is usually spongy, so note the manner in which the chip breaks when crushed on the microscope stage.

Rock salt (halite) can usually be identified by its taste, solubility in water and perfect cubic cleavage. When it occurs in finely disseminated crystals it might be confused with barite, which has a prismatic cleavage. Barite is insoluble in cold or hot hydrochloric acid, whereas halite is soluble in all three.

5.29 Hydrocarbon Evaluation: The standard mineralight for the investigation of fluorescence in minerals has a wavelength of 2700A ("A" is the angstrom unit and is equal to a ten-millionth of a millimeter). In petroleum work, however, the long ultraviolet (3600 A) source used by Exploration Logging is much more effective in producing fluorescence in the visible region since it is itself on the threshold of visibility; even a small change in the energy of the fluorescing radiation will bring it into the visible region. Ultraviolet light of this wavelength causes fluorescence of almost all crude oils over the entire range of gravities. The color of fluorescence is generally characteristic of the gravity of the crude oil, as shown below and in Figure 5-21.

Gravity, API	Color of Fluorescence
Below 15	Brown
15 - 25	Orange
25 - 35	Yellow to Cream
35 - 45	White
Over 45	Blue-white to Violet

Very-low-gravity oils are difficult to see because they fluoresce very little, probably because of molecular decomposition. The high-gravity oils are difficult to see because (1) some of their fluorescence occurs in the ultraviolet region, and (2) they have high concentrations of the low-molecular-weight paraffin hydrocarbons which do not fluoresce at all. The presence of refined rig oils, including pipe dope, complicates detection of the light crudes because refined oils may fluoresce white or blue-white. Each sample is examined under ultraviolet light for evidence of fluorescence. Representative chips which do exhibit fluorescence should be picked out, placed in a spot dish and examined under the microscope. Chlorothene, as a leaching agent, is placed on these cuttings under ultraviolet light to determine whether the fluorescence is oil or mineral. The extent to which the cuttings will "cut" in the leaching agent is the prime consideration in evaluating the show in the cuttings. If mineral fluorescence is seen, mention it in the description.

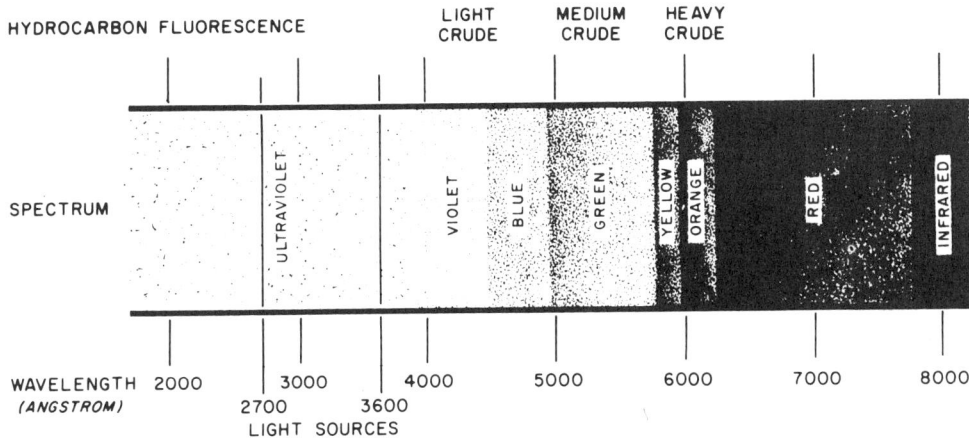

Figure 5-21. Electromagnetic Radiation Spectrum

One of the biggest problems facing a new geologist when evaluating a show is how to recognize and eliminate pipe dope. Pipe dope, which fluoresces, is applied to each tool joint when making a connection and when making up pipe while tripping-in. By keeping a sample in the UV box of the dope used on the rig, comparison can be made when examining cuttings. Also, periodic checks must be made of the chlorothene, for when it is contaminated or old it sometimes gives a cut and therefore leads to false interpretations. If fluorescence or oil staining is seen, or if an oil show is expected, perform tests as described under paragraphs 5.30 and 5.31

5.30 Oil Determination from the Drilling Mud: If any producible oil is left in the formation after flushing and at the time of drilling, it may be produced from the cuttings and found in the mud. The mud, therefore, is the place to look for evidence of producible liquid hydrocarbons from the formation. The mud sample for study must be collected at the point of access nearest the wellhead, usually at the end of the flowline.

Place 200 ml of mud in a small dish, examine it under UV light and observe the color and intensity of oil fluorescence floating on the surface. If none is present add 100 ml of water, stir, and observe again. Let this set for five minutes, then observe it again. Any increase or decrease in fluorescence helps you to assess the type of oil present. If sufficient oil occurs it may be scooped from the surface of the mud and saved in a sealed cut bottle.

Oil shows in the mud are then evaluated on the basis of the percentage of the mud surface covered by oil fluorescence and plotted on the basis of a rating of trace, fair, good, or very good. Check the mud periodically for fluorescence caused by mud additives.

5.31 Oil Determination from Formation Samples: Both washed and unwashed cuttings are observed, as described below.

Unwashed Cuttings

- Place 100 ml of unwashed cuttings (with minimum possible mud) in the blender jar, then add water to bring the volume to 700 ml.

- Agitate the mixture for 30 seconds, then let it set for 30 seconds.

- Take a reading for TG and PV. Compare this reading with the ditch gas reading to get a qualitative estimate of the permeability by indicating to what extent gas is retained in the cuttings.

- If necessary, you may manually inject the blender gas into the sample injection port of the chromatograph for a full analysis.

- Also with unwashed cuttings, place a 200-ml sample in a dish and add 100 ml of water. Check for fluorescence as described under paragraph 5.30.

- Record all findings.

Washed Cuttings

There are two main factors you must consider in evaluating an oil show in the cuttings: (1) the type of fluorescence, its color, intensity and cut, and (2) the percentage of the total cuttings sample which exhibits such fluorescence and cut.

- Check for visible oil staining under the microscope and select the visibly stained cuttings for a "cut" test.

- Note the percentage of staining and type of porosity, and check for any petroleum odor.

- Check the sample under UV light and note the percentage, color and intensity of fluorescence. Pick out fluorescent grains and reexamine them under the microscope for oil staining. Be careful not to confuse oil staining with contamination of the sample by pipe dope or fluorescent mud additives.

The type of fluorescence and cut gives some indication of the quality of the show within the section. The percentage of cuttings containing show in the sample should be some reflection of the massiveness and extent of the formation from which they came. Should oil be contained in the rock cuttings, it sometimes can be leached out with a solvent and observed with the naked eye under an ultraviolet light or an ordinary light. The procedure is described below.

Place individual stained grains in a cut dish, or 25 ml of representative sample in an evaporating dish. Place the dish under UV light and add the solvent. Note the rate of cut: instantaneous, fast, slow, streaming; and note the color and intensity of cut fluorescence. Remove it from UV light and note the color of solvent cut. If fluorescence is seen but no cut, this may be due to low permeability -- and a cut test should be tried on a dried sample. If this is still unsuccessful, crush the sample or add a drop hydrochloric acid.

Show descriptions should be written in the following order:

1. Free oil in mud
2. Odor
3. Visible staining — color and percentage
4. Fluorescence — color, intensity and percentage
5. Cut — rate, color, intensity and natural color

The chart in Figure 5-22 outlines the procedure to be followed when a show is encountered.

5.32 Secondary Equipment and Services

The following discussion pertains to equipment that is secondary to the job of basic mud logging. However, a standard logging unit usually contains numerous pieces of secondary equipment to aid the logging geologist in observing the engineering and pressure evaluation parameters encountered while drilling. A GEMDAS service unit is equipped with many of the pieces of equipment that are in a standard unit plus additional equipment not normally found in a standard unit. The size of the standard unit and its electrical power capabilities restrict the installation of the additional GEMDAS equipment, except for the ALFA/GEMDAS X unit configuration.

The following is a list of equipment other than that used for basic mud logging, which Exploration Logging can supply as an extra service to the client. The flow diagram in Figure 5-23 shows the relationship of one piece of equipment to another.

<u>Secondary Equipment and Services Most Commonly Used</u>

- Engineering and Pressure Evaluation "Tools"
  Mud Weight
  Mud Temperature
  Mud Resistivity (Conductivity)
  Mudflow Monitor
  Drill Monitor (System and Panel)
  Computational System (configured as 1, 2 or 3 below)
    (1) Computer/calculator with no interface to equipment (basic programs are run)
    (2) Computer/calculator with interface, data storage and printing capabilities
    (3) Computer to perform (2) above and to conduct many technical functions
  Shale density
  Shale factor
- Pit Volume Totalizer (PVT)
- Hydrogen Sulfide and Carbon Dioxide Detection
- Sample Chemical Analysis
  Stain kit
  Calcimetry
- Core Analysis

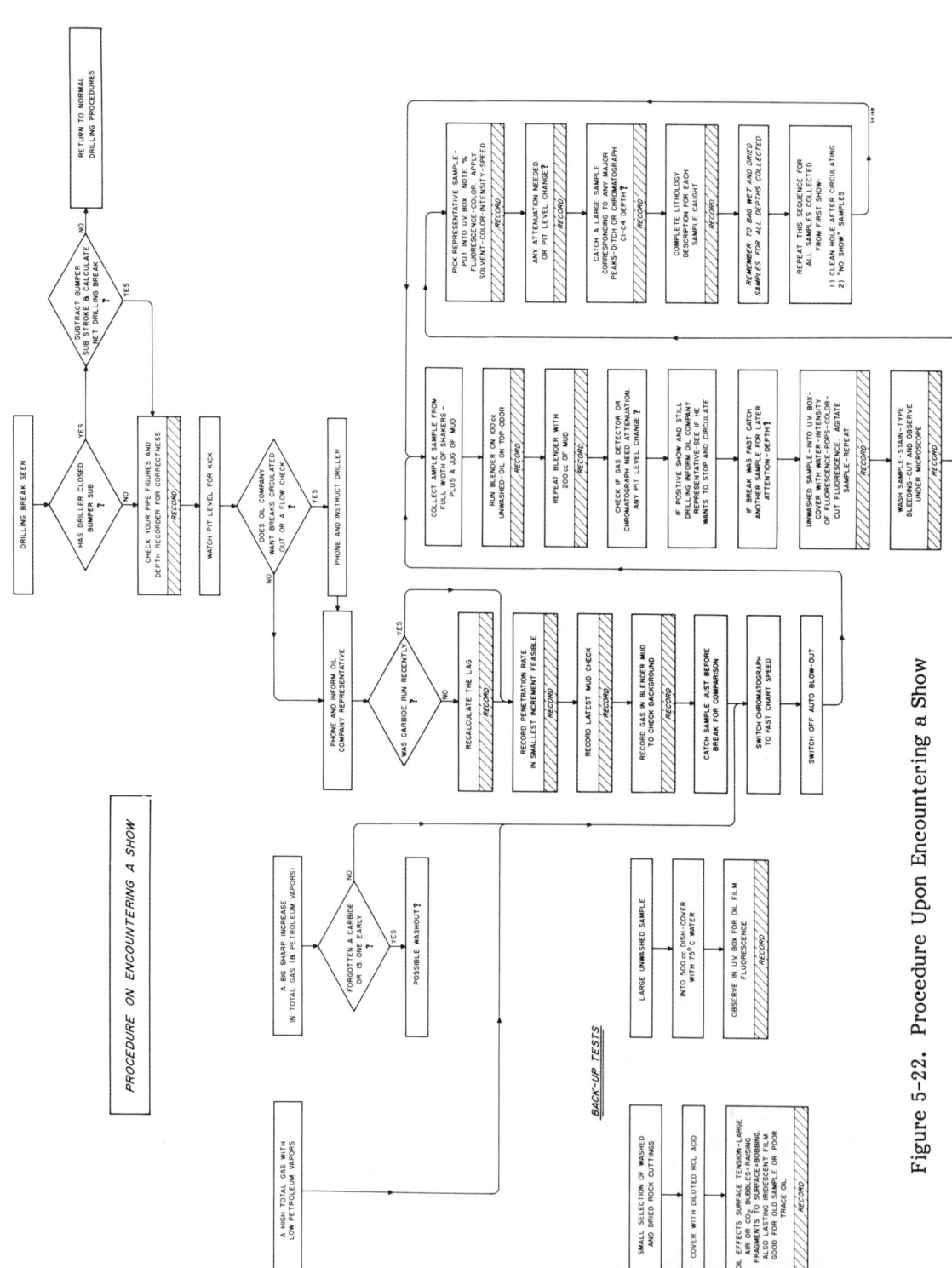

Figure 5-22. Procedure Upon Encountering a Show

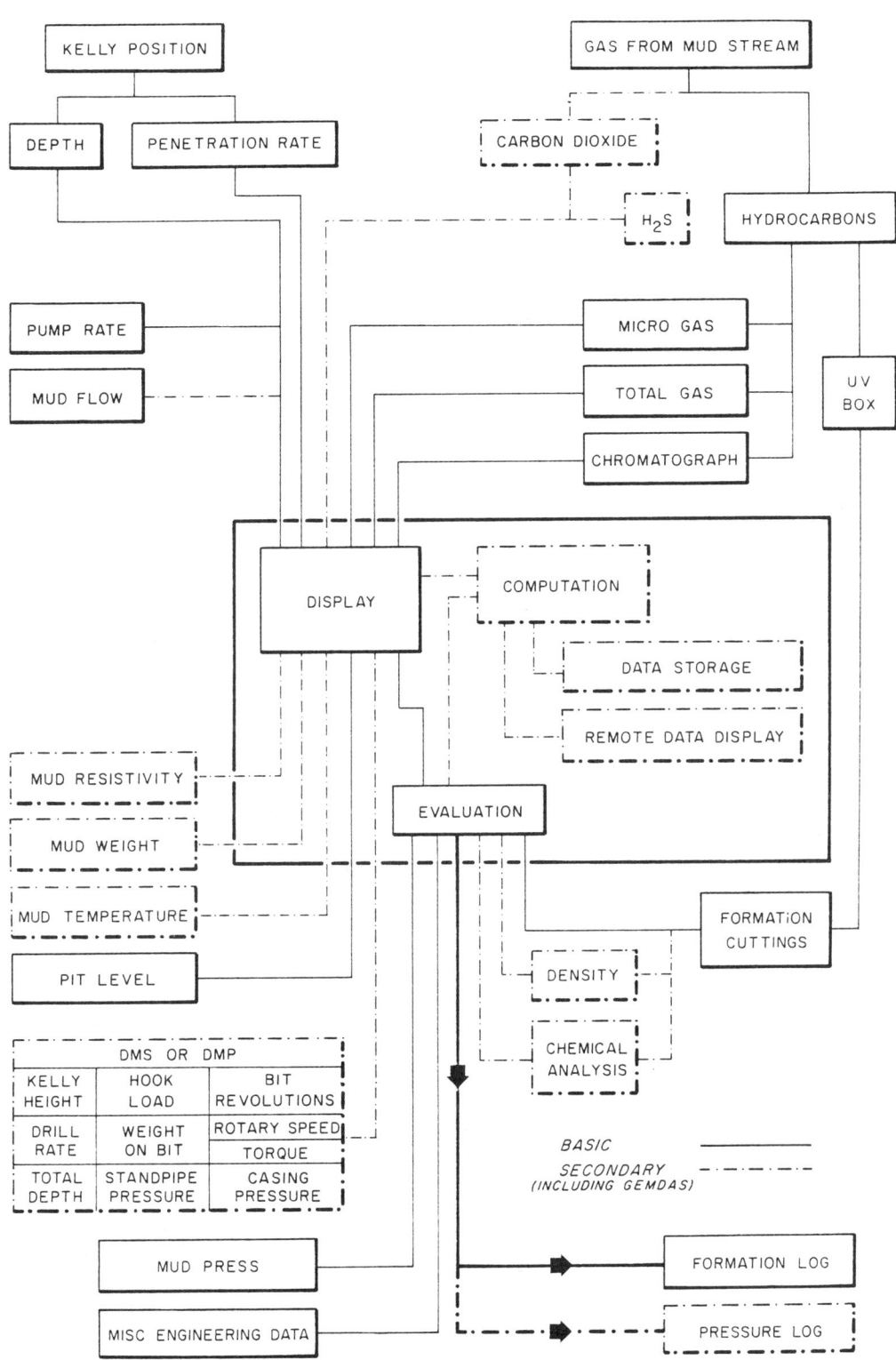

Figure 5-23. Basic and Secondary (Including GEMDAS) Data Inputs

Secondary and Supplementary Equipment and Services
Standard with a GEMDAS Unit

- Mud Weight
- Mud Temperature
- Drill Monitor (System and Panel)
- Mudflow Monitor
- Pit Volume Totalizer (PVT)
- Computation
   (1) Computer/calculator with interface, data storage and printing capabilities
   (2) Computers to perform (1) above and to conduct many technical functions
- Hydrogen Sulfide Detection

Detailed descriptions of individual equipment are not included here since they can be found in relevant operations and/or applications manuals.

5.33 Mudweight Recording: Controlling the mudweight is essentially a process of maintaining the mud column at a density which exerts a hydrostatic pressure that is greater than the formation pore pressure at any point in the open hole. If the hydrostatic pressure exerted by the mud column exceeds either of these limits, then hole problems can arise. Excessive departures from these values can mean severe or catastrophic hole problems. Current balanced-drilling techniques often leave only a fine margin between effective pressure control and a possible kick. Consequently, close and accurate monitoring of mudweight into the hole and mudweight out of the hole (in the ditch) is important for safe and efficient drilling.

Detecting and interpreting mudweight changes constitute an important role for the GEMDAS operator, Pressure Evaluation Geologist (PEG) or logging geologist. Equally important is the communication of such changes to the appropriate rig personnel.

The presentation of mudweight data is most useful on a real-time basis. Any changes usually require immediate response, and you must keep a record of such changes. This can be (1) noted on the GEMDAS hardcopy printout, (2) plotted on the Exlog pressure log sheets, or (3) typed on the mud log. For further information on mudweight changes, refer to the Mudweight Measurements manual.

5.34 Mud Temperature Recording: The measurement of downhole temperatures while drilling is a valuable aid to formation pressure evaluation. The most common method of estimating downhole temperatures is to continuously monitor the temperature of the circulating mud as it enters and leaves the well. This is done by positioning sensors in the suction pit and in the ditch. Additional methods of determining downhole temperatures include the use of electric logging techniques and temperature plates (Temp Plates).

Downhole temperatures which follow the general trend of the geothermal gradient are subject to subtle changes caused in certain cases by the existence of abnormal formation pressures. In order to recognize these temperature changes as they are reflected by the mud temperature, close monitoring of the mud system must be maintained so that the effect of external influences on temperature can be eliminated. Flowline mud temperature is significantly affected by changes in pump rate, change in hole size, rpm, length and diameter of riser, and ambient sea and air temperatures. Temp Plate and electric log methods do not suffer from these disadvantages but are not continuous. See the <u>Pressure Log</u> manual (MS-156) for more information.

5.35 Mud Resistivity Recording: Resistivity measurements of the drilling fluid reflect the ionic concentration of the drilling fluid, the interstitial fluid of drilled cuttings, and, where there is communication into the wellbore, the ionic concentration of the formation fluid. The complex mixing of the formation fluids with the drilling fluid makes it possible to detect subtle changes in resistivity of the formation fluid and thus the salinity. In order to remove the effect of surface salinity changes which are due to adding water or chemicals to the drilling fluid, differential resistivity is plotted. This is made possible by Exlog's dual probe system.

The plotted differential resistivity shows a general increase with depth in a normally pressured section. A pressure transition zone, if present, is indicated by a further sharp increase in resistivities closely followed by a sharp decrease as the low salinity levels in the overpressured formations are encountered.

5.36 Mudflow Monitor: The mudflow monitor calculates the flowrate of the mud through the drillstring. It monitors and counts the strokes per minute (spm) of the pumps and multiplies that figure by the number of gallons of mud pumped by each stroke. The results are displayed on the front panel of the monitor and are multiplexed if a computer system is available. The flow of the return mud is monitored directly at the return line and displayed on the front panel. The monitor then compares the return flow to that pumped, and if there is a gain or loss of mud in the well that is greater than a predetermined level, it activates audible and visible alarms.

5.37 Drill Monitor (System and Panel): The Drill Monitoring System (DMS) is the heart of Exploration Logging's GEMDAS service. Its primary function is to continuously monitor various sensors located on the drilling rig (Figure 5-24), display the data received from the sensors, and convert the data to suitable format to interface with various external computing and recording devices. The ALFA/GEMDAS-X series logging units operate a similar system called the Drill Monitoring Panel (DMP). The DMP monitors six parameters of the well and displays them in digital format. The six parameters monitored are:

- Hook load
- Weight-on-bit
- Standpipe pressure
- Rotary speed
- Torque
- Casing pressure

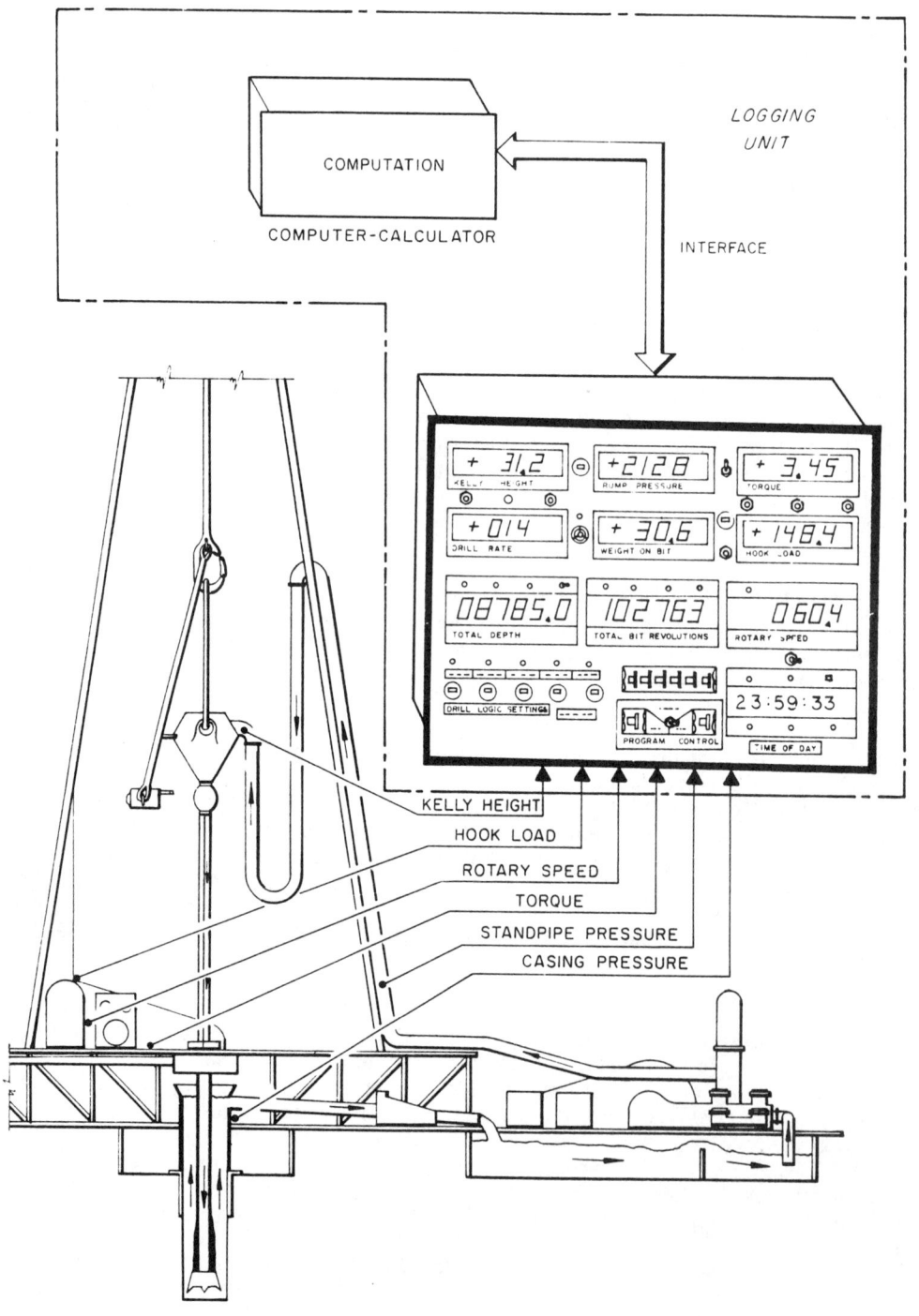

Figure 5-24. Drill Monitor System

5.38 Computational System: Exploration Logging uses a computer system on drilling rigs for the purpose of data acquisition and storage. Sensors placed at strategic points on a drilling rig send analog signals into the logging unit where they are converted to a digital format and, in most cases, displayed on digital meters. The computer has access to these readings at this point. The raw data is read by the computer, and from these raw data readings other values such as D-exponent are computed. The data is then recorded on tape at predetermined depth intervals and printed out by the high-speed printer. It is also possible to have the data displayed on CRT screens in various locations on the rig. All this is done in real time, i.e., while the drilling is taking place. There are many offline functions of the computer, too, such as obtaining additional printouts of the data recorded on tape or graphically plotting the recorded data. Hydraulic and E-log analyses can also be performed.

5.39 Shale Density Determination: During the course of normal sedimentation, interstitial water is squeezed from shale as it compacts. Under these conditions shale porosity should decrease with depth, and density should increase.

In the case of "overpressure," the shales exhibit higher-than-normal porosity with correspondingly lower density. Shale density has often proved to be very effective in determining the degree of undercompaction and consequent abnormal pore pressure in shale bodies. For further information about overpressure and shale density determination, refer to the <u>Pressure Log</u> manual (MS-156).

The shale density kits provided by Exploration Logging are to help rapidly determine the density of drilled shale cuttings. Three methods of shale density determinations are available:

- Single-Solution Shale Density Kit
- Multi-Solution Shale Density Kit
- Mercury-pump measurement of bulk density

5.40 Shale Factor Determination: Shale factor reveals the diagenetic state (or maturity index) of the clays. If methylene blue solution is titrated with a crushed slurry of the shale, the dye will be adsorbed onto the available sites by cation exchange mechanisms. The amount of dye required to saturate the cation exchange capacity of the shale depends up on the latter's geologic maturity.

If the principle primary clay is montmorillonite which undergoes compaction to "mixed-layer" clays and illite, the shale factor values are initially high and show a steady decline with diagenesis. Overpressured, undercompacted sections theoretically show an increase of the shale factor due to the increased porosity and hence larger surface areas for cation exchange. If illite and kaolinite are the primary clays, the shale factor is low initially. In the case of an overpressured section of such clays, the shale factor may show no increase whatever. Hence, in sections of mature, reworked clays, shale factor may be of little use in the detection of overpressure.

If the geopressures are generated by tectonic forces rather than by abnormal compaction through gravity loading, shale factor ceases to be useful as a pressure indicator. The procedure for shale factor determination is included in Appendix B.

5.41 H2S and CO2 Detection: Anybody who works with or in the vicinity of hydrogen sulfide must understand the hazardous nature of this highly toxic gas which, when inhaled in <u>any</u> concentration is <u>extremely</u> dangerous, and can cause death. The danger of exposure to this gas even for short periods, and the <u>unreliability</u> of odor as a means of detection, must be understood. Because H2S is heavier than air it can accumulate in "low" areas, so be particularly alert on calm days when there is no wind and the atmosphere is heavy. Make it your personal business to learn the locations of all emergency equipment and how to use it — including all fire alarms, fire-fighting equipment, and emergency shutoff valves (H2S is highly flammable). Certain high concentrations in air can explode upon ignition. Although it has a characteristic odor of rotten eggs, it rapidly deadens the sense of smell; therefore, odor cannot be depended upon as a means of detecting the gas. More detailed information, including emergency treatment, is outlined in <u>Hydrogen Sulfide Detection: Occurrence & Hazards</u>, MS-3016.

Carbon dioxide is not commonly encountered, but its presence in typical concentrations (unlike H2S) has no adverse effect on health. However, both gases do affect the gas-detection instruments in such a way as to render the catalytic Total Gas readings unreliable. For this reason it is desirable to know of their presence. Many logging units are equipped with either the basic Mine Safety H2S-Indicating Tubes or with the General Monitors H2S Monitor. The Beckman 100 CO2 Detector is also optional and is usually used only in areas of known CO2 occurrence.

5.42 Stain Kit: The identification of rock types does not present any problems except in the case of carbonates (see paragraph 5.27). The Exploration Logging Stain Kit has been produced to help with these carbonate determinations. It is capable of differentiating thirteen carbonates and two of the common sulfates. Refer to Appendix B for detailed staining procedures.

5.43 Calcimetry: Another useful tool when working with carbonates is calcimetry. It is used to determine percentage calcium carbonate and dolomite in a sample. The procedure for this determination is included in Appendix B.

5.44 Core Analysis: From analysis of cores it is possible to measure some of the physical properties of the rock and its fluids to help determine productive possibilities, fluid contacts, effective pay thickness, and the economical feasibility of testing and completing the well. Routine core analysis also includes the determination of fluid saturation, porosity and air permeability. Porosity and fluid saturation measurements are important in determining the type of production and potential reserves. Permeability measurements provide an indication of the productive capacity of a well and are important in understanding reservoir behavior. However, in each analysis the particular procedures and equipment vary with the size and nature of the core. We may classify core analysis into one of the following three types:

- Conventional core analysis
- Sidewall core analysis
- Whole core analysis

The procedures and equipment covered in the Exploration Logging Core Analysis manual are for conventional or sidewall core analysis. A flow diagram outlining core analysis procedures is included in Appendix B. General information about core analysis is included in Section 4 (4.39).

5.45 Mud Log Presentation and Standardization

Before drafting the log, be sure the drafting table is clean. All hand-drawn plots should be completed in ink <u>unless specified by the Exlog area office</u>. If orange carbon is used, smearing occurs when the log is run through the typewriter, so try to avoid doing this too often. If the log on a slow drilling section requires several runs through the typewriter, it is best to redraft the whole sheet when it is completed. Experiment with rotolite speeds when drafting in ink; you may find that typing can be accomplished without using orange carbon paper, providing blueline copies remain clearly legible. Refer to Appendix A for examples of prepared sheets.

5.46 **Drafting:** Follow the instructions below for drafting the log.

- Penetration Rate Column. Plot this over 5-ft or 1-metre intervals unless shorter significant drilling breaks occur. All corners should be rectangular with no overlap, and parallel with printed lines. Extend the lines over the top and bottom of the log sheet to ensure continuity when the log is spliced. The checkmark for a casing point should extend two squares vertically and one and one-half horizontally. Mark visual porosity by diagonal strokes of appropriate length.

- Core Column. Block-in cored intervals on the left side, in a line approximately 2mm wide for the whole core recovered. Continue the line, but do not block-in for any core not recovered; this is always assumed to be from the bottom of the core. Sidewall cores are also entered in this column, at the appropriate depths. Triangles representing these cores are available in Letraset form.

- Test Column. Entry is similar to that of cored intervals. However, the whole interval is blocked-in on the right-hand side of the column. Wireline tests (FITs, RFTs, etc.) should be annotated as crosses (also available in Letraset form).

- Cuttings Lithology Column. Hand-draw lithology symbols only upon special instructions from the client, because normally they are typed. When hand-drawn symbols are requested they should be done in ink (without orange carbon). Great care is required to achieve consistency. The letters "NR" should be typed in this column when there are no returns. Use two diagonals to cross-out this area.

- Oil Evaluation Column. Block-in the appropriate evaluation, using an ink pen. It is acceptable to use part divisions to indicate degrees of show.

- Continuous Total Gas and Cuttings Gas Column. The continuous total gas curve needs to be evaluated before it is plotted on the log.

Consider a drilling break — the formation at this break contains hydrocarbons. Ten feet of this formation is cut; circulation time is two hours. During the time the mud is carrying this show to the surface, the cuttings, oil and gas are becoming separated and being distributed over an area wider than the 10-ft interval in the borehole. Also, the gas is expanding because of the pressure reduction and is in fact getting slightly ahead of the cuttings. In two hours this "stretching" of the show can be quite marked, such that the equipment detects slow increases in gas which peak sometime after the beginning of the show has been detected. If the time to drill the 10 feet is computed and the duration of the show on the recorder is compared to this, it is usually found that the latter endures quite a while longer due to the "stretching" effect. Therefore, the gas curve must be condensed. It is reasonable to assume that the bulk of the gas entered the borehole as the bit cut into the top of the formation, so the plotted curve will have a steeper initial gradient than the curve on the recorder, commencing at the drilling break. During the show, continue the curve at the appropriate unit values and modify the chart record to represent a more realistic picture of the gas relative to the drill rate and the lithology. Try to incorporate on the log as much of the peak and trough detail as possible. However, rapid changes in background should be averaged out.

Cuttings gas is graphically represented as a square plot. Take care to ensure that the line is joined at right angles at each reading. A 2mm broken line drawn with a .25mm pen should be used to represent cuttings gas values.

- Chromatograph Column. This is a square plot as indicated in Figure 5-25. Take care in joining the lines and in keeping them parallel. Think, before plotting, so that it will be possible to plan where lines cross, thus avoiding too much confusion. When values almost coincide, you may either slightly increase or reduce the values to obtain more clarity and definition. It is generally acceptable to plot on a 10-ft interval, but do show smaller variations when they occur.

  Occasionally it may be necessary to change the style of the square plot in order to induce more clarity into the graphic representation. When plotting gas values over small depth increments, with large fluctuations in gas magnitude, you should consider using a ladder plot, as indicated in Figure 5-26. Careful labeling is a must to ensure clarity of this plot. It must be stressed that whenever the use of this plot is contemplated, chromatographic analysis for the whole well should be plotted in this manner.

- Cut Column. As with the oil evaluation, this is a "quick look" evaluation. Details are found in the description column, and these columns allow easy identification of zones of interest from a rapid scan of the whole log. Block this column the same as for the Oil Evaluation column.

- Miscellaneous Column. This column can be used for plotting calcimetry, shale density, carbon dioxide, etc. The E-log flash can also be entered in this column.

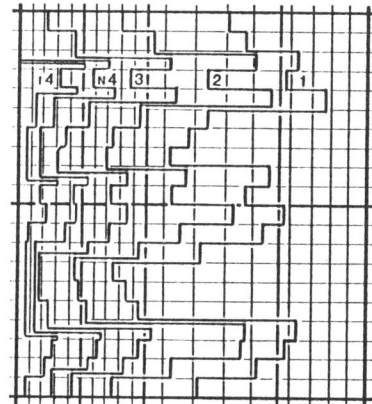
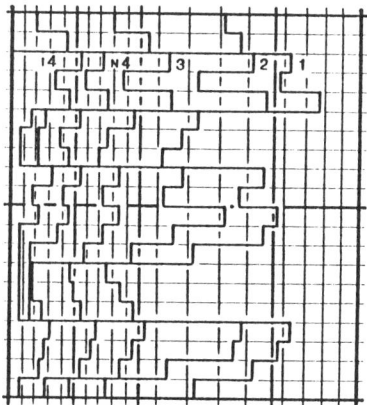

Figure 5-25. Standard Recessed Plot

Figure 5-26. Ladder Plot

- Interpreted Lithology Column. The option of including an interpreted formation log is provided in the "Lithology Description and Remarks" column. Two alignment ticks are provided — one in the sheet heading, and one located at the base of the sheet. When required, these marks should be joined using a .35mm ink pen. This line should be drawn on the reverse side of the log sheet. The title of the column, "Interpreted Lithology," can be entered on the sheet heading. In the absence of preprinted Letraset titles, it is recommended that this title be typed.

  All lithology entered in this column should be ink-drafted. Use the recommended hand-drawn Exlog lithology symbols or, if so requested, those used by the client. Do not forget to enter these symbols in the lithology legend of the master log heading.

5.47 Typing: An office typewriter is used in all logging units; it has a standard keyboard, but geologic symbols have been added for typing the Master Log. A basic rule is to become thoroughly familiar with a new typewriter before attempting to type the log.

Never send in incomplete log prints unless a process is not yet completed; e.g., bit still running, casing not yet on bottom, etc. When typing, always keep within the column in which you started. Try not to type over gas curves.

- Log Heading. A provisional log heading should be sent in with the first prepared sheet. Be sure that all typing is aligned. Keep the log head up-to-date throughout the well, even though this may lead to misalignments and smudging. A completely new heading should be made at the end of the well.

- Sheet Heading. The top of each sheet should include the following information:

  OIL COMPANY   WELL/CONCESSION   SHEET NUMBER

  (National Oil Co.)   (21/6-3)   (Sheet #6)

- Data and Penetration Rate Column. Type "X" in the correct Penetration Rate box. Type the Drill Rate Scale as close to the top of the page as possible. Take care not to let numbers overlap the top line as they will then be incomplete when the log is cut and spliced. The scale chosen should be that which makes best use of the whole column without recourse to scale changes or back-up scales. The client's wishes must be respected in this, but you should suggest a new scale should the one you use prove to be unsatisfactory. Some suggested scales are included in Appendix A.

  Record dates at the left-hand edge of the column, using the pattern month/day (6/28). If the date coincides with a bit change, record the date first, the new bit underneath it and, under that, the bit data. For example:

  ```
   - 3/29
  N B #9 HTC OWV-J
  356 ft/21.5 hr
  ```

  Bit data should begin one space in from the left edge. For bit type use only uppercase letters. The amount and method of bit data recording must be in accordance with the client's wishes, but a consistent pattern should be used. The standard Exlog practice is

  ```
  NB #10 HTC  OSCIG-J
  329 ft/14 hr
  14 T/80 RPM
  2500 #/60 SPM
  ```

  If desired, strokes-per-minute can be replaced by gallons-per-minute. Bit grade can also be included below the bit information. Bit grading is described in Section 3, paragraph 3.22. When working with metric systems, carefully note in the master log heading the abbreviations used to denote metric units.

  Bit data can also be entered on the Bit Data Record (see Appendix A, Figure A-7) if it is felt that entry in the Rate of Penetration column will cause the drill rate curve to become obscured. When using this method only a simple notation, including bit number and type, is entered in the Rate of Penetration column.

Any drilling parameter changes which may lead to a change in drill rate should be recorded. For example:

    Incr. Wt to 20T
    Decr Pump Press

Intervals logged after a trip (LAT) and intervals circulated out of the hole (CR) should be noted on the left-hand side of the column, one space in.

Mark the Casing Record and change in hole size a little above and below the casing point, centered in the column and boxed. Similarly, note Depth Corrections as + or - x feet (centered and boxed).

No notes other than those discussed should be included in the Drill Rate column. Try not to obscure the curve. If this happens, go over it again so it will be visible.

- Depth Column. Numbers should be aligned so as to be exactly bisected by the depth line. (See Appendix A, Figure A-5.) For example:

```
═══╪════════════╪═══
   98 00        10800
═══╤════════════╤═══
```

Avoid depth markings which overlap at the top and bottom of the sheet, but if this is necessary, be sure that they will align properly when the log is spliced.

- Cuttings Lithology Column. This column is used to give rapid visual indication of lithology; hence we present a fixed set of symbols in a fixed order. The standard Exlog symbols must be presented on the log in the same order in which they appear on the master log heading (the order is from left to right, horizontally), an expanded version of which is illustrated in Figure 5-27.

Enter the following accessories to the right of the lithology in which it is an accessory. If the association is unclear, enter in the right-hand side of the column:

)   coquina, shell fragments

    G   glauconite

    P   pyrite

    S   siderite

    K   potassium

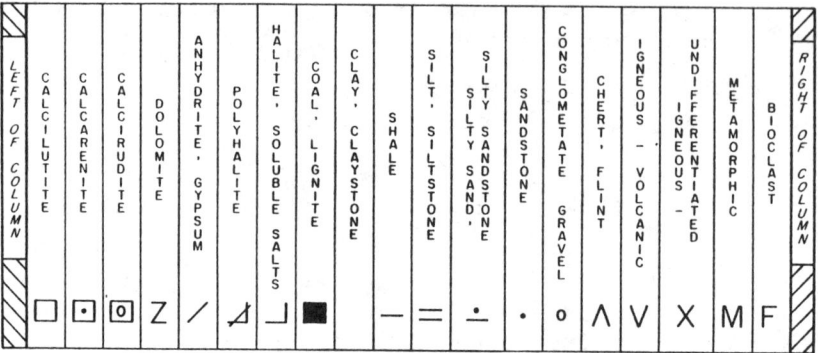

Figure 5-27. Exlog's Standard Lithology Symbols

It is the convention that sand and conglomerates be separated from other lithologies by a line. You may deviate from this system and order <u>only</u> at the request of the client.

- Continuous Total Gas and Cuttings Gas Column. Do not record trip gas on the total gas curve, but as a typed notation just in from the left-hand margin. For example:

    TG 25

Also make a notation if there is no gas

    TG 0
    CG 4   (connection gas)

Note scale changes by placing the appropriate attenuation (e.g. X1, X2, X5, X10, etc.) adjacent to the curve at the changeover point. Whenever information is available from adjacent wells, it may be possible to predetermine the attenuations necessary to accommodate the entire range of expected gas magnitudes. However, if there is any uncertainty, go from a X1 to a X10 attenuation. It is recommended that a maximum of only two attentuations be used. This ensures some continuity of the continuous total gas curve. Presentation of a curve unbroken by changes in attenuation not only presents a more readable plot, but is also pleasing to the eye. Therefore, whenever gas readings fluctuate between attenuations over short distances, keep the curve on the upper attenuation to retain continuity.

Take care to ensure that the attenuation of the cuttings gas is on the same attenuation as the continuous total gas. This allows for easier comparison of the two plots.

- Chromatograph Column. Identify the hydrocarbon curves at least once every sheet and wherever there is a crossover of curves, as illustrated in Figures 5-25 and 5-26.

- Miscellaneous Column. In the event of a full description column (e.g., during shows or caving), carbides and deviations may be recorded here. But be careful not to obliterate any plots within the column.

- Lithology Description and Remarks Column. The following should appear on sheet 1 of each log, suitably spaced and boxed:

  | Using 100 ml unwashed sample | All grain sizes are Wentworth unless otherwise stated |

- Remarks and Lithology Description Column. Descriptions must follow a logical order as outlined below:

    rock type (and depth at which it was observed)
    color
    hardness
    grain size ⎫
    grain shape ⎬ or texture, as appropriate
    sorting ⎭
    cementation
    porosity
    accessories
    show

    All abbreviations must conform to the standard Exlog Abbreviations List (Appendix illustration A-3). Any word not listed there should be written out in full. Punctuation is also standardized, as below:

    -- Capitalize the initial letter for each rock type
    -- Put a comma after each item of description
    -- New line for each rock-type
    -- Use no full stops (periods) at the end of a description

    You may deviate from these rules only at the request of the client.

    Record mud checks and carbides on the log once per tour or whenever a notable change occurs. When penetration is fast, record weight (and at least viscosity) on every sheet. Deviation surveys should also be recorded here. These and any other remarks should be spaced, boxed and recorded in a regular format as indicated by the examples in Appendix A.

    Whenever a core is taken or a drillstem test is conducted, include information about them in this column. Type detailed core descriptions and test results on a separate sheet and attach it to the end of the log as shown in Appendix A, "Supplemental Log for Core Analysis and Test Results."

5.48 Copying: Copies of the mud log are produced using a Rotolite blueline printer. Gear cogs provide control of the light and dark quality of the copy. ALFA/GEMDAS X units have a Rotolite copier with a rheostat for controlling the speed.

5.49 Routine Responsibilities of the Logging Geologist

Before arriving at the wellsite or as soon as possible thereafter, the logging crew should seek out the client's representative who has jurisdiction over the logging operation and get clear-cut instructions in writing as to the responsiblity and authority the logging crew is expected to exercise. These instructions should include the following:

- Whether drilling breaks are to be circulated out and, if so, how far the break is to be penetrated before circulating.

- What conditions of gas readings, fluorescence, drilling break, and type lithology constitute sufficient show to warrant calling the wellsite geologist for coring or testing as the case may be.

- Names and addresses of persons to receive daily logs and number of copies to send to each.

- Any other factors the client deems necessry to a well-coordinated mud logging job.

Ordinarily, wellsite geologists who have previously used well logging units will have very definite instructions regarding the above points. If the wellsite geologist or engineer has not had such prior experience, the senior member of the Exlog crew should explain this and work out with him the details of the instructions and put them into writing. These instructions should be posted in a prominent place for ready reference by the Exploration Logging crew.

5.50 Well Initiation: It is imperative to have detailed logging instructions when starting a well. There are three primary means:

- A completed logging instructions form is sometimes provided by the Exlog office. It is the responsibility of the logging geologists assigned to the well to clarify or supplement these instructions (if they are not clearly understood) by contacting the client or his representative.

- Frequently the client will provide a complete drill returns logging program (note the example in Appendix A). Normally, however, certain things will need clarification, or maybe something has inadvertently been omitted. Again, it is your responsibility to acquire supplementary instructions.

- The client will occasionally telephone or radio the initial or supplemental instructions to the logging geologists at the wellsite. This verbal instruction must <u>always</u> be written down and communicated to the other Exlog personnel.

    If a situation or problem arises that is not specifically covered in the client's instructions, use your own best judgement and experience in arriving at a solution — or, in more important circumstances (or if in doubt), contact the client's representative for further instructions. The logging geologists working the first shift must send an Initial Well Report form to the Area Office upon start-up of a new well.

5.51 Specific Responsibilities: The routine responsibilities of the logging geologist are controlled to a large degree by the instructions furnished by the client and therefore change from job to job, but the following are responsibilities specific to every job.

- Obtain and plot all necessary data for a complete graphic well log.

- Evaluate all oil and gas shows, using factual data and keeping in mind the prevalent conditions.

- Immediately notify the drilling contractor and client of any unsafe operating conditions.

- Maintain and repair all logging equipment and fixtures.

- Clean and maintain the interior and exterior of the logging unit.

5.52 Routine Reports: During the course of a well, certain reports must be filled out on a regular basis.

- Time and Expense Account Form. Every employee must fill this out and submit it to the local office at the end of each month, keeping a copy for his own records.

- Personnel Status Report Form. Fill this out at the end of each month. It is a record of crew movements (hours worked, standby and time off). It is sent to the responsible Exlog office and a copy kept in the unit files.

- Monthly Well Report Form. Submit this with the Personnel Status form. It is a record of materials (sample bags, core boxes, etc.) used during that month and includes an estimated completion date of the well, made by the logging geologist.

- Formation Logging Report. This report, used for communicating with the client's engineer, lists a broad spectrum of geological and engineering data. It is used for routine morning, phone, or radio interim reports concerning specific shows or lithology changes, and may be mailed or delivered daily (depending on circumstances). A more detailed geological format will have to be provided for the client's wellsite geologist. Draw and type a format to suit the client and make blueline copies for each morning.

- GEMDAS Logging Report. This fulfills the reporting responsibilities of the GEMDAS operator and should be handed in on the same basis as the Formation Logging Report and Geological Morning Report.

- Supply Inventory List and Equipment Operating Form. Use this form to keep a record of supplies and materials used in the course of logging a well. Toward the end of a well (or bi-monthly on long wells), take a complete inventory and note any items lacking. Also, any repairs or damage should be described in detail. When completed, mail the form to the local office.

5.53 Well Completion: Fill out a Final Well Report Form at the end of the well. The logging geologist working the last shift is responsible for completing the form and ensuring that it gets to the Area Office immediately after the completion of the well.

Type the total depth and date of completion on the Master Log heading, and the total depth and last suite of electric logs on the bottom of the Master Log. Make the required number of prints of the last sheets and distribute as before. It is the Trailer Captain's responsibility to ensure that the Master Log is delivered to Exploration Logging's Area Office for brown-line copies (a continuous sepia of the log) or given to the client's representative at the wellsite. The client should also receive all logs, worksheets, kelly height and chromatograph charts, and any other recorded information unless specified otherwise.

Any remaining wet and dry samples must be boxed, labeled and distributed according to the client's instructions.

Finally, you must complete a Supply Inventory List and Equipment Operating Form and send it to the Area Office.

Consult the appropriate Care & Operations manual and MS-3009, <u>Logging Geologist's Routine Checklist</u>, whenever a unit is likely to be inoperative for awhile or when it is to be released from an operation. These manuals contain detailed shut-down and tear-out procedures.

In addition to the duties and responsibilities described here, a logging geologist should do whatever else he feels necessary to give the client a superior service.

5.54 Wellsite Relationships: In any business environment, people tend to communicate what they as individuals are <u>aware</u> needs to be communicated. The relationships you develop with wellsite personnel depend almost entirely on your communication skills. One of the things you must convey to other wellsite personnel is that <u>you are aware</u> of what is needed and that <u>you are concerned</u> about the safety and success of the well. You will receive many forms of feedback: questions, smiles of approval, or possibly language that irritates. Be sure you understand <u>why</u>, in each case; if you misunderstand <u>any</u> of this feedback it's your fault — not theirs. The burden of communication lies with the service company!

- During critical drilling periods, tensions and tempers can rise quickly. Show a genuine interest in what's happening, and offer whatever assistance you can. (Sometimes that means staying out of the way!)

- There may be people of different nationalities who are not fluent in the language spoken at the wellsite, which can cause "language barriers" and lead to misunderstandings due to misinterpetation.

- Poor relationships stemming from misunderstandings are costly. If you receive new instructions, repeat back your understanding of the new information rather than assume you thoroughly understand. <u>Always write down</u> any change of instructions for yourself and for the rest of the Exlog crew. Before a well is "spudded" (i.e., drilling begins), details have been negotiated and instructions have been prepared by the client. Upon reaching the well you must become thoroughly familiar with the planned drilling program. (An example program may be seen in Appendix A.) Discussing the program with the drilling engineer communicates to him that you care about helping to make the well a success.

The Logging Geologist's job encompasses far more than accurately logging lithologies and reporting changes in pit level!

5.55 PRESSURE LOG

The Pressure Log is a graphic display which is not only a permanent record of the data generated by the equipment but also forms a sound basis for evaluation of abnormal formation pressure <u>while drilling</u>. The procedures used are well established and accepted by the petroleum industry. The pressure evaluation service offered by Exlog is not performed on every well. When a Pressure Log is kept, however, it is the responsibility of the GEMDAS Operator, the Pressure Evaluation Geologist, and occasionally the Logging Geologist in a standard logging unit.

Abnormal formation pressure can be evaluated graphically by recording the gas shows, their magnitudes, appearance and behavior, associating this with other factors, and then linking all these factors with their geological indicators (formation type, size of cuttings, etc.). Thus the Pressure Log qualifies as a stand-alone tool for abnormal pressure evaluation, rather than relying on petrographical or minerological changes. Exploration Logging's <u>Pressure Log</u> manual (MS-156) outlines the major pressure indicators and is a useful introduction to pressure evaluation.

Formations exhibiting high pressures at a particular depth are usually zones of abnormally high porosity. This is based on the compaction principle that, as burial depth increases, overburden pressure increases, thereby compacting the rock bulk. For compaction to occur fluid must be removed from the pore space, allowing the pore space to decrease to the point where the matrix and fluid are supporting their normal share of the overburden. If for some reason the bulk rock is buried and the fluid in the pore space cannot be removed, then the additional overburden load is transmitted to the fluid; the fluid will be abnormally high-pressured and the porosity will be abnormally high. This may occur if the deposition rate is high or the shale-to-sand thickness is high. (Sand beds act as pipelines for liquid removal.) Under normal conditions of burial, shale porosity decreases as a function of depth.

A number of items of specialized equipment systems with surface sensors have application in the evaluation of downhole pressure while drilling. These specialized systems, in addition to gas monitors, are:

- Continuous mud weight recording
- Continuous mud temperature recording
- Continuous mud resistivity recording
- Bulk density determination kits
- Shale Factor determination kits
- Drill Monitoring (System and Panel)

Pit Volume Totalizer (PVT)
Differential mudflow comparator
Computational hardware and software

The Pressure Log was introduced to maximize usage of the data generated by this equipment. Field testing of various formats plus years of developing the techniques of pressure evaluation have produced a useful pressure data log.

Before it is transferred to a Pressure Log, the raw data must be processed by an experienced field geologist who may be a Pressure Evaluation Geologist (PEG), a GEMDAS operator, or occasionally a Logging Geologist, so that a correct interpretation can be made. The experienced field geologist is aware of factors that might influence evaluation parameters and produce false pressure indicators. Experience of this nature can be gained only from field operation of the relevant equipment and from observing the behavior of the data produced by each under various circumstances.

5.56 CORE LOG

The Core Log is a record of core analysis data and lithology versus depth (see Appendix illustration A-6). Core analysis is used for exploring and evaluating the productive possibilities of edge wells and wildcats at the time of drilling. In the exploitation of discovered reserves, the uses to which core analysis information is put ranges from guiding the completion of the wells and then preliminary evaluation of the oil property through all the the production problems and remedial operations, and finally to its use in pressure maintenance or in deciding the probable effect and results of secondary recovery operations near or at the end of the primary production of the field. Another important factor is the value of core analysis in the evaluation and calibration of other logging methods. It is increasingly realized that such data, along with exact knowledge of porosity, permeability and fluid saturations, provides the basis for good reservoir operations and prediction of performance. Exlog's core analysis service is performed only at the request of the client and when the necessary equipment is available.

Core analysis is discussed briefly under paragraph 5.44, procedures are explained with the aid of a flow diagram in Appendix B, and core handling is described in Section 4, paragraph 4.39.

5.57 WIRELINE LOGS

After each section of the hole is drilled and before casing is run into the hole, it is necessary to "log" the hole. This involves running one or more wireline logging tools, either singly or in combinations.

The procedure for wireline logging is shown in Figure 5-28. The sonde contains one or more measuring devices and a cartridge which contains electronic control and transmission circuitry. It is lowered into the hole on an armored steel cable. This cable contains seven separate conductors within its core which transmit power and signals between the surface and the tool. The sonde is usually 3-5/8 inches in diameter and may be up to several feet long.

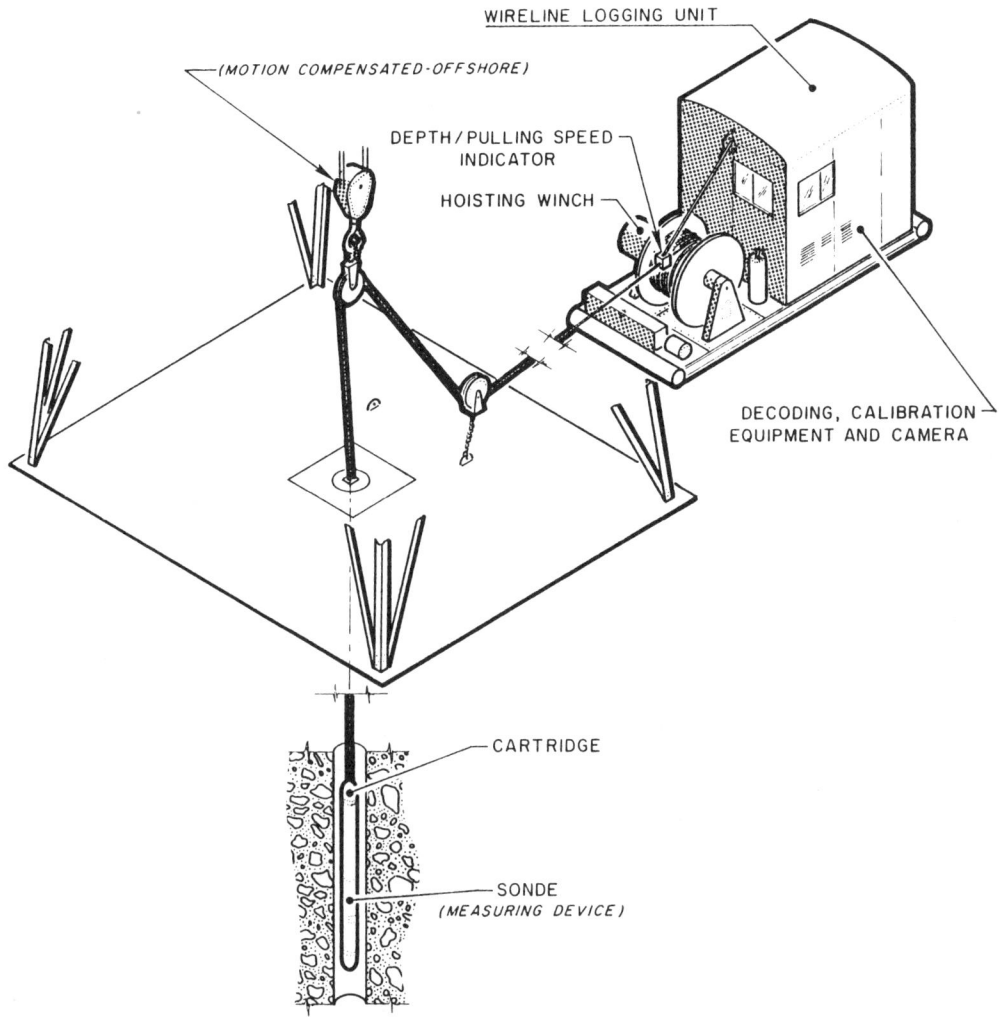

Figure 5-28. Wireline Logging Operations

After reaching bottom, wireline logs are run by pulling the sonde and cartridge up the hole at a fixed speed which is determined by the type of measurement to be made. As the tool is withdrawn from the hole, a continuous measurement signal is sent to the surface (via electrical conductors in the cable) where the raw data is processed in a control panel and recorded in the proper log format on film by an optical recorder.

These signals may also be recorded by a digital tape recorded which allows computer processing of all logs seen, either at the wellsite or in a critically located computing center.

Wireline logs are used for:

- Correlation
- Bed thickness
- Porosity
- Water saturation

These measurements enable conclusions to be drawn on:

- Lithology
- Permeability
- Presence and type of hydrocarbons
- Mobility of hydrocarbons

5.58 Classification

Wireline logs can be separated into four functional groups which measure the following:

- Resistivity
- Porosity/lithology
- Production characteristics
- Miscellaneous parameters

5.59 Resistivity: Depending on the type of resistivity measurements required, various tools are used. These are discussed below.

- Direct Resistivity Measurements. Tools of this type measure the electrical resistivity of the formation by passing an electric current out into the formation. Since the current must pass through the mud, mud cake and invaded zone (see Figure 5-29), the resultant measurement is a combination of their resistivities (R_m, R_{mc} and R_{xo}) with the true resistivity of the undisturbed formation (R_t). The standard unit of measurement of resistivity used in well logging is ohm.metre^2/metre. Commonly this unit is expressed in the simpler form of ohmmetres.

  It is possible to make correction for these effects or to minimize their importance by forcing the electric current deeper into the formation. Older logging tools (e.g., 16-inch or "short" normal) achieved greater depth of investigation by increasing the spacing between electrodes. This led to longer tool length and loss of vertical bed definition. Modern resistivity tools (e.g., Laterologs, Spherically-Focused Log) achieve deeper investigation by means of a "focused beam" without excessive electrode spacing.

  Even focused tools are affected by the mud and filtrate. If the resistivity of the mud is much greater than that of the formation water, e.g., the mud is less saline, these effects will be too great and the tools will be unreliable. If the mud is nonconductive (for example, an oil-based mud; see Appendix C), no conductive path will exist into the formation and the tools cannot be used.

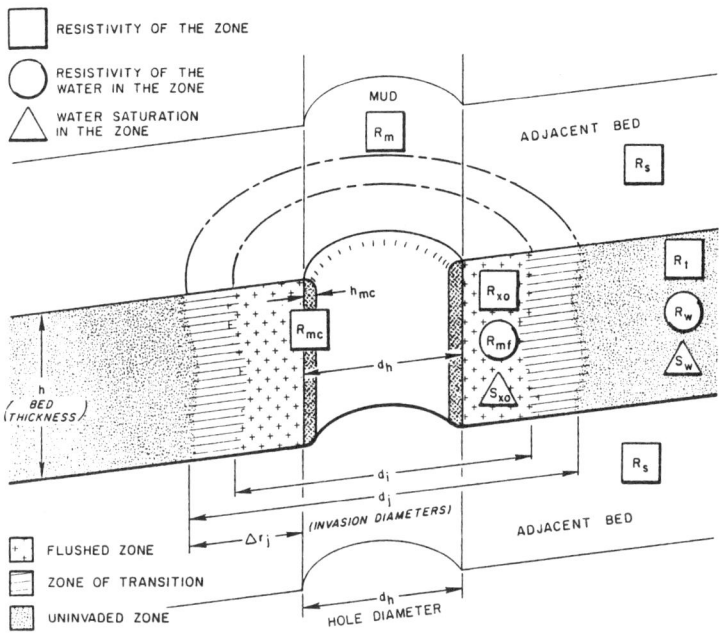

Figure 5-29. Symbols Used in Wireline Log Interpretation

- Micro-Resistivity Measurements. These tools (Micro Laterolog, Proximity Log, Micro-SFL) are direct resistivity devices which are run with the tool forced against the borehole wall and are referred to as "pad" tools. Being adjacent to the borehole wall and having only a very shallow depth of investigation, these tools measure the resistivity of the invaded zone (Ri) with some slight effect from mud cake resistivity (Rmc), and approximate the resistivity of the flushed zone (Rxo).

  It is sometimes the practice to leave these tools turned on when going into the hole, with the pad retracted. This gives a reading of mud resistivity (Rm) in the hole and may confusingly be referred to as a "mud log."

- Induction Measurements. In this type of tool an alternating electric current is passed through transmitter coils. This induces secondary eddy currents to flow in the formation which in turn induces signals in the receiver coils. The receiver coil current is proportional to the conductivity of the formation. Since the tool is investigating the formation's ability to conduct an electric current rather than to resist one, it is the convention to scale the log in conductivity, the standard unit in well logging being millimho/metre (i.e, 1000/ohmmetre), thus avoiding the almost continuous use of fractional numbers.

Induction logs are focused and have deep investigation and so give a good estimate of true formation resistivity (Rt). There is, however, some interference caused by eddy currents in the mud and invaded zone. These tools therefore work best in nonconductive (e.g., oil-based) muds, and work worst where the mud is more conductive (e.g., saline) than the formation water. This is the opposite of the Laterolog response. Thus, the two logs will work best at opposite extremes of logging conditions. Under many common logging conditions (of mud and formation water salinity), both logs may be applicable but the results may require some correction.

Combinations of direct resistivity and induction (conductivity) devices with varying depths of investigation are commonly run together. Comparison of the various signals can yield informaion about invasion profiles, diameter of invasion, better values of Rt (true formation resistivity), and ratios of Rxo/Rt for water saturation determination.

- Spontaneous Potential (SP). This is the measurement of the electrical potential between a point in the borehole and a grounded electrode at the surface. This potential is caused by electromotive forces in porous and permeable formations which are of electrochemical and electrokinetic origins. Where formations are impermeable no potential will exist, and a baseline will be developed which is commonly referred to as a "shale baseline." Any deviation from this baseline will be an indication of the presence of permeability. In fact, there is a slight potential adjacent to shales due to the migration of sodium ions, but this is constant and is ignored.

  Where a porous and permeable zone is invaded by mud filtrate, ions migrate between the mud filtrate and formation water when there are differences in salinity. There is also a flow of positive ions from the formation water to the adjacent shale due to the inherent negative charge of the shale particles. These two electrochemical actions form in essence a battery, causing a potential to be developed which is measured as a small voltage in the borehole.

  Since mud and formation water have similar electrolytes (predominately sodium chloride), it can be said that SP deflection is dependent on permeability and relative salinities of mud filtrate and formation water. This is shown in Figure 5-30 in which each of the sand stringers have equal permeability. Stringers A and B deflect the SP to the left in proportion to the salinity. Stringer C gives no deflection since no ionic migration is taking place. Stringers D and E deflect to the left in proportion to their salinity although, since the ratio is now inverted, their deflection is not so great as that seen in A and B (e.g., 5/2 is greater than 2/5).

  The SP can be used to detect permeable beds, define bed boundaries, determine formation water resistivity (and hence salinity), and give qualitative indications of shaliness in sandstones. Although rarely used quantitatively, the SP gives an excellent, easily read correlation log with good bed definition. The deflections, when caused by thin sands of

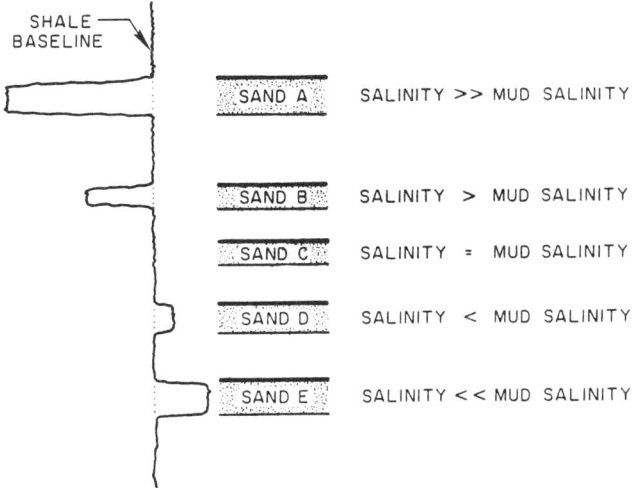

Figure 5-30. The Effect of Salinity on SP Response

varying shaliness in a massive shale section, give good correlation with drilling breaks and can be a useful aid to the geologist. Note that Figure 5-30 is for demonstration only; salinities do not normally change this rapidly within a section.

5.60 Porosity/Lithology: The tools used for these measurements are discussed below.

- Sonic. This tool measures the time required for a sound pulse to travel a fixed distance between a transmitter and receiver. The tool is centered in the borehole and uses multiple transmitters and receivers to remove the effects of travel through mud and mud cake. In this way, only the time taken to pass through a fixed length of formation is recorded (Interval Transit Time, microseconds/ft).

  Transit time (the reciprocal of velocity) is directly related to porosity in a known rock matrix. Charts or computer programs can be used to carry out this conversion. Because interstitial clay has an interval transit time similar to that of formation water, the sonic response to clay-filled porosity is the same as if the pore space contained all water. Therefore, porosities derived from the sonic in a porous sandstone with some clay filling will be optimistic compared to the effective liquid-filled porosity of that same sandstone.

  Gas in the mud, unconsolidated formations, formation fractures, gas saturation in a porous reservoir rock, aerated muds and highly rugose holes will tend to attenuate the sonic signal and cause "cycle skipping." Because of the multiple transmitters and receivers and centralization of the tool in the borehole, hole irregularities and sonde tilt have little effect on the sonic signal.

- Formation Density. This is a pad tool containing a high-energy gamma-ray source. The gamma rays "collide" with electrons, lose energy and are detected at the tool. The degree of scattering of gamma rays is proportional to the electron density and hence to the bulk density of the formation (gm/cc):

$$\rho_b = \phi \rho_f + (1 - \phi) \rho_m$$

where

ρ_b = bulk density
ρ_f = fluid density
ρ_m = matrix density
ϕ = porosity

hence

$$\phi = \frac{\rho_m - \rho_b}{\rho_m - \rho_f}$$

The porosity may be calculated if the rock matrix is known. Since the tool has a very shallow depth of investigation, the fluid is assumed to be mud filtrate with a density of 1.0 (fresh) or 1.1 (salt).

The presence of a mixed matrix leads to possible errors in the assumption of matrix density. Low-density interstitial clays will especially result in overestimates of porosity. By filling the pore space, the clay reduces the volume of liquid-filled porosity — but its low density decreases the bulk density of the rock, indicating that a portion of that porosity remains. In this instance, i.e. the clay has a lower density than the matrix but higher than the fulid, the porosity shown by the tool will be higher than the actual liquid-filled volume in the rock but lower than the total pore volume (the total void space between the matrix grains). If the void space is filled with interstitial material having a density similar to that of the matrix, the tool will indicate a correct value of liquid-filled porosity, although this will be less than the total void space in the rock. Similarly, the presence of gas in the pore space, with a density much lower than that assumed, will produce a slightly optimistic apparent porosity.

- Neutron. This tool, which may be a pad-type or centralized, bombards the formation with high-energy neutrons. These neutrons rebound from heavy nucleii with high energy but lose energy when colliding with light hydrogen nucleii. Low-energy neutrons are detected at the tool. Capture rate is proportional to the hydrogen content of the formation, i.e., the water and hydrocarbons contained in porosity. The log is normally calibrated in limestone porosity units (i.e. porosity (%) in an assumed limestone matrix). Conversion charts are used to determine actual porosity in other rock types. On more recent logs, the log may be rescaled for the matrix types of major interest in the section being logged.

Clay contains bound or interlayer water which, when interstitial clay is present in a reservoir, contributes to the hydrogen index of the formation, while the clay reduces the porosity. Interstitial clay in a formation will therefore result in an optimistic apparent porosity. Gas which has a lower hydrogen index than water or oil gives a lower neutron response and hence pessimistic apparent porosity.

Although the direct response of the porosity tools may be misleading, each of the tools responds differently in different lithologies and to the presence of interstitial clay and gas. For this reason it is possible to crossplot the various results to assist in the determination of the composition of mixed lithologies, clay content, gas, and true porosity.

- Gamma Ray. A scintillation detector measures the natural radiation of the formation. Since shale contains potassium which has radioactive isotopes, it establishes a relatively consistent maximum. Normal sandstones and carbonates have little or no radioactivity and therefore will be offset. Micaceous or argillaceous sandstones or carbonates show an intermediate response. Gamma ray is therefore a useful tool for correlation and a quick assessment of potential payzones. Certain evaporites (e.g., sylvite) have high radioactivity but can usually be discriminated from shales by their much higher density.

5.61 Production Characteristics: There are various logs which can determine the fluid type production rate, productivity and current status of a reservoir. They will not be studied here.

5.62 Miscellaneous Parameters: These parameters are related to the physical characteristics of the hole. Hole size, shape and deviation can be measured using the following tools:

- Caliper. This measures hole diameter and is useful for sidewall sample selection and calculating cement volume. It will indicate permeable formations by reduced diameter due to filter cake development. The caliper can read borehole diameters ranging from 5 to 20 inches.

- Borehole Geometry Tool (BGT). This records hole deviation, azimuth and relative bearing. Two calipers measure hole volumes in cubic feet. It can measure an opening that ranges from 5 to 40 inches, and is useful to detect eccentricity and calculate cement volumes.

- High-Resolution Dipmeter Tool (HDT). There are four pads, each of which has micro-resistivity electrodes. Depth differences between signals give angle and direction of formation dip. Also, there are two calipers measuring two diameters in two vertical planes 90° apart. It records deviation, azimuth, relative bearing, and (consequently) hole deviation (inclination and direction). All data is sent uphole on FM mode and appears on film and tape. Interpretation is carried out with the help of computers.

5.63 Interpretation

The analysis of wireline log data depends upon one fundamental assumption which is that the only conductive medium present in the formation is water in the pore space, and that matrix material and hydrocarbons are essentially nonconductive. Hence the electrical conductivity (or resistivity) of the formation will be a function of:

- Porosity — the percentage of bulk volume available for fluid.

- Water saturation — the percentage of pore volume filled with water. The remainder will be filled with oil and/or gas. Empty porosity does not exist!

- Salinity — assuming the only electrolyte present is NaCl, the resistivity can be directly converted to salinity.

- Temperature

- Shape, size and communication of the pore spaces

It can therefore be said that the resistivity (Ro) of a porous and permeable rock material which is 100 percent saturated with an aqueous solution is directly proportional to the resistivity of that solution (Rw). Thus

$$\frac{Ro}{Rw} = \frac{Ro'}{Rw'} = \frac{Ro''}{Rw''}, \text{etc.}, = F$$

where Ro, Ro', Ro'' are the resistivities of a formation (at constant temperature) when flushed with various aqueous solutions of resistivities Rw, Rw', Rw''. The constant of proportionality, F, is called the Formation Factor and is a function of porosity, permeability, and the shape, size and distribution of the pore spaces. Archie (1942) proposed the relationship

$$F = \frac{a}{\phi^m}$$

where m is the "cementation factor," a function of the type and degree of consolidation of the rock. The constant "a" is empirically derived and varies from section to section.

The Humble Formula proposed by Winsauer et al (1952) provides general values which are suitable in most rocks:

$$F = \frac{0.62}{\phi^{2.15}}$$

For ease of calculation it is common to use a value of 2 for cementation factor. Satisfactory results can then be obtained with

$$F = \frac{.81}{\phi^2} \quad \text{in clastic formations (e.g., sandstones)}$$

$$F = \frac{1}{\phi^2} \quad \text{in consolidated formations (e.g., carbonates and indurated sandstones)}$$

Although departing at extremes, all three formulae give very similar results in the commonly encountered porosity range.

Since oil and gas are nonconductive, the Ro/Rw relationship can be modified as follows:

$$\frac{Ro}{Rw} = \frac{Rt}{Rw}(Sw)^n = \frac{Rxo}{Rmf}(Sxo)^n = F$$

where n is the "saturation exponent" and is generally taken to equal 2 (see Figure 5-29 for other nomenclature).

In a clean formation (i.e., none shaley) with regularly distributed porosity and where porosity, (Rt Laterolog or Induction), Rxo (Micro-Laterolog), Rmf (from a surface filtrate measurement) and temperature (from a maximum reading thermometer on the tool) are known, these formulae can be used to determine:

- Hydrocarbons in place (1 - Sw)
- Mobility of hydrocarbons (Sxo - Sw)
- Salinity of the pore water (from Rw and temperature)

In more complex reservoirs, some corrections are required and more complicated mathematics, sometimes using a computer. The same basic principles are nevertheless used.

5.64 Abnormal Pressure Evaluation

Formations exhibiting high pressures at a particular depth are zones of abnormally high porosity. It is possible, therefore, to plot parameters which are a function of porosity and observe the deviation from the normal trend. Thus, plots of shale acoustic transit time, bulk density, neutron response and resistivity-versus-depth should deviate from the normal trend when abnormally high-pressure zones are encountered.

These methods are practiced by Exploration Logging and are explained in detail in the <u>Pressure Log</u> manual (MS-156).

5.65 FORMATION TESTS

Formation testing is a direct means of obtaining information concerning the liquids and pressures in a formation open to a borehole. The traditional method of achieving this is by way of a temporary completion with a drillstem test (DST). For an outline of Exploration Logging's role in drillstem testing, refer to Appendix E.

5.66 DRILLSTEM TESTING

A drillstem test is made by lowering a valve, a packer, and a length of perforated tailpipe on the end of the drillpipe to the level of the formation. The packer is set against the wall of the borehole so that it seals off the test interval from the mud column above. The valve is then opened. This procedure effectively reduces the pressure opposite the formation to atmospheric pressure, and the formation fluids can flow into the hole and be produced through the drillpipe. It amounts to a temporary completion of the well, and the produced fluids are therefore representative of the fluid production that may be expected if the well is eventually completed.

The basic formation test tool assembly (shown in Figure 5-31) consists of (1) a resilient sealing element or packer which can be expanded against the hole to segregate the formations above and below the packer, (2) a tester valve to exclude mud when the tool is being run into the hole and to allow formation fluids to enter the drillstem during the test, and (3) a bypass valve to allow equalization of pressure across the packer after a test is completed. The figure also shows the various types of formation tests usually used.

Drillstem testing is the most hazardous of all drilling operations and is therefore conducted with utmost care. Before beginning a test, it is determined that the hole and mud are in good condition. Mud is circulated for at least one cycle to be sure that all cuttings have been removed. The mudweight is measured during this period so that the hydrostatic pressure, which will be indicated by pressure recorder charts, can be checked. The general test procedure is explained below for the "open hole single packer test" and by reference to Figure 5-32. This diagram also contains typical pressure recorder charts.

(1) While going into the hole, the rubber packer is at its maximum length and minimum diameter.

(2) After the tool reaches bottom and the necessary preparations are made on the surface, the packer is set or collapsed by slacking off weight of the drillstring upon the anchor pipe resting on the bottom of the hole. This shortens, compresses and expands the packer against the wall to isolate the lower zone from the rest of the open hole. Further slack-off on the drillstring will open the tester valve which is the main valve of the tool. When going into the hole the upper part of the drillstring is empty (or nearly empty), and the tester valve is closed to prevent mud entry into the drillpipe. Bypass ports below and above the packer are open, thus drilling fluid can bypass the packer without pressure buildup below the packer if a restriction in the hole is encountered. These bypass ports are closed as weight is slacked off on the tool to open the tester valve, and, simultaneously, formation fluids below the packer can enter the drillstring.

(3) At the completion of the flow the shut-in valve is closed and pressure is allowed to build up in the wellbore opposite the formation. Quantitative information about the formation can be obtained from this portion of the pressure record if the shut-in period is sufficiently long to allow the pressure in the wellbore to approach the static reservoir pressure. (It is common for this cycle to be repeated to record two or more flow and shut-in periods.)

(4) At the end of the test the tester valve is closed, the drillstring is raised to retain fluid in the drillpipe, and the bypass ports are opened to equalize the pressure across the packer.

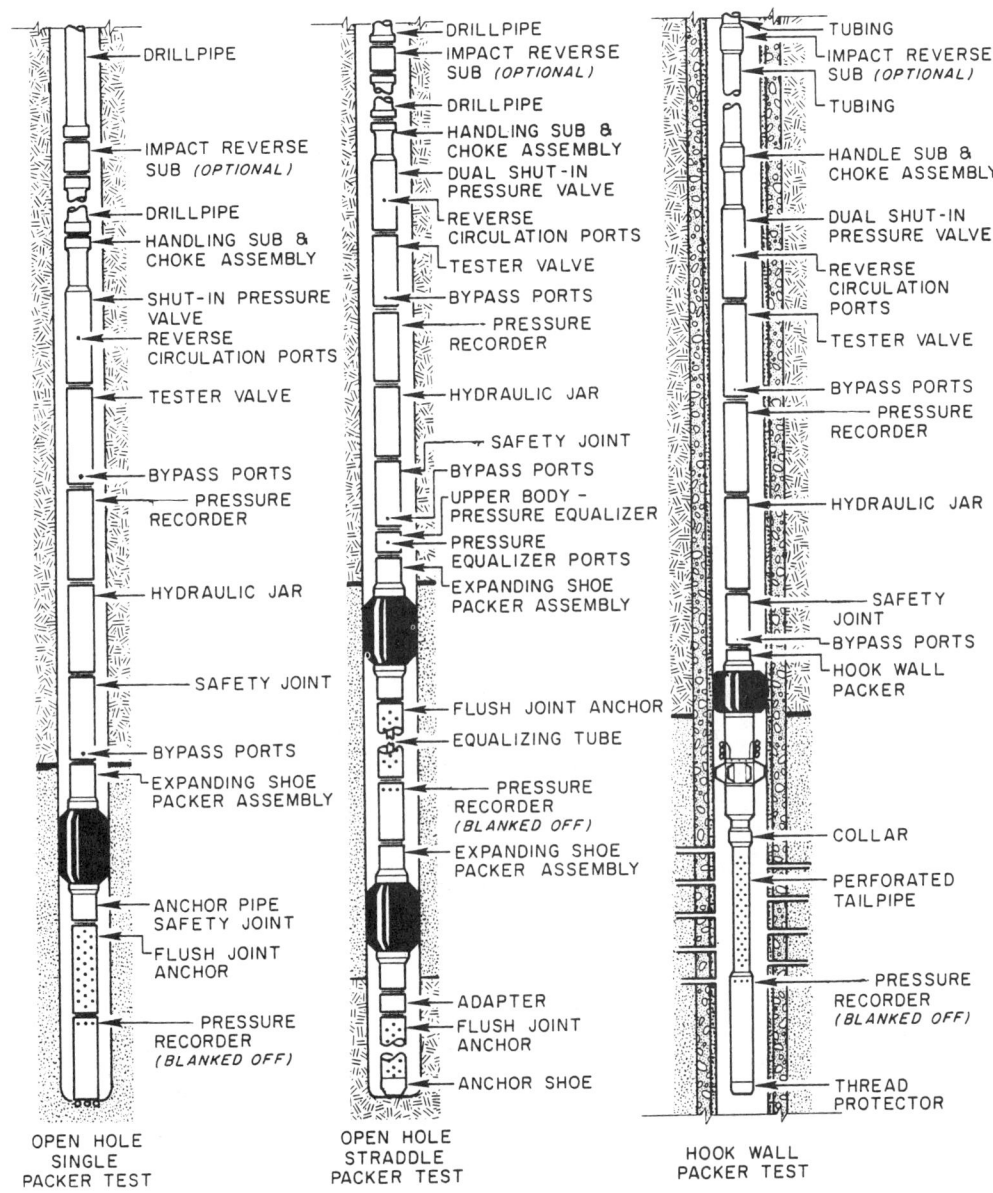

Figure 5-31. Principal Types of Drillstem Tests

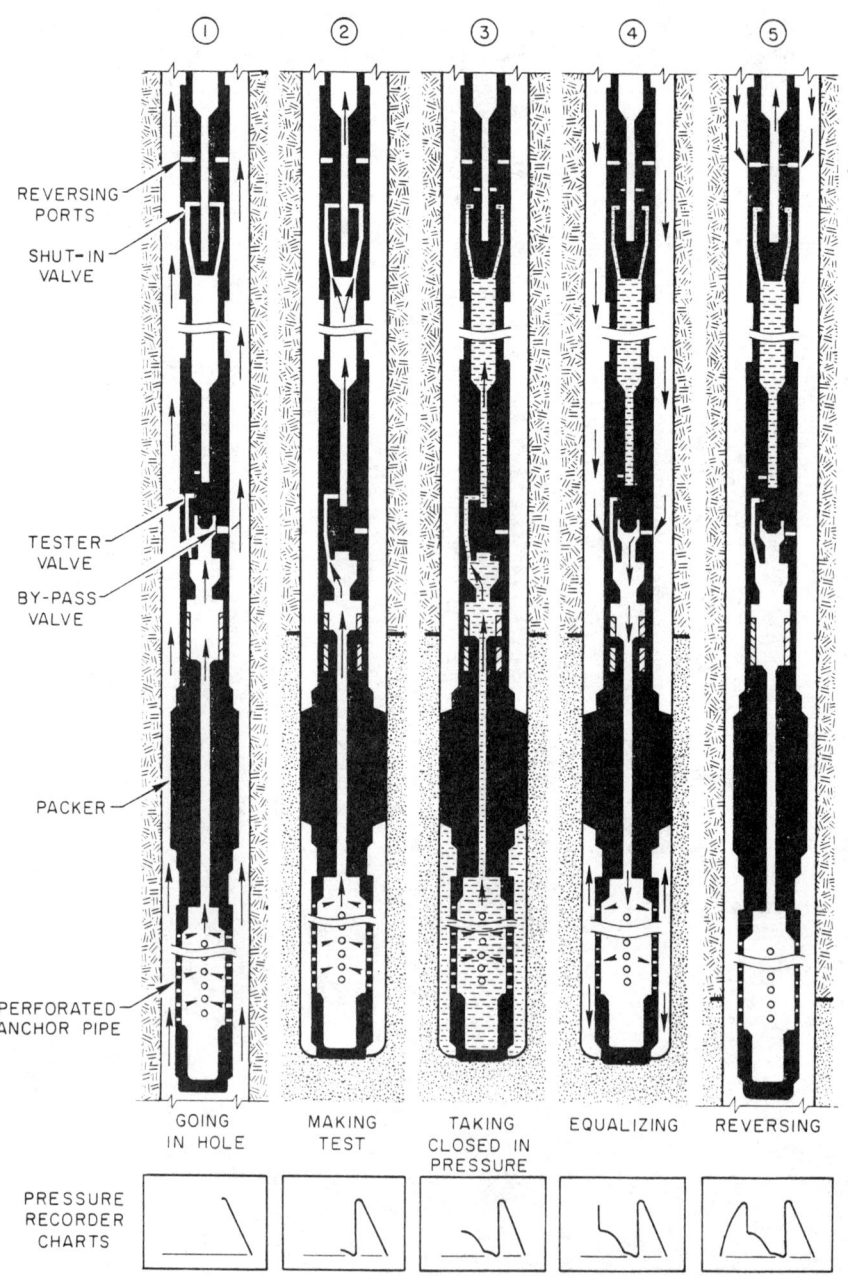

Figure 5-32. DST Procedure

(5) Finally, the setting weight is taken off the anchor pipe and the packer is freed. The drillstring is then pulled from the hole until the section containing fluid reaches the surface. The fluid content is examined as each successive stand is broken out. Frequently such stand-by-stand sampling is neither necessary nor desirable; therefore, the fluid recovery may be pumped out of the drillstem by reverse circulation before pulling the drillstem. This is accomplished by opening ports in the DST assembly and by pumping fluid into the annulus at the surface with the blowout preventer (BOP) rams closed, forcing it to circulate down the annulus and up the drillstem. The reverse circulation ports may be opened by turning or raising the drillstem at the surface, but are usually opened by means of a drop bar.

Variations for drillstem testing include the open-hole straddle packer test, illustrated in Figure 5-31 (center panel). This arrangement is used whenever it is desired to isolate formations both above and below the zone to be tested. Straddle packer testing is more subject to failure than conventional testing because both packers must hold pressure-tight to obtain a successful test, and the double packers increase the chance of sticking. Sometimes the zone of interest is a considerable distance offbottom and a long anchor assembly is needed, further enhancing the chance of stuck pipe.

Another formation test procedure, the hook wall packer test (Figure 5-31, right panel), is employed inside the casing. A DST inside casing is frequently used on a workover because a production test can be quickly obtained and such tests usually eliminate the need for swabbing. Except for packers, the general arrangement of valves and accessory items is nearly the same for single packer, straddle packer, and hook wall packer test assemblies. Drillstem tests are often made with hook wall packers inside the casings of individual zones scheduled for completion in a new well when a multizone completion is contemplated.

A variation of the hook wall packer test, used in open hole when the test interval is a considerable distance above bottom, is to employ a sidewall anchor tool similar to the hook wall tool for casing. This device will set against hard formation and, provided a smooth, strong section of the hole is selected, will hold almost as well as the hook wall tool inside the casing. The static load on the sidewall tool can be reduced by employing straddle packers.

The most important information obtained from a DST is the amount and composition of the recovered fluids, for it is on the basis of this information that completion of the well will be decided. Water recovered from a formation or drillstem test cannot always be identified as invasion filtrate or natural connate. The usual method is to determine the salinity, or resistivity, of the mud filtrate from a filter press. However, in some areas and in some circumstances, the salinity of the naturally occurring formation (connate) water may be indistinguishable from that of the mud filtrate.

The natural occurrence of the nitrate ion (NO3) in connate water is very rare. Therefore, nitrate is sometimes added to the mud in order that the filtrate will contain nitrate and carry these ions into the formation during invasion. On a subsequent formation test, if the water recovered contains the same concentration of nitrate ion, the fluid recovered is solely filtrate and not formation water. Conversely, if the nitrate ion concentration is reduced, it can be concluded that the ion concentration has been diluted by the entry of formation water into the test tool.

From this type of information the evaluation of a drillstem or wireline test is made quickly and confidently at the wellsite. The method may also be used upon well completion for determining the oil-water contact and thus help in the setting up of production zones.

Exploration Logging often performs the nitrate ion test, the procedure for which can be found in Appendix B. We can also be called upon to collect regular samples of formation fluid as the test progresses and to perform chromatograph gas analysis on samples from the separator. The procedure for this analysis is described in Appendix E and additionally applies to analysis of gas samples taken during Wireline Formation Testing, described under paragraph 5.67.

5.67 WIRELINE FORMATION TESTING

Three different types of testers are available to provide testing for almost every type of hole condition. Wireline testing is faster and safer than drillstem tests because pressure data is recorded at the surface and the full fluid column remains in the hole.

- Repeat Formation Tester -- for any number of formation pressure tests and two fluid samples.

- Formation Interval Tester -- for single tests in open or cased hole.

- Multiple Fluid Sampler Tool -- for multiple tests in open or cased hole, this tool is essentially a number of Formation Interval Tester tools linked one above the other.

The aim of these tools is to provide a sample of formation fluids (generally between one and five gallons) to enable the following evaluations:

- Fluid Identification
  - Oil gravity determination
  - Gas analysis
  - Gas/oil ratio determination
  - Water cut

- Pressure Determination
  - Flowing pressure
  - Formation shut-in pressure
  - Hydrostatic pressure

- Permeability Good indication of zones of permeability. Approximate determinations of midrange permeabilities. Positive indications of high permeabilities.

Wireline formation testing tools are run with either an SP tool or a gamma ray tool to provide correlation, or in cased hole they may be run with a collar locator. When these tools are run in open hole, care should be taken to seat the tool in an area where there is a smooth response on the caliper log.

5.68 Repeat Formation Tester

The repeat formation tester, illustrated in Figure 5-33, is run into the hole and a continuous digital readout of hydrostatic pressure is obtained. At any point in the open hole, the tool may be hydraulically actuated to force a rubber pad against the wall of the hole, and a tube in the center of the pad is forced hard against the formation. As the tool is actuated a low pressure area is created in a chamber behind the central tube, and an extension of the tube against the formation connects this chamber with formation pressure. A readout of formation pressure can thus be obtained a number of times, and at anytime two chambers in the tool may be filled from the tube forced against the formation. Both the small area of this tool in contact with the formation and the positive hydraulic retraction reduce the possibility of differential sticking.

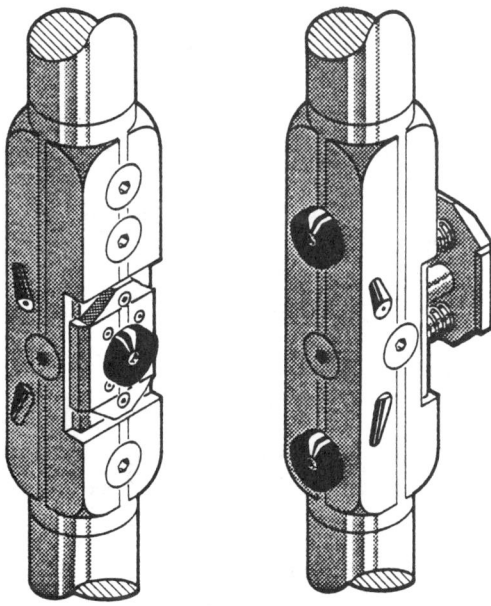

Figure 5-33. Repeat Formation Tester

5.69 Formation Interval Tester

After positioning the tool in the hole, the tool is hydraulically actuated, forcing a rubber pad or pads tightly against the formation to form a seal against the hydrostatic pressure of the fluid in the hole. A shaped charge (or charges) is then fired into the formation, opening a passageway for formation fluids to flow into a chamber in the tool. A surface readout of pressure enables the monitoring of the sampling, shut-in and hydrostatic pressures. After the sample chamber is sealed and a shut-in pressure has been obtained, the tool may be retracted and returned to the surface. If the tools are used in cased hole it is possible to provide a chamber containing four gallons of cement which may be injected into the perforations through the casing prior to retracting the tool.

Testing through casing reduces the possibility of sticking this type of testing tool and increases the likelihood of obtaining a successful test. If the tests are made a sufficient time after casing has been set, and with a good cement bond behind the casing, the effect of invasion is reduced and interpretation of the recovered fluids is easier.

5.70 Evaluation of Recovered Sample

When the tool is recovered at the surface, the chamber pressure is recorded (1) as a check against the value for formation pressure obtained downhole and (2) to indicate any leaks in the sample chamber valves. The chamber is then discharged through a simple separator and any gas is passed through a positive displacement flowmeter. The liquids recovered are separated into oil and water, and the resistivity of the water will be recorded. If nitrate tracer was run in the mud system, the nitrate ion test will be run on any recovered water (paragraph 5.19). If the sample contained oil or gas, the oil gravity and gas analysis may be obtained and the ratios of gas, oil, formation water and mud filtrate enable an empirical estimate of the quantity and quality of fluids that may be produced from the interval tested.

5.71 WELL STIMULATION

Well stimulation treatments were originally developed to rejuvenate old oil and gas wells by improving the porosity and permeability of the producing formations. As techniques have improved, however, they have been used more and more to initiate acceptable producing rates from new wells.

The first stimulation method, nitro-shooting, started about a hundred years ago to liven up wells that had almost ceased to produce. Sometimes the improvement after shooting was spectacular. Other techniques have almost completely taken the place of shooting, although it is still employed on a limited scale. During the 1930s, acid stimulation for limestone and dolomite formations became commercially available. The first treatments were with hydrochloric acid; by 1940 mud acid mixtures of hydrofluoric and hydrochloric acids were being used. In 1948, hydraulic fracturing (hydrafrac) was developed as a specific stimulation process. The original procedure, after breaking down the formation by applying pressure, was to pump mixtures of gelled oil and sand into the induced fractures to hold open the fissures after hydraulic pressure had been released.

There are many variations of the original process, the most important being the use of acids on calcareous formations and large-volume jobs with high rates of injection, employing water instead of refined or crude oil. "Frac" techniques have been developed to apply fracture pressure at specified points in a well. Fracture treatments are expensive, but the method is frequently employed. The process does for tight sandstone reservoirs what acid treatments do for limestone or dolomite reservoirs. Many wells that would not have otherwise justified their costs have done so with fracture treatments.

APPENDIX A
EXAMPLES RELATING TO MUD LOGS

Appendix A contains examples of various types of information referenced in the text of MS-178. A brief description of the examples follows.

Figure A-1 is an example of typical instructions issued at the wellsite to operating company personnel and the mud logging crews. This information is to make all personnel aware of the communications interaction required during operation of the well and of the followup action and reporting procedures. The instructions will often contain specific orders to the toolpusher and Wellsite Geologist.

Figure A-2 shows four scales that can be used for rate of penetration. The two top scales indicate different options, depending on the range of drill rate. The second scale from the bottom is used only occasionally, while the bottom scale is in standard usage. These scales are in the Drill Rate column on the Formation Evaluation Log.

Figures A-3 and A-4 list the standard abbreviations to use on Exlog's logs and reports. Most of these are industry standard.

Figures A-5a, -5b and -5c are examples of logs formulated to standard preparation in various operating areas. Although the requirements and conventions of specific areas or clients may call for variations and modifications to the log, always remember that the prime rule in preparing any log is consistency —

- on any individual log

- on all logs prepared for the client (unless the client specifies otherwise)

- on all logs prepared in any given area (unless the client specifies otherwise)

The client's wishes are of course paramount. However, as you gain experience and knowledge of the area in which you work, you can recommend the format that the company has found to be most successful in the area. Any major change in logging format and procedure must first be discussed with your local office.

Figure A-6 illustrates a supplemental log for reporting core and test results. It is added to the bottom of the Formation Evaluation log.

Figure A-7 illustrates the bit data record which also can be added to the bottom of the Formation Evaluation Log.

1. Start logging at the shoe of the surface casing.

2. Take two sets of washed samples and two sets of unwashed samples every 30 feet in top hole and at 20 ft and 10 ft thereafter when the drilling rate permits.

3. Circulating Instructions: Do not circulate out drilling breaks above 5100 ft. Below 5100 ft the following circulating instructions apply.

   A. When a drilling break at least 5 feet thick is encountered after drilling through at least 15 feet of overlying shale, stop and circulate. Do not drill over a total of 15 feet from the top of the drilling break.

   B. When a series of thin sand stringers aggregating at least 5 feet is encountered, stop and circulate after drilling 5 feet beyond the base of the lowest sand stringer; however, do not drill over a total of 25 feet from the top of the highest sand stringer of the series.

   C. After drillstem tests, connections after tests, etc., circulate out for <u>a sufficient period of time</u> so that new shows can be properly evaluated.

   D. Circulate up bottom samples prior to EVERY trip below 5100 feet.

   E. <u>Testing</u>: Circulate at least 10 feet above the top of potential test zones (even if the driller has to lay down a single).

4. Record bottomhole trip gas and connection gas on the log if 0.2 percent or greater.

5. Carbide (record with viscosity):

   A. Run twice a day.

   B. Run before testing (approximately 10 minutes after circulating, following hole-wiping trip).

6. Salinities to be measured and recorded on the log:

   A. Base sample immediately prior to gas show.

   B. Gas shows.

   C. Wiping-hole gas prior to testing.

   D. Bottomhole trip gas after testing.

   E. Record with viscosity and mudweight every six hours below 3500 ft.

7. If evidence of light mudweight is noted (high connection gas or trip gas, persistent gas shows, anomalous shale shows, etc.):

   A. <u>Immediately</u> instruct the driller to stop drilling and build up mud weight.

   B. Inform company man and wellsite geologist.

8. Stop drilling and notify wellsite geologist or operator's office if a significant gas increase is encountered <u>(even if not associated with a drilling break)</u>.

9. This program is to serve as a guide only, and loggers should use their own good judgement in situations not spelled out above.

10. Daily mud logging reports should be ready by 7:00 a.m.

11. Prints of the daily mud log should be mailed to the responsible people or offices.

12. <u>No information is to be released to any other party</u>.

Figure A-1. Typical Mud Logging Program

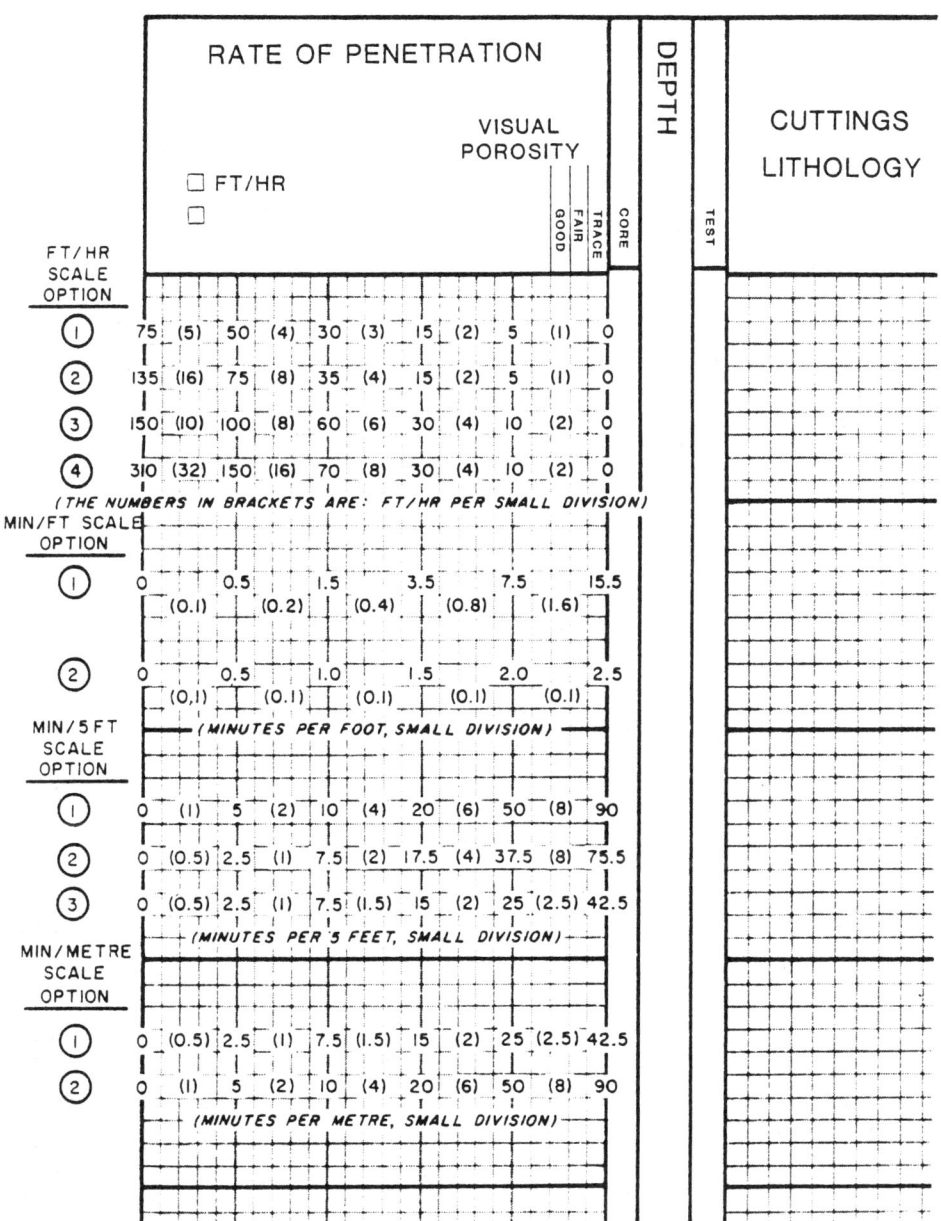

Figure A-2. Suggested Scale for Rate of Penetration

@	at	chit	chitin (ous)	drlg	drilling
abnt	abundant	Chk	Chalk (y)	drsy	druse (y)
abt	about	chl	chlorite (ic)	dtrl	detrital (us)
abv	above	choc	chocolate	dtrm	determine
accum	accumulation	chk	choke		
acic	acicular	cht	chert (y)	Ech	Echinoid
aft	after	circ	circulate (ing)	elg	elongate
agg	aggregate	c-in-c	cone-in-cone	elip	elliptical
aglm	agglomerate	clas	clastic	embdd	embedded
alg	algae (al)	cln	clean	Endo	Endothyra
alt	altered (ing)	clr	clear	enlrg	enlarged
amb	amber	clus	cluster	est	estimated
amor	amorphous	Cly	Clay	ethy	earthy
Amph	Amphipora	Clyst	Claystone	equiv	equivalent
amt	amount	cmt	cement (ed)	euhed	euhedral
andes	andesite (ic)	cncnc	concentric	evap	evaporitic
ang	angular	cntr	center (ed)	exp	expose
anhed	anhedral	Coal	Coal	extr	extrusion (ive)
anhy	anhydrite (ic)	col	color (ed)		
apr	apparent	com	common	f	fine (ly)
aprox	approximate (ly)	conc	concretion (ionary)	fac	facet (ed)
aprs	appears	conch	conchoidal	fau	fauna
arag	aragonite	Cono	Conodont	Fvst	Favosites
aren	arenaceous	cons	considerably	Fe	Iron-Ferruginous
arg	argillaceous	cont	contact	Fe-st	Ironstone
argl	argillite	contm	contaminate	FFP	final flowing pressu
ark	arkose (ic)	contrt	contorted	FHP	final hydrostatic pr
asph	asphalt (ic)	Coq	Coquina	fib	fibrous
av	average	Cor	Coral	fig	figure
		corr	correct (ed)	fis	fissile
B	bottom of	cov	covered	fl	fill (ed)
bar	barite (ic)	cpct	compact	flat	flattened
bas	basalt	cren	crenulated	fld	feldspar (athic)
bcm	become (ing)	crev	crevice	flgy	flaggy
bd	bed	crg	coring	flk	flake (y)
bdd	bedded	Crin	Crinoid (al)	flnt	flint (y)
bdg	bedding	crm	cream	flor	fluorescence
Belm	Belemnites	crnk	crinkled	fls	flesh
bent	bentonite (ic)	crpgr	cryptograined	flt	fault
bf	buff	crpxl	cryptocrystalline	fltg	floating
bioc	bioclastic	csg	casing	fm	formation
biot	biotite	ctc	contact	fnt	faint (ly)
bit	bitumen (inous)	ctgs	cuttings	fol	foliated
bl	blue (ish)	cvg	cavings	Foram	Foraminifera
bldg	bleeding	Cyp	Cypridopsis	fos	fossil (iferous)
bldr	boulder (256 mm +)	xbdd	crossbedded	fr	fair
blk	black	xl	crystal	frac	fracture
blky	blocky	xlam	crosslaminated	frag	fragment (al)
bnd	band	xln	crystalline	fri	friable
bndd	banded	xstrat	cross-stratifield	fros	frosted
Brac	Brachiopod			frs	fresh
brec	breccia (ed)	DC	drill collar	FSIP	final shut-in pressur
bri	bright	dd	dead	Fus	Fusulinid
brit	brittle	DD	drillers depth		
brkn	broken	deb	debris	g	good
brn	brown	decr	decreasing	Gab	Gabbro
Bry	Bryozoa	degr	degree	Gast	Gastropod
btm	bottom	dend	dendrite (ic)	gil	gilsonite
btry	botryoidal	dia	diameter	gl	glass (y)
		diat	diatoms	glau	glauconite (ic)
c	coarse (ly)	dif	difference	Glob	Globigerina
ć	core	dism	disseminated	glblr	globular
calc	calcite (areous)	dk	dark	glos	gloss (y)
carb	carbonaceous	dns	dense (er)	gn	green
cav	cavernous	DP	drillpipe	Gns	Gneiss
cbl	cobble (64-256 mm)	dto	ditto	gr	grain (ed) (y)
Ceph	Cephalopod	Dol	Dolomite (ic)	gran	granular
cgl	conglomerate	dolc	dolocast (ic)	Grap	Graptolite
Chaet	Chaetetes	dolmd	dolomold (ic)	grd	grade (ed) (ing)
chal	chalcedony	Dolst	Dolostone		

Figure A-3. Standard Abbreviations for Descriptions and Reports

grnl	granule (2-4 mm)	lt	light (er)	perm	permeability
Grnt	Granite	ltc	lithic	pet	petroleum
grsy	greasy	ltl	little	phos.	phosphate (ic)
grty	gritty	lyr	layer	piso	pisolite (ic)
Gvl	Gravel			pit	pitted
gy	gray	m	medium	pk	pink
gyp	gypsum (iferous)	macfos	macrofossil (s)	pkr	packer
Gywk	Graywacke	mag	magnetite	plag	plagioclase
		magn	magnetic	plas	plastic
hd	hard	mar	maroon	Plcy	Pelecypod
hem	hematite (ic)	mas	massive	pl fos	plant fossils
hex	hexagonal	mat	material	plty	platey
hi	high	max	maximum	pol	polish (ed)
hrznl	horizontal	mbr	member	por	porous (osity)
hvy	heavy	Mdst	Mudstone	porc	porcelain (ous)
hydc	hydrocarbon	meta	metamorphic	pos	possible (ility)
		msm	metasomaticous	p-p	pin point
ig	igneous	mica	mica (ceous)	pred	predominate (ly)
IHP	initial hydrostatic pressure	micfos	microfossil (iferous)	pres	preserved (ation)
imp	impression	micgr	micrograined	prim	primary
imprg	impregnated	mic-mica	micro-micaceous	pris	prism (atic)
incl	included (sion)	micxl	microcrystalline	prly	pearly
incr	increase (ing)	mid	middle	prob	probable (ly)
ind	indurated	min	minimum	prom	prominent (ly)
indst	indistinct	mky	milky	psdo	pseudo
Inoc	Inoceramus	mnr	minor	pt	part (ly)
ISIP	initial shut-in pressure	mnrl	mineral (ized)	ptg	parting
intbdd	interbedded	mnut	minute	purp	purple
intcl	interclast (s)	mod	moderate	pyr	pyrite (ic) (tized)
intfrag	interfragmental	Mol	Mollusca	pyrbit	pyrobitumen
intfm	interformational	mot	mottled	pyrclas	pyroclastic
intgran	intergranular	Mrl	Marl (stone)		
intgwn	intergrown	mtx	matrix	qtz	quartz
intlam	interlaminated	musc	muscovite	qtzc	quartxitic
intpt	interpretation			qtzs	quartzose
intr	intrusion (ive)	n	no, non	Qtzt	Quartzite
intstl	interstitial	nac	nacreous		
intv	interval	nod	nodule (ar)	r	red
intxl	intercrystalline	num	numerous	rad	radiate (ing)
invrtb	invertebrate			rand	random
ireg	irregular	oil	oil	rbl	rubble (y)
irid	iridescent	obj	object	rnd	round (ed)
Fe	iron	occ	occasional	rec	recovered
Fe-st	ironstone	och	ochre	reg	regular
		od	odor	repl	replaced (ing) (ment)
Jasp	Jasper	OH	open hole	resd	residue (al)
jtd	jointed	olv	olive	rhmb	rhomb (ic)
jts	joints	ooc	oolicast (ic)	rhmbl	rhombohederal
		ool	oolite (ic)	rk	rock
kao	kaolin	oom	oomold (ic)	RKB	rotary kelly bushing
KB	kelly bushing	opq	opaque	rmn	remains (nant)
ker	kerogen	opp	opposite	rmg	reaming
KO	kick off	org	organic	rng	rang (ing)
		orig	origin (ate)	ro	rose
lam	laminated	orng	orange	rr	rare
lav	lavender	orth	orthoclase	rsns	resinous
lch	leach (ed)	Ost	Ostracod	RU	rig up
ldg	ledge	ox	oxidized	run	run (ning)
len	lentil (cular)				
lg	long	p	poor	s	small
Lig	Lignite (ic)	P&A	plugged & abandoned	sa	salt
lith	lithographic	PB	plugged back	S	Sulfur
lmn	limonite (ic)	Para	Paraparchites	sach	saccharoidal
lmpy	lumpy	pbl	pebble (y) (4-64 mm)	sal	salmon
Lngl	Linguloid	pchy	patchy	sat	saturate (d)
low	lower	PD	present depth	sb	sub
lrg	large (er)	Pdct	Productids	sbang	subangular
Ls	Limestone	pel	pellet	sbhed	subhedral
lse	loose	perf	perforate (ion)	sbrnd	subrounded
lstr	luster			sc	scales

scat.	scattered	srtg	sorting	TVD	true vertical depth
Sch	Schist	Ss.	Sandstone		
Scol.	Scolecodonts	st.	stone	unconf.	unconformity
scs	scarce	stip	stippled	uncons.	unconsolidate
Sd	Sand (1/16-2 mm)	stn	stain (ed) (ing)	uni	uniform
sdy	sandy	str	streak	up	upper
sec	secondary	strat	strata (ified)		
sed	sediment (ary)	strg	stringer	v	very
sel	selenite	stri	striated	var	variable
sft	soft	strmg	streaming	vcol.	varicolored
Sh	Shale	Strom.	Stromatoporoid	ves	vesicular
shad.	shadow	struc	structure	vgt	variegated
SICP	shut-in casing pressure	styl	stylolite (ic)	vit	vitreous
shy	shaly	suc	sucrosic	vn	vein
sid	siderite (ic)	sug	sugary	Volc.	Volcanic
SIDP	shut-in drillpipe pressure	surf	surface	Vrtb.	Vertebrate
sil	silica (eous)	srct	sericite	vrt	vertical
sks	slickensided	swb	swab (bing)	vrvd.	varved
sl	slight (ly)	sx	sacks	vug	Vug (gy) (ular)
slky	silky	sz.	size		
slo	slow			w	well
slt	silt	T/	top of	wh	white
Sltst	Siltstone	tab	tabular	wk	weak
slty	silty	TD	total depth	wo	waiting on
sm	smooth	Tent	Tentaculites	wthr	weather (ed)
smpl	sample	tex	texture	wtr	water
sol	solution	tgh	tough	wvy	wavy
solb	soluble	thk	thick	wxy	waxy
sp.	spot (ted) (ty)	thn	thin	w/	with
spec.	speck (led)	thru	throughout	W/C	water cushion
Spfr.	Spirifers	tr.	trace		
Spg	Sponge	Trilo	Trilobite	xbdd	crossbedded
sph	spherules	trip	tripolite (itic)	xbdg	crossbedding
sphal	sphalerite	trnsl	translucent	xl	crystal
spic	spicule (ar)	trnsp	transparent	xlam	crosslaminated
splty	splintery	Troc	Trochiliscus	xln	crystalline
spr	spore	tt.	tight (ly)		
spy-ca.	sparry calcite	tstg	testing	yel	yellow
squ	squeeze	tub	tubular		
srt	sort (ed)	tuf	tuff (aceous)	zeo	zeolite
				zn	zone

```
                    L E G E N D
              B:   Birdwell           Mc: McCullough
              D-A: Dresser Atlas      S:  Schlumberger
              G-O: Gearheart-Owen     W:  Welex
```

a	apparent (subscript)	FIT . . .	Fluid Identification Tool (G-O)
A-BHC .	Acoustic Amplitude Log (S)	FPC . . .	Free Point Indicator (B)
AVL . . .	Acoustic Velocity Log (W)	FPI . . .	Free Point Indicator (G-O)
		FPK . . .	Flo-Pak (B)
b	bulk (subscript)	FPP . . .	Free Point Indicator (B)
BATS . .	Borehole Audio Tracer Survey (G-O)	FPT . . .	Free Point Indicator (B)
BC . . .	Borehole Camera (B)	FT . . .	Formation Tester (D-A, G-O, S)
BGT . . .	Borehole Geometry Tool (S)		
BHC . .	Borehole Compensated Sonic Log (D-A, S)	g	gas (subscript)
BHC-VD .	Variable Density Log (S)	GDS . . .	Guard Log (B)
BL . . .	Beta Log (B)	GL . . .	Guard Log (G-O)
		GM . . .	Gradiomanometer (S)
C	Conductivity	GR . . .	Gamma Ray Log
CA . . .	Caliper Log (B)		
CAL . .	Caliper Log (B, S)	h	bed thickness
CBL . . .	Cement Bond Log (D-A, G-O, S)	h	hole (subscript)
CBL-VD .	Cement Bond - Variable Density Log (S)	HDT-D .	High Resolution 4-Arm Digital Dip Log (S)
CCL . .	Continuous Collar Log (B, G-O)	HRDIP .	High Resolution 4-Arm Dip Log (D-A)
CDI . . .	Continuous Directional Inclinometer Log (B)	HRT . .	High Resolution Thermometer (S)
CDL . .	Borehole Compensated Density Log (D-A, G-O, W)		
CDR . .	Continuous Directional Survey Log (S)	i	invaded zone (subscript)
CDS . . .	Continuous Directional Survey Log (G-O)	IEL . . .	Induction Electric Log (D-A, G-O, W)
CFM . .	Continuous Flowmeter (S)	IES . . .	Induction Electric Log (B, S)
CICL . .	Casing Inspection Caliper Log (G-O)	IFL . . .	Interface - Density Log (G-O)
CL . . .	Caliper Log (G-O)	IL . . .	Induction Log (G-O)
CNL . .	Compensated Neutron Log (D-A, S)	IS	Induction Log (B)
C/O . . .	Carbon/Oxygen Log (D-A)	ISF . . .	Spherically Focused Induction Log (S)
CPP . . .	Casing Potential Profile (D-A)		
CSC . . .	Sonar Caliper (B)	j	zone of transition (subscript)
CST . . .	Sidewall Core Gun (S)		
CSVL . .	Borehole Compensated Sonic Log (G-O)	L	Laterolog (S)
CTN . .	Nuclear Cement Top Locater (B)	LL . . .	Laterolog (D-A)
		LR . . .	Last Reading
d	diameter		
DCS . . .	Core Slice Tool (S)	m	cementation factor
DF . . .	Fluid Density Log (B)	m	mud (subscript)
DIFL . .	Dual Induction Focused Log (D-A)	ma . . .	matrix (subscript)
DIGL . .	Dual Induction Guard Log (W)	mc . . .	mud cake (subscript)
DIL . . .	Dual Induction Laterolog (S)	mf . . .	mud filtrate (subscript)
DL . . .	Density Log (G-O)	MFS . .	Multiple Fluid Sampler (S)
DLL . . .	Dual Laterolog (S)	MGL . .	Micro-Guard Log (G-O)
DNLL . .	Dual Space Neutron Lifetime Log (D-A)	ML . . .	Micro-Electrical Log
DTL . . .	Differential Temperature Log (G-O)	MLL . .	Micro-Laterlog (D-A, S)
ΔT . . .	Delta T, Interval Transit Time	MSFL . .	Micro Spherically Focused Log (S)
		MSG . .	Acoustic Cement Bond Log (W)
e	bed thickness		
EL . . .	Electrical Log (D-A, G-O)	NL . . .	Neutron Log (B, G-O)
EMC . . .	Micro-Contact Caliper Log (B)	NLL . . .	Neutron Lifetime Log (D-A)
ENP . . .	Epithermal Neutron Log (B)	NNL . . .	Neutron Log (B)
ES . . .	Electrical Log (B, S)	NML . .	Nuclear Magnetism Log (D-A, S)
ES-ULS .	Saltdome Profile Electrical Log (S)		
		o	oil (subscript)
F	Formation factor		
f	fluid (subscript)	PAL . . .	Pipe Analysis Log (S)
FDC . .	Compensated Formation Density Log (S)	PDC . .	Perforating Depth Control (S)
FDIP . .	Focused 3-Arm Dip Log (D-A)	PFM . .	Packer Flowmeter (S)
FDL . . .	Formation Density Log (G-O)	PL . . .	Pressure Log (B)
FIT . . .	Formation Interval Tester (S)	PLL . . .	Proximity-Microlog (S)

Figure A-4. Standard Abbreviations for Wireline Logs

PML	. . Proximity-Microlog (D-A)	SVL . . .	Sonic Log (G-O)
Ø	Phi, Porosity	SWAN . .	Sidewall Acoustic Neutron Log (D-A)
		SWC . .	Sidewall Core Gun (G-O)
QPGS . .	Quartz Pressure Gauge Service (G-O)	SWD . .	Sidewall Core Gun (D-A)
		SWN . .	Sidewall Neutron Log (D-A, W)
R	Resistivity		
RFT. . .	Repeat Formation Tester (S)	t	thickness
RFVL . .	Radioactive Fluid Velocity Log (G-O)	t	true (subscript)
RTL. . .	Radioactive Tracer Log (G-O)	T	Temperature
RTP. . .	Radioactive Tracer Log (S)	TDT. . .	Thermal Decay Time Log (S)
ρ . . .	Rho, Density	TDT-K .	Dual Space Thermal Decay Time Log (S)
		TL . . .	Temperature Log (B, G-O)
S	Saturation	TLD. . .	Differential Temperature Log (B)
SC . . .	Signature Curve (G-O)	TRP. . .	Radioactive Tracer Log (B)
SFAL . .	Sonic Formation Amplitude Log (G-O)	TVT . . .	Borehole Television (S)
SFL. . .	Spherically Focused Log (S)		
SFVS . .	Spinner Fluid Velocity Survey (G-O)	V	Volumetric Percent
Sign. . .	Signature Log (D-A)	VDL. . .	Variable Density Log (D-A)
SL . . .	Salinometer (B)	V2S . . .	Continuous Velocity Log (B)
SNP . . .	Sidewall Neutron Log (S)	V3D . . .	3D Velocity Log (B)
SP . . .	Spontaneous Potential Log		
SS . . .	Seismic Spectrum (G-O)	w	water (subscript)
		WSS . . .	Wellbore Sibilation Survey (B)
		xo . . .	invaded zone (subscript)

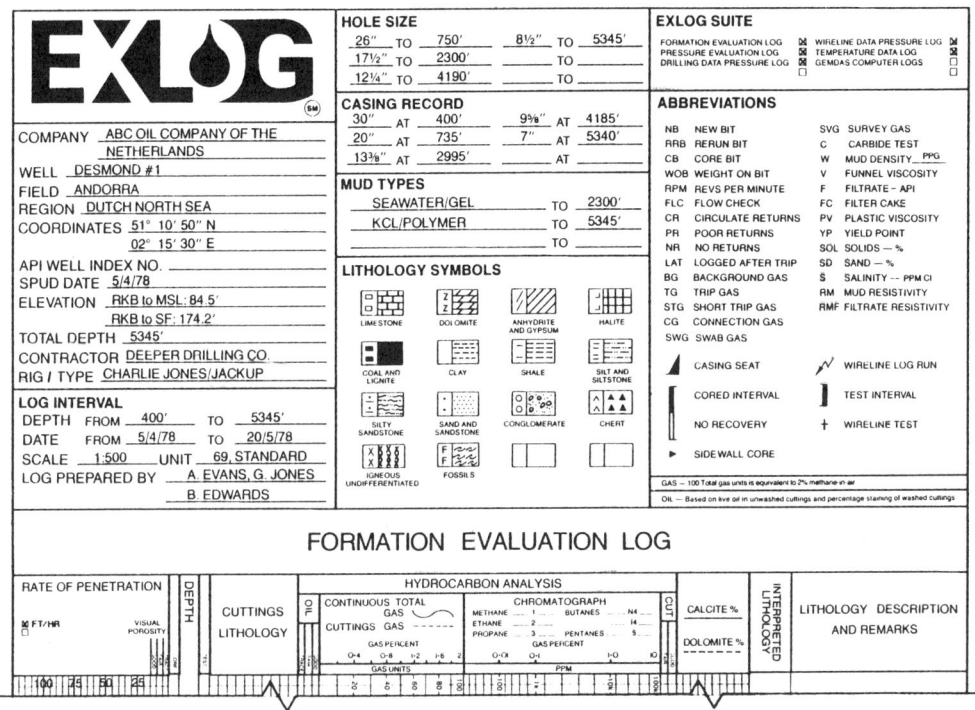

Figure A-5a. Sample Mud Log Heading

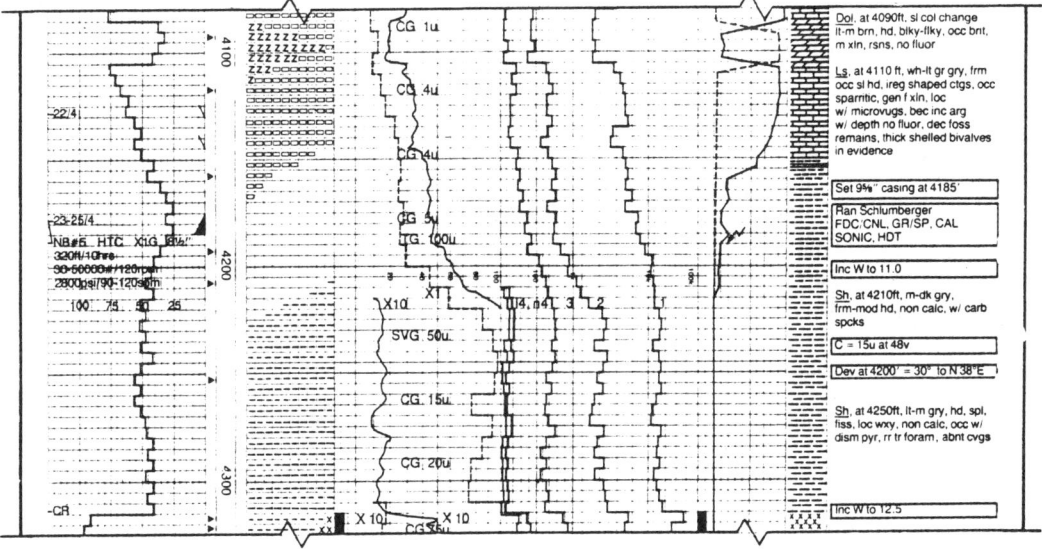

Figure A-5b. Sample Mud Log

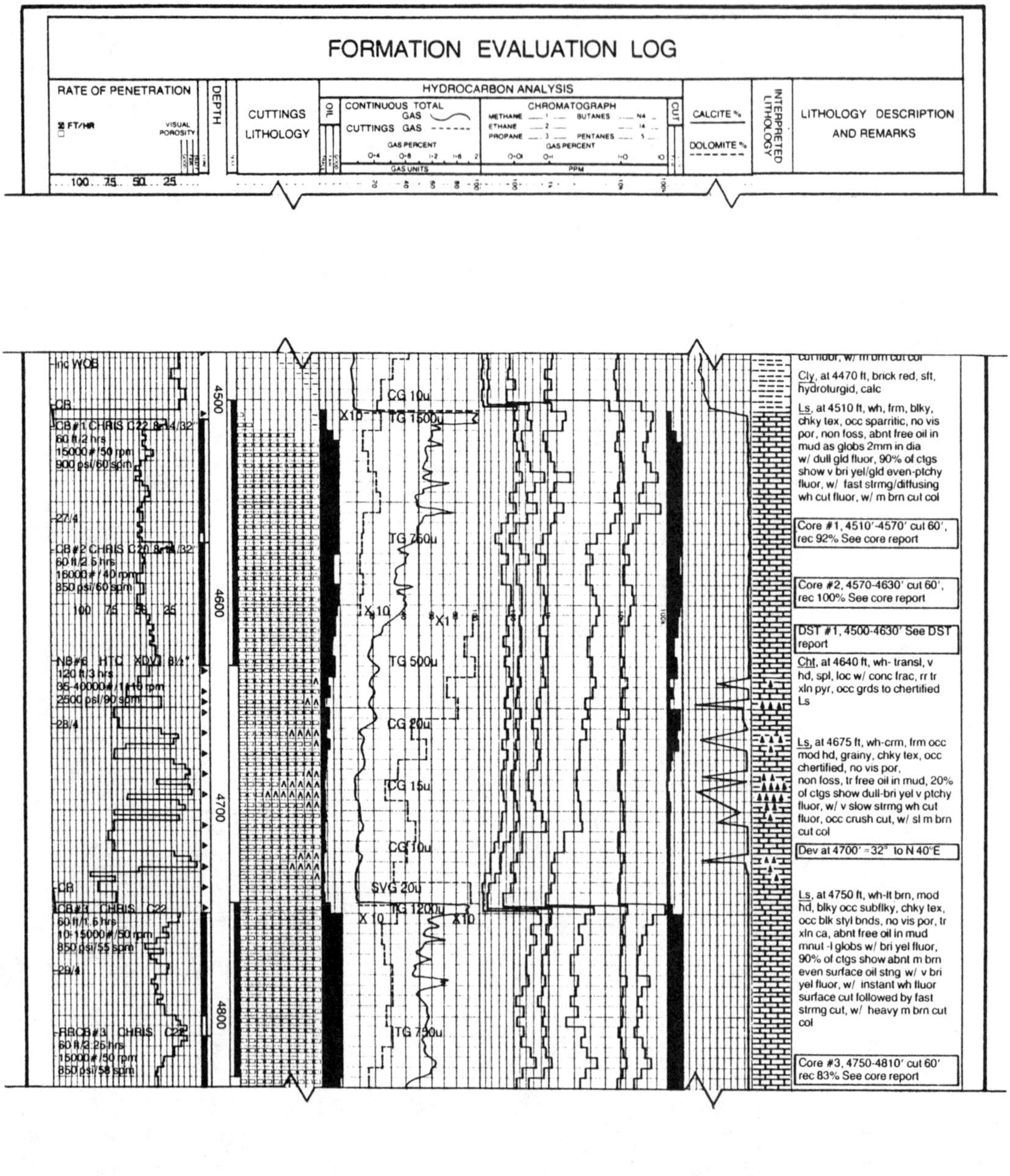

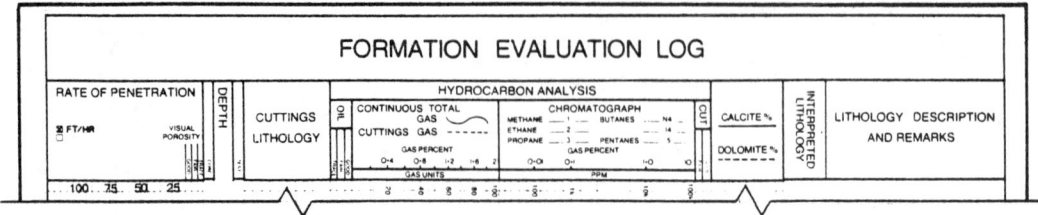

(Sample Mud Log, Cont'd)

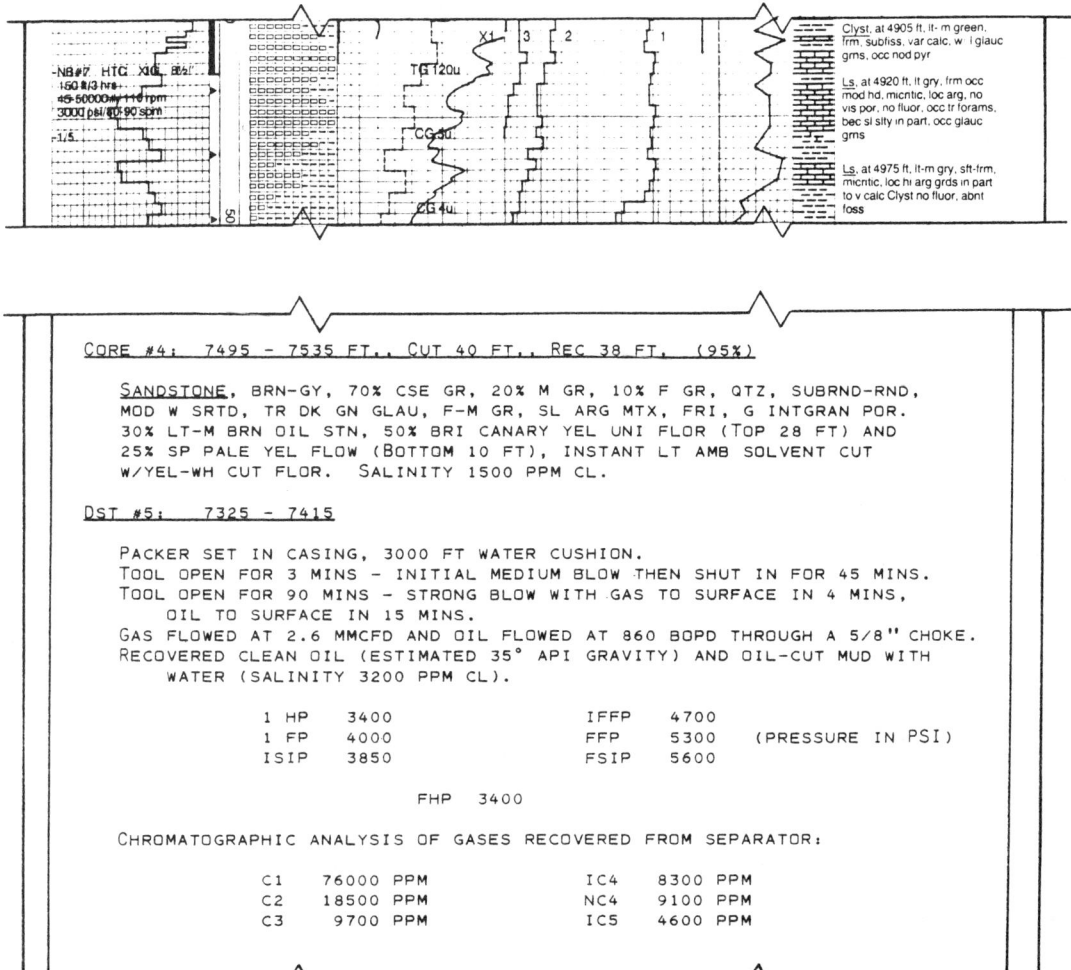

Figure A-6. Supplemental Log for Core Description and Test Results

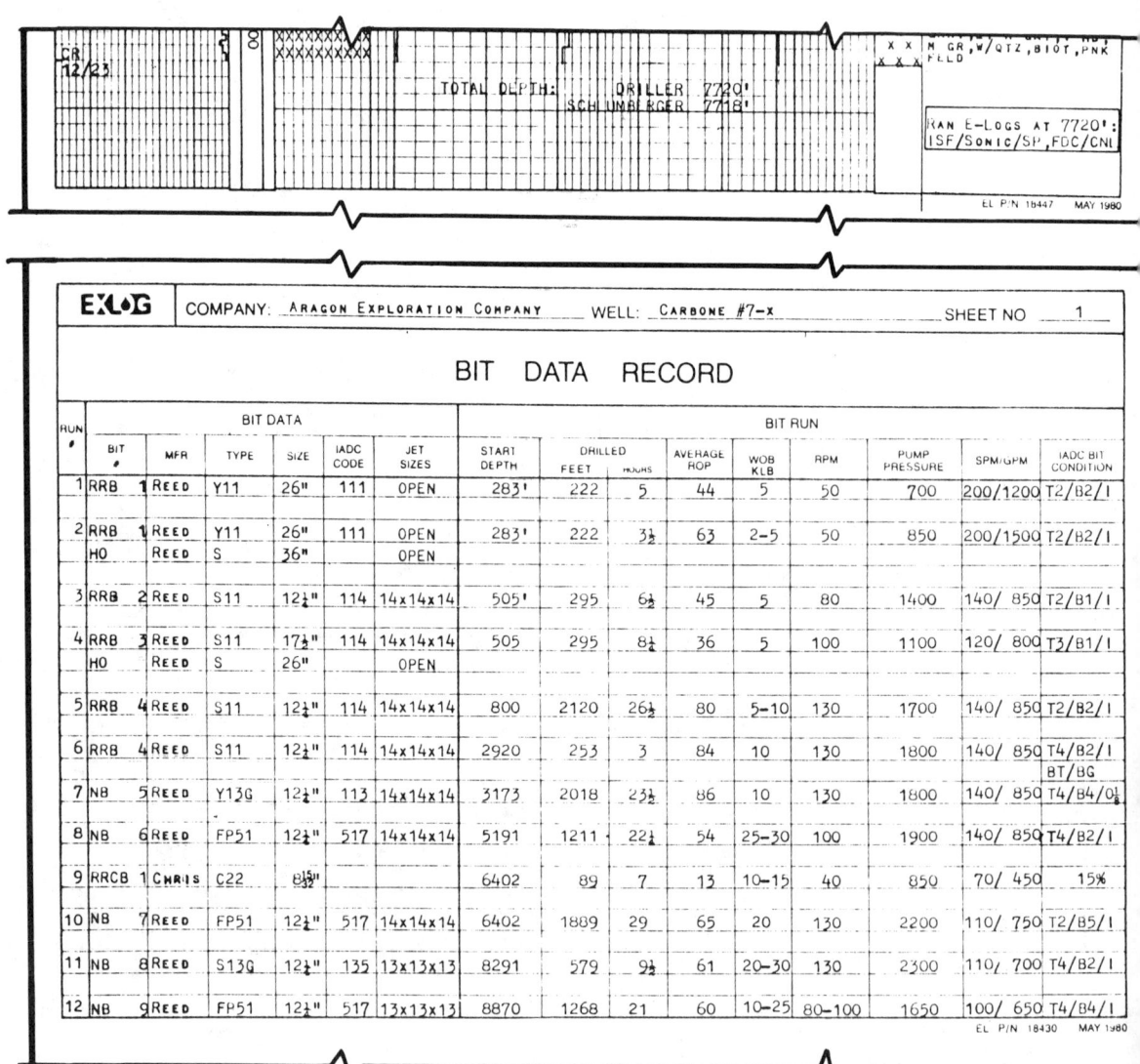

Figure A-7. Bit Data Record

APPENDIX B
TEST PROCEDURES

The following procedures (except for that of chloride) refer to tests using pieces of secondary equipment available only at the client's request.

As these procedures are only for quick reference and serve as a memo of standardization, for further description of procedures and theory of operation, refer to the appropriate equipment manual.

The differentiation of carbonates by normal wellsite petrographic methods is extremely difficult and sometimes even impossible. The Exploration Logging Stain Kit has been produced explicitly to help with these carbonate determinations. The following is a brief description of its use.

The cutting samples, specimen, etc., to be stained must first be thoroughly washed in water to remove any drilling mud or other contaminants, then etched with 10 percent hydrochloric acid and washed again in distilled water.

The first flowchart illustrates the standard procedure. The second flowchart is mainly used for confirmation, and the third is used to determine the reaction produced by dissolved sulfate ions on a solution of barium chloride (Figure B-1b, 1c and 1d).

Figure B-1a. Carbonate Staining

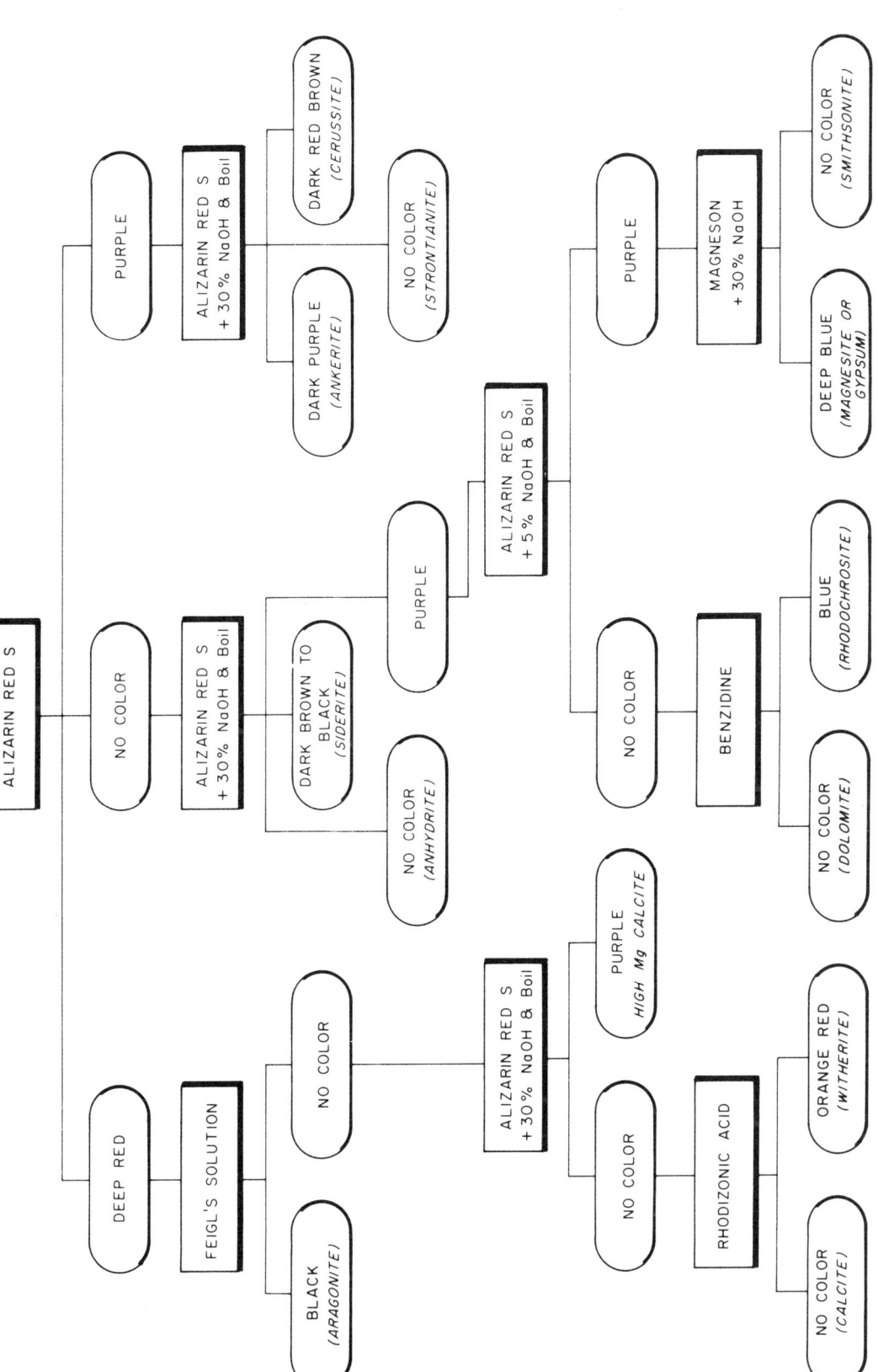

Figure B-1b. Flowchart for Stain Procedure, Using Alizarin Red S

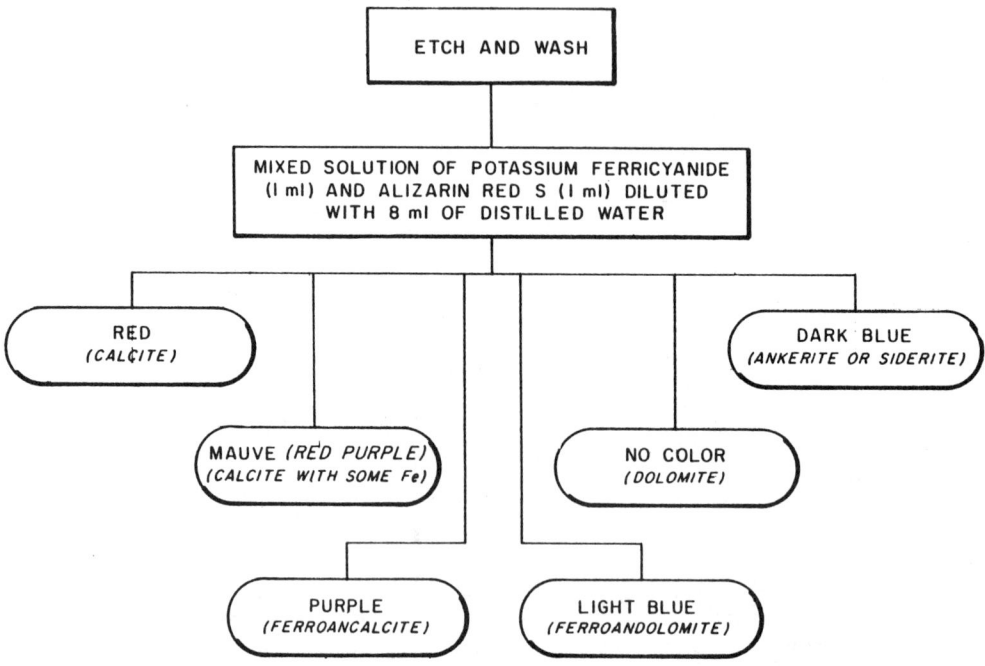

Figure B-1c. Flowchart for Stain Procedures, Using Distilled Water

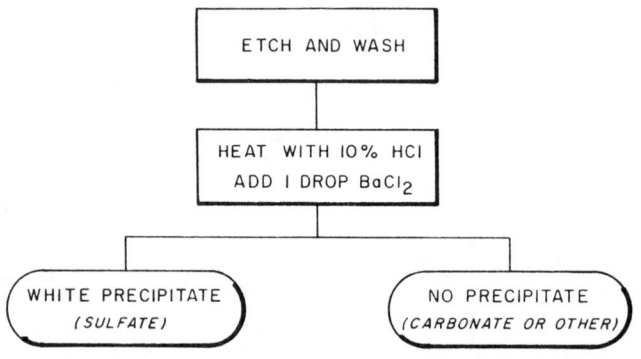

Figure B-1d. Flowchart for Stain Procedures, Using Heat

PURPOSE

The salt or chloride test is very significant in areas where salt can contaminate the drilling fluid. This would include the majority of the world's oil fields. The salt may come from make-up water, salt stringers or beds, or from saltwater flows.

EQUIPMENT

The following materials are required to determine the chloride ion concentration in the mud filtrate:

1. Silver nitrate solution containing 4.7910 gm/litre (equivalent to 0.001 gm chloride ion per millilitre), preferably stored in an amber bottle.

2. Potassium chromate indicator solution (5 gm/100 ml of water).

3. Acid solution 0.02 normal sulfuric or nitric acid.

4. Phenolphthalein indicator solution (1 gm/100 ml of 50% alcohol).

5. Calcium carbonate, precipitated, chemically pure grade.

6. Distilled water.

7. Two graduated pipettes, one 1-ml and one 10-ml.

8. Titration vessel, 100 to 150 ml, preferably white.

9. Stirring rod.

PROCEDURE

Measure one or more millilitres of filtrate into the titration vessel. Add 2 or 3 drops of phenolphthalein solution. If the indicator turns pink, add acid drop by drop from pipette, with stirring, until the color is discharged. If the filtrate is deeply colored, add an additional 2 ml of 0.02 normal sulfuric or nitric acid and stir. Then add 1 gram of calcium carbonate and stir.

Add 25 to 50 ml of distilled water to 5 to 10 drops of potassium chromate solution. Stir continuously while adding standard silver nitrate solution drop by drop from the pipette, until the color changes from yellow to orange-red and persists for 30 seconds. Record the number of millilitres of silver nitrate required to reach the end point. If over 10 ml of silver nitrate solution are used, repeat the test with a smaller sample of filtrate.

Note: If the chloride ion concentration of the filtrate exceeds 10,000 ppm, a silver nitrate solution equivalent to 0.01 gram chloride ion per millilitre may be used. The factor 1000 in the equation in the following paragraph is then changed to 10,000.

Report the chloride ion concentration of the filtrate in parts per million, calculated as follows:

$$\text{chloride, ppm} = \frac{(\text{ml of silver nitrate})(1000)}{\text{ml of filtrate sample}}$$

Figure B-2. Chloride Test Procedures

This procedure is illustrated on the following flowchart.

1. Collect 10 ml of mud filtrate. If the filtrate is clear, proceed to step (3). If not, it must be treated to remove the color.

2. Dilute a 5-ml sample to 30 ml with distilled water (see notes). Add one tablespoon of Ca(OH)2, shake well and allow to settle. Filter the solution. If it's clear, continue with the next step. If not, repeat this step. Make a note of the number of treatments. Each treatment involves a six-times dilution (see step 6).

3. Dilute a 5-ml sample of the original solution (if clear) or the treated solution from step (2) so that the developed color may be read on the colorwheel scale. Normally, a reference background of 150 to 200 ppm nitrate is run in the mud, and to read this value it is necessary to make a ten-times dilution (5 ml made up to 50 ml). The answer read in step (6) must be multiplied by a factor of 2 for this dilution (e.g., scale reading 95; 95 x 2 = 190 ppm). If your answer exceeds 200 ppm, take a 5-ml sample of the X10 diluted filtrate and repeat step (3). Multiply the answer by a factor of 4.

4. Measure 5 ml of the diluted filtrate into a test tube, and add the contents of the grey ampules. Shake for three minutes (<u>no less</u>). Stand and allow to settle, then decant the solution to a clean, dry test tube. Add the contents of the pink ampule, and shake to dissolve the crystals. Stand for ten minutes (time is critical for full color development.)

5. Next make up the colored filtrate to 10 ml (by adding distilled water), and stand in the clear side of the colorwheel. Dilute another 5 ml of the diluted filtrate to 10 ml and place in the other side of the colorwheel. Compare the color intensities. Read the scale.

6. Multiply the scale reading by the original dilution. If Ca(OH)2 was used, you must also multiply further by the dilutions used; e.g., Ca(OH)2 used for one treatment (6 x dilution)

    A. scale reading 18
    B. 10 x dilution, therefore, 2 x 18 = 36
    C. 6 x 36 = 216 ppm

7. Record ppm (NO3) ion on log sheet

<u>Notes</u>

1. Check distilled water periodically for ppm NO3, and deduct this from the value.

2. Ca(OH)2 and filter paper contain nitrates. Check Ca(OH)2 bottle for concentration of nitrates. Also, subtract 5 ppm NO3 for each filter paper used (including filter press).

3. The test chemicals are carcinogenic. Shake the tubes with a cork in the top and spare your fingers.

Figure B-3a. Nitrate Ion Test Procedure

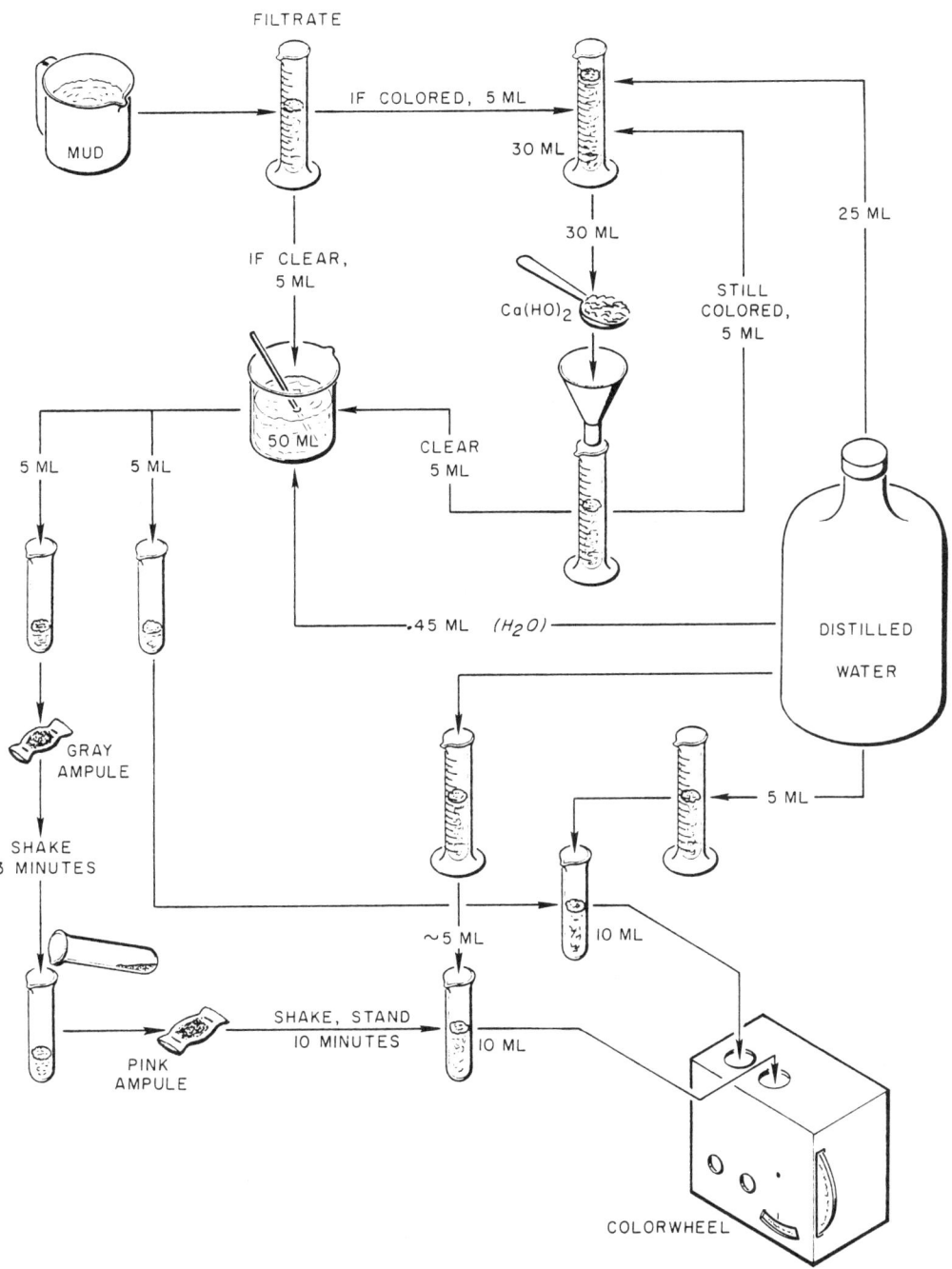

Figure B-3b. Nitrate Ion Test Procedure Flowchart

The following procedure is for the Bernard (manual-type) calcimeter only. There is also an Autocalcimeter available but, because it is easier and more convenient to operate, it is not described here.

1. Crush a few cuttings from the washed and dried sample. Try to avoid selecting obvious cavings. Seive.

2. Weigh out 0.5 gram of the sample retained in the fine seive and place it in the reaction flask.

3. Fill a 5-ml test tube with diluted HCl and place it in the same reaction flask. TAKE GREAT CARE. Do not spill any acid.

4. Place the bung firmly in the flask. Equate the liquid levels in the tube and reservoir by lowering the reservoir. Write down the first reading from the graduated burette.

5. Tilt the flask to spill all the acid into the flask. Hold the flask by the neck to avoid absorbing any heat of reaction. Shake gently, and stand.

6. After one minute, take a reading from the graduated burette (don't forget to equate the levels in the glass reservoir and burette). Subtract this second reading from the first. This value is the amount of CO_2 evolved due to calcite reaction.

7. After ten minutes, equate the levels and take a third reading, from which the second reading must be subtracted. The value of the subtraction is due to the slower reaction of the magnesium ion, and represents the dolomite content.

8. The gas volumes are converted to percentages by reference to the conversion chart prepared during calibration.

9. If the dolomite percentage is high, the reaction can be quickened by using muriatic acid (30% HCl). Take care - concentrated HCl is extremely corrosive.

10. Check your results frequently by running two or more calcimetry tests on the same sample.

Figure B-4a. Calcimetry Procedure

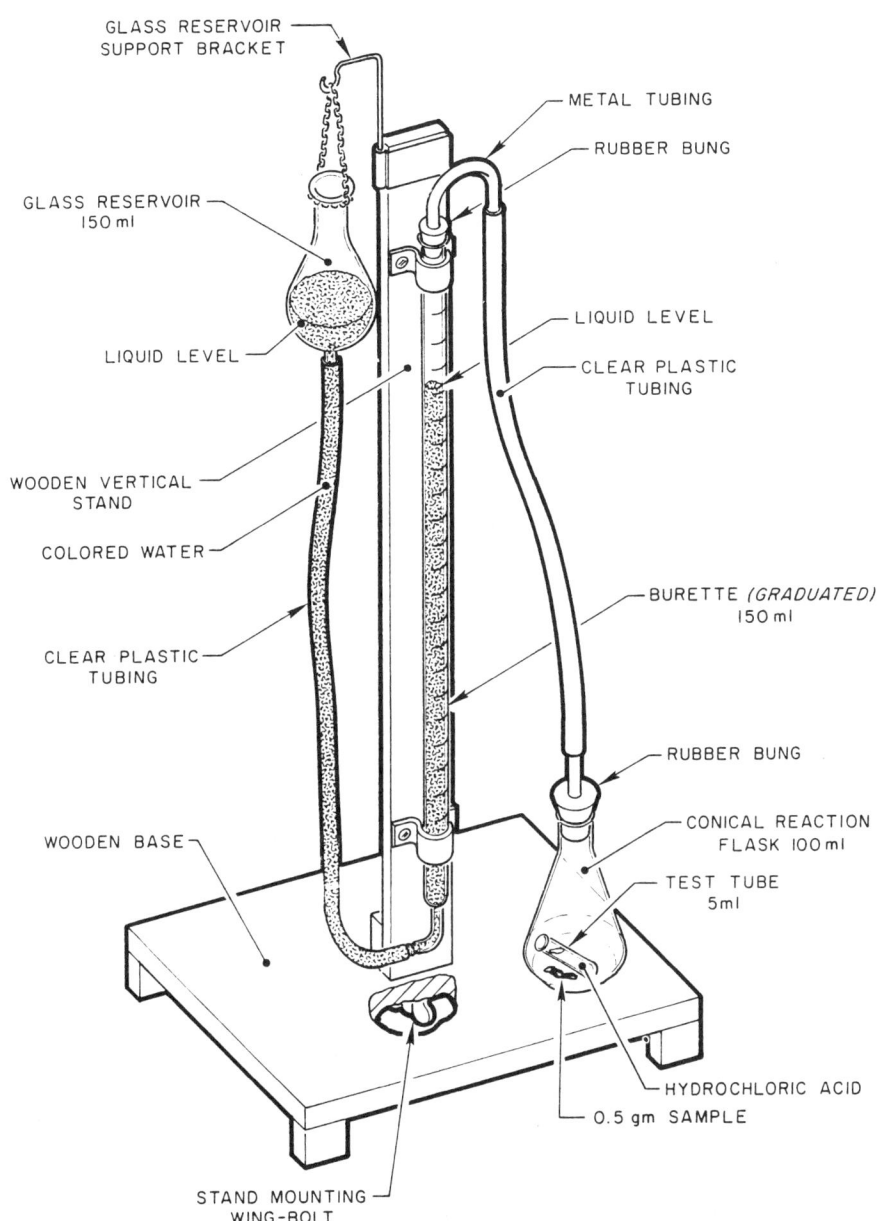

Figure B-4b. Bernard Calcimeter

1. Select shale cutting wherever possible.

2. Crush gently (rotational movement) in mortar and pestle so as not to break down any "gangue" quartz or calcite.

3. Seive and throw away coarse fraction.

4. Weigh out 0.5-gram of the crushed sample and place in mud cup. Add a couple of drops of 0.02N H_2SO_4 to acidize, then add distilled water (a few squirts).

5. Heat the crushed slurry to boiling point (stirring occasionally), then remove from heat.

6. Titrate 0.01N methelene blue solution slowly, checking frequently for the end point by placing a drop of the titrate on filter paper. The end point is shown as a greenish ring around the blue spot.

7. When the end point is reached, warm the solution again but do not boil. If the end point still shows, the titration is finished. If not, continue titrating until the new end point is reached.

8. Check results by performing two or more titrations of the same sample.

Shale Factor Value

$$\frac{100}{0.5 \text{ gm}} \times 0.01 \times \text{millilitre of titration} = \text{milliequivalents shale factor}$$

i.e., Shale Factor = 2 x millilitre of methelene blue titration

Figure B-5. Standard Procedure for Shale Factor

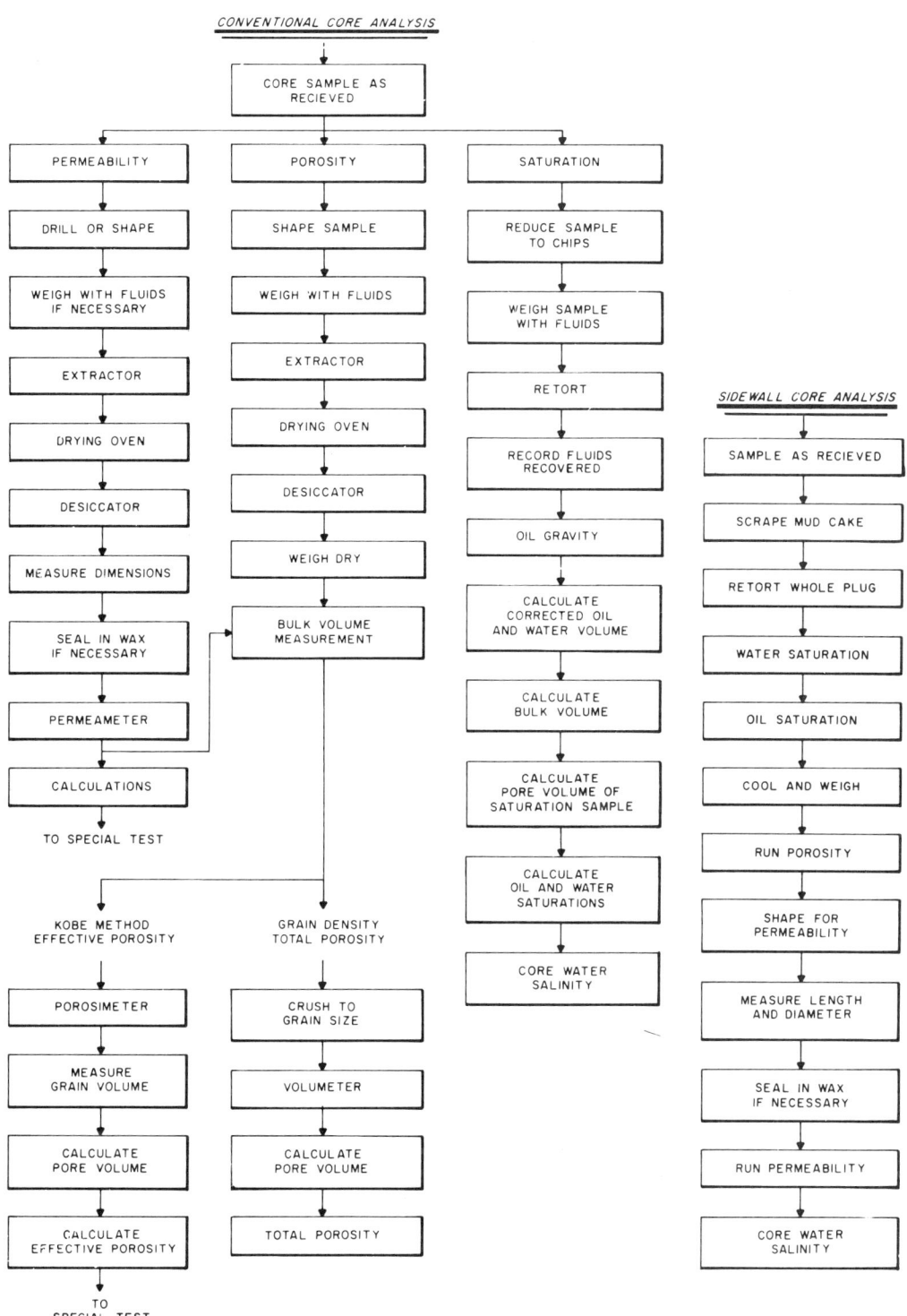

Figure B-6. Flowchart for Core Analysis Procedures

APPENDIX C
MUD LOGGING TECHNIQUES WITH OIL-BASED DRILLING FLUIDS

INTRODUCTION

An oil mud is defined as a mud with a continuous liquid phase of oil. These include

- Oil-based muds
- Inverted emulsion oil

The following material applies to most types of oil muds, and the techniques are applicable to any mud system which contains an appreciable amount of diesel or oil.

PREPARING CUTTINGS SAMPLES

Cuttings samples removed from the hole are in excellent condition when using an oil mud, especially when drilling soft gumbo or bentonite montmorillonite shales. With oil muds it is possible to obtain cuttings from formations which, due to their swelling nature, no cuttings could be obtained had a water-base mud been used.

All samples are coated and contaminated by the oil drilling fluid, and can be cleaned as follows:

1. Improvise suitable diesel-bath containers by cutting an oil drum in half. Ensure that they are clean and free from any contaminating fluid and debris. Fill both with clean refined diesel.

    - Diesel bath (A): use for initial washing and sieving of the cuttings
    - Diesel bath (B): use for final washing of the sieved cuttings

    Periodic renewal of the diesel in the baths is imperative. At this stage take a representative sample on a sample tray and leave the sieves outside the unit.

2. A further diesel wash may be needed. Use a mud-jug full of diesel, pour it over the sample, and wash and drain. Generally at this stage the sample is adequately free from drilling fluid to enable identification and description. However, it may be necessary to wash the sample again.

3. Use drilling detergent degreaser (obtainable from the rig) to wash the sample and then rinse off with water. Use a fine mesh screen. Experience has shown that drilling detergent clouds the sample and that water enhances the muddy appearance. Remember that oil muds are used in anticipation of "swelling shales." Washing the sample in water at

this stage may cause the shales to swell and hydrate and further enhance the turbid sample appearance.

Compare a representative sample from each stage of the washing procedure. With experimentation, the washing procedure may be adjusted to suit the degree of contamination and type of formation being drilled. Usually, diesel baths A and B are adequate.

Washed and Dried Samples

Dry the samples outside the unit to avoid the diesel fumes. It may be necessary to remove the unit oven from the wall to a safe area and use an extension lead for the power supply. (Check with responsible personnel for a safe area.) The local Exlog Service Department may provide additional drying ovens for oil-mud logging operations.

When gas is present, the ovens outside the unit must be turned OFF as they are not explosion-proof. Samples needing to be dried will accumulate, so when there are not enough sample trays or space to tolerate accumulation, washed samples should be placed in cloth bags and hung on a rail in order to partially air dry. The samples must be hung under shelter to prevent rain contamination. During periods of non-logging or when there is no gas hazard, the samples may be oven dried. Label each bag as described in the next paragraph.

Wet Samples

The waterproof ink pens are not entirely oil- or diesel-proof; after sustained contamination by diesel, the ink is obliterated. Ensure that the samples are hung on the rails in proper order so that, if relabeling is required, there will be no confusion as to which sample is which. The labeling should be done heavily on both sides of the bag and on the yellow tag. A ball point pen can be used on the yellow tag and the diesel will not impair legibility. A good method is to use a piece of elastoplast or cloth tape labeled and stuck to the sample bag string. To facilitate legibility, take care when filling the bag to keep the outside of the bag and the label clean. Do not leave it on the ground in a pool of oil mud. The wet samples may be given a cursory wash in diesel prior to bagging just to remove some of the contaminating mud.

When hanging wet samples on rails, ensure that they do not hang over each other and drip fluid on the sample below. Work from the top rail to the bottom rail or use parallel rails (in a horizontal plane).

Similar precautions apply to the collection and preparation of samples to be canned for geochemical analysis. A mud sample must be collected and canned every 500 feet or whenever there is a change in mud composition. (These samples are not washed.)

It is preferable to prepare all samples outside the unit. This is possible only if space permits, so advise the client that you will need diesel, detergent, two oil-drum halves and a dry, protected work area close to the logging unit.

EXAMINATION FOR FLUORESCENCE AND CUT

Take samples of the mud and any mud additives to make comparisons for fluorescence and cut. A sample of the diesel used in the washing procedure must also be used for comparison. Initially, examine unwashed cuttings and compare them with washed cuttings, then compare further with washed and dried cuttings.

Fluorescence

Light oils from formation will probably be masked by the fluorescence of the drilling fluid and washing diesel. Light oils are difficult to see even under normal conditions due to their volatility. Heavy oils, though, should show a significant difference in fluorescent color from that of the mud.

The intensity of the fluorescence is critical. The formation oil may have a fluorescent color that is similar but more intense. In unwashed cuttings, any live formation oil will usually have a duller fluorescence than that of the mud. Do not overlook the possibility of mineral fluorescence.

Test for live oil by covering washed cuttings with water heated to $70°C$ ($158°F$). Under UV light, highly volatile live oil breaks out of the cuttings with rapid popping, with a bright yellow to blue-white fluorescence. It may be possible to see viscous low gravity, or dead oils. Also, place some washed cuttings in hydrochloric acid and check for live oil fluorescence.

Cut

The washed sample may be rinsed with chlorothene before a cut test is made. A slight difference in fluorescence color or intensity of oil bleeding from aggregate (sandstone) may indicate formation oil.

Take care in evaluating the fluorescence and cut observed when using oil muds, and remember that formation oil may be broken down by the emulsifier used in the mud make-up. High porosity reservoirs will release their oil which will disperse in the mud system. Usually only "dead oils" or "low porosity reservoir oils" will be seen.

GAS ANALYSES

DUAL DITCH GAS ANALYSER (TG and PV)

The diesel in the mud causes a continuous high background reading. The addition of oil to maintain the mud system is done gradually to avoid sudden increases in background gas. Slugging the mud with diesel should be avoided to prevent questionable gas peaks. Peaks from diesel additions are sluggish or rounded, while those from formation gas are sharp and peaked. Diesel is relatively low in the light end (aromatics) associated with crude "live" oil. Petroleum vapors tend to be high in relation to total gas. Small shows tend to be masked by the diesel.

The diesel added to the mud gives a reading both on the Petroleum Vapor and Total Gas readouts. Allow for this by subtracting the number of background units from both PV and TG. This should give the number of gas units from the formation. The background diesel units constantly change with reference to the last addition of diesel to the system. The number of units subtracted should be mentioned in the daily morning report to the operator. The Wellsite Geologist may even request that the oil-mud background units be mentioned on the log.

Oil muds tend to contaminate the Gas Detector System, which causes a loss of sensitivity of the filaments. This is because diesel is a complex hydrocarbon which may not completely combust at normal voltage settings. This results in the filament being coated and inhibiting the catalytic oxidation effect. It is very important that frequent calibration checks are run and that the filaments are burnt-off at a higher voltage during trips. Further, diesel vapors cause a rapid build-up in distillate, thus contaminating the filters and condensate bottles. Lines must be flushed regularly with compressed air. The placement of a second distillate bottle in the system may be of some help.

MICROGAS ANALYZER

Compare blender analyses of the mud, unwashed cutitngs, and washed cuttings. Also, run the gas samples through the chromatograph and note the relative compositions. The blender TG and PV readings are affected; they are usually erratic and too high, depending on the amount of residue oil mud present on the cuttings. It may be necessary to neglect blender readings.

BLENDER SAMPLE CHROMATOGRAPHIC ANALYSIS

The following procedure applies only to the Standard (catalytic) Chromatograph as the F.I.D. Chromatograph allows only manual injection of the blender gas sample. Arrange the connections between the blender and chromatograph and the Microgas Detector and chromatograph as shown in Figure C-1. Then follow these instructions:

1. Place the chromatograph in the Test Sample mode.
2. Stop the chromatograph cycle (watch the TIMER dial) by pressing the HOLD button.
3. Run the blender.
4. Set the SAMPLE/AIR valve on the Microgas Detector to the SAMPLE position.
5. Press the AUTOMATIC button on the chromatograph.

The sample is then drawn from the blender, through the chromatograph and Microgas Detector. Both chromatograph and microgas readings may be taken from the same blender sample using this method.

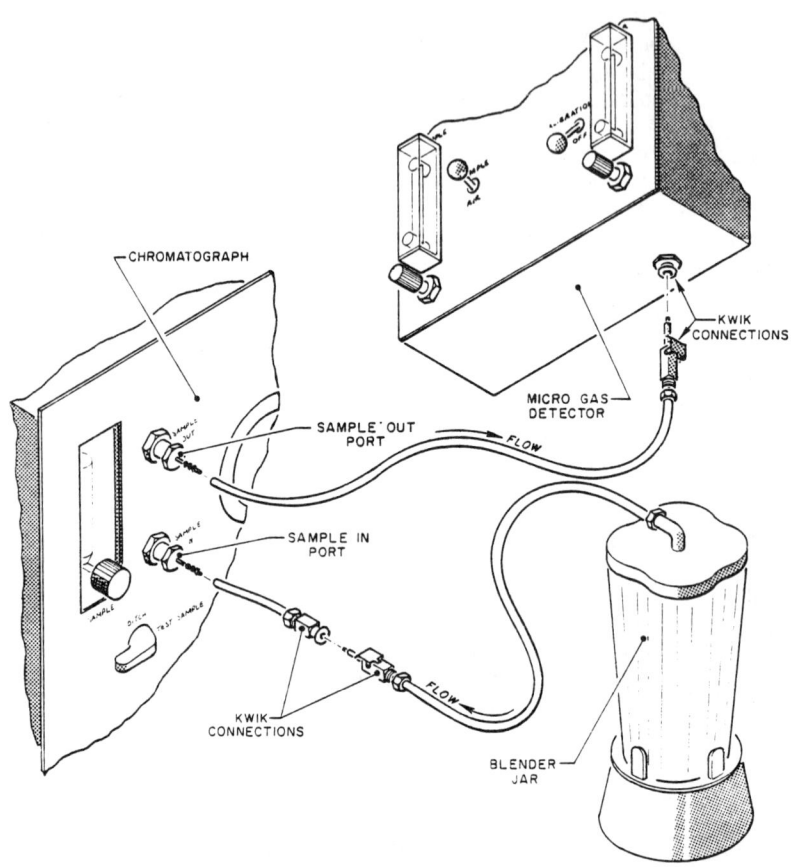

Figure C-1. Set-Up for Chromatographic Analysis of Blender Samples

CATALYTIC AND F.I.D. CHROMATOGRAPHS

These are the best tools for evaluating shows with an oil-mud system. With a newly made mud there is a significant increase in C1-C4. With time, however, after many circulations the background will stabilize (as for the Ditch Gas Analyzer). As diesel consists of higher molecular weight hydrocarbons, sudden increases of C1-C4 will be indicative of formation gas. The advantage of the chromatograph is that it selectively analyzes low molecular weight hydrocarbons and flushes out the heavier hydrocarbons, which constitute diesel wet gases, from the chromatograph columns. This decreases the possibility of logging false shows. The catalytic chromatograph baseline though, does drift, and will show a loss of sensitivity. This is due to resin build-up, and periodically the blockage will have to be cleared. This is necessary throughout drilling, but especially so in a prospective payzone if a geological column is known for the area.

APPENDIX D
IADC BIT
CLASSIFICATION

With the following charts, any tri-cone bit can be described using a four-digit code, as shown below:

    1st digit: Series (type of information for which the bit is suitable).

    2nd digit: Type (subdivision of the series).

    3rd digit: Feature (mechanical or metallurgical design variations).

    4th digit: Vendor. This digit is an Exploration Logging code, not an industry standard, and is used to define the manufacturer of a bit by adding an additional number at the end of the three-digit IADC bit code.

This system of classification permits comparison of the bit types offered by various manufacturers. With reference to the following charts, to find a comparable bit to the Hughes ODG, look up its IADC code on the Hughes chart (1331):

    1333 -- Reed Tool Company   Y13G

    1335 -- Smith Tool Company   DGH

    1336 -- SMF (Creusot-Loire)   TS5K

SERIES	FORMATIONS		TYPES	FEATURES								
				STANDARD (1)	"T" GAUGE (2)	GAUGE INSERT (3)	RLLR. SEAL BEARING (4)	SEAL. BRG. & GAUGE (5)	FRICTION SEAL BRG (6)	FRICTION BRG & GAUGE (7)	DIREC- TIONAL (8)	OTHER (9)
MILLED TOOTH BITS	1.	Soft Formations Having Low Compressive Strength and High drillability	1.	OSC-3AJ			X3A		J1			
			2.	OSC-3J			X3		J2	JD3		
			3.	OSC-1GJ		ODG	X1G	XDG	J3			
			4.									
	2.	Medium to Medium- Hard Formations with High Compressive Strength	1.	OWV-J/OW4-J		ODV/OD4	XV	XDV	J4	JD4		
			2.	WO								
			3.	OWC			XC					
			4.									
	3.	Hard Semi-Abrasive or Abrasive Form- ations	1.	W7-J/W7C		WD7	X7	XD7	J7			
			2.	W7R-2J								
			3.									
			4.	WR		WDR	XWR	XDR	J8	JD8		
	4.	For Future Use	1.									
			2.									
			3.									
			4.									
INSERT BITS	5.	Soft to Medium Formations with Low Compressive Strength	1.							J22		
			2.					X33		J33		
			3.									
			4.									
	6.	Medium Hard Form- ations of High Com- pressive Strength	1.			HH44		X44		J44		
			2.					X55R		J55R		
			3.			HH55		X55		J55		
			4.									
	7.	Hard Semi-Abrasive and Abrasive Formations	1.			HH77				J77		
			2.			A89/RG7AJ		RG7XJ		J88		
			3.					R91XJ				
			4.									
	8.	Extremely Hard & Abrasive Formations	1.									
			2.					RG2BXJ		J99		
			3.									
			4.									

Figure D-1. Tri-Cone Bit Code: Hughes Tool Company, Vendor Code 1

261

SERIES	FORMATIONS		TYPES	STANDARD (1)	"T" GAUGE (2)	GAUGE INSERT (3)	RLLR. SEAL BEARING (4)	SEAL. BRG. & GAUGE (5)	FRICTION SEAL BRG (6)	FRICTION BRG & GAUGE (7)	DIRECTIONAL (8)	OTHER (9)
								FEATURES				
MILLED TOOTH BITS	1.	Soft Formations Having Low Compressive Strength and High Drillability	1.									
			2.	V3S/V3								
			3.	V3M/VH3G								
			4.									
	2.	Medium to Medium - Hard Formations with High Compressive Strength	1.	V2								
			2.		VH2							
			3.									
			4.									
	3.	Hard Semi-Abrasive or Abrasive Formations	1.	V1								
			2.	VH1								
			3.			VQM						
			4.									
	4.	For Future Use	1.									
			2.									
			3.									
			4.									
INSERT BITS	5.	Soft to Medium Formations with Low Compressive Strength	1.			QMC9						
			2.									
			3.									
			4.									
	6.	Medium Hard Formations of High Compressive Strength	1.			QMC7						
			2.			QMC77						
			3.									
			4.									
	7.	Hard Semi-Abrasive and Abrasive Formations	1.			QMC6						
			2.			QMC6S						
			3.									
			4.									
	8.	Extremely Hard & Abrasive Formations	1.			QMCH						
			2.									
			3.									
			4.									

Figure D-2. Tri-Cone Bit Code: Varel, Vendor Code 2

SERIES	FORMATIONS	TYPES	STANDARD (1)	"T" GAUGE (2)	GAUGE INSERT (3)	RLLR. SEAL BEARING (4)	SEAL BRG. & GAUGE (5)	FRICTION SEAL BRG (6)	FRICTION BRG & GAUGE (7)	DIREC-TIONAL (8)	OTHER (9)
1. (MILLED TOOTH BITS)	Soft Formations Having Low Compressive Strength and High Drillability	1.	Y11			S11					
		2.	Y12	Y12T		S12		FP12			
		3.	Y13	Y13T	Y13G	S13	S13G	FP13			
		4.									
2.	Medium to Medium-Hard Formations with High Compressive Strength	1.	Y21		Y21G	S21	S21G	F21			
		2.	Y22				S23G	F22			
		3.				S23					
		4.									
3.	Hard Semi-Abrasive or Abrasive Formations	1.	Y31		Y31G	S31	S31G		F31G		
		2.									
		3.									
		4.	Y34		Y34G			F34			
4.	For Future Use	1.									
		2.									
		3.									
		4.									
5. (INSERT BITS)	Soft to Medium Formations with Low Compressive Strength	1.							FP51		
		2.					S52		FP52		
		3.					S53		FP53		
		4.							FP54		
6.	Medium Hard Formations of High Compressive Strength	1.									
		2.			Y62/Y62B		S62		FP62	FP62X	FP62B
		3.			Y63		S63		FP63		
		4.					S64		FP64		
7.	Hard Semi-Abrasive and Abrasive Formations	1.									
		2.			Y72		S72		FP72		
		3.			Y73		S73		FP73		
		4.					S74		FP74		
8.	Extremely Hard & Abrasive Formations	1.									
		2.									
		3.			Y83		S83		FP83		
		4.									

Figure D-3. Tri-Cone Bit Code: Reed Tool Company, Vendor Code 3

263

SERIES	FORMATIONS	TYPES	STANDARD (1)	"T" GAUGE (2)	GAUGE INSERT (3)	RLLR. SEAL BEARING (4)	SEAL BRG. & GAUGE (5)	FRICTION SEAL BRG (6)	FRICTION BRG & GAUGE (7)	DIREC-TIONAL (8)	OTHER (9)
MILLED TOOTH BITS											
1.	Soft Formations Having Low Compressive Strength and High Drillability	1.	S3S			S33S		S33SF		S3SJD	S3SJ4
		2.	S3	S3T		S33		S33F		S3JD	S3J4
		3.	S4	S4T		S44		S44F		DS	
		4.	S6	S6T		S66					
2.	Medium to Medium-Hard Formations with High Compressive Strength	1.	M4N			M44N		M44NF			
		2.	M4							DM	
		3.	M4L			M44L		M44LF			
		4.									
3.	Hard Semi-Abrasive or Abrasive Formations	1.	H7	H7T		H77					
		2.	H7U			H77U					
		3.			H7SG		H77S				
		4.				H77C		H77CF			
4.	For Future Use	1.									
		2.									
		3.									
		4.									
INSERT BITS											
5.	Soft to Medium Formations with Low Compressive Strength	1.					S84		S84F		
		2.					S86		S86F		
		3.			S8		S88		S88F	DS88	
		4.									
6.	Medium Hard Formations of High Compressive Strength	1.							M84F		
		2.			M8		M88		M88F/M89TF		
		3.							M89F		
		4.									
7.	Hard Semi-Abrasive and Abrasive Formations	1.							H84F		
		2.			H8		H88		H88F		
		3.			H9		H99		H99F		
		4.									
8.	Extremely Hard & Abrasive Formations	1.			H10		H100		H100F		
		2.									
		3.									
		4.									

Figure D-4. Tri-Cone Bit Code: Security, Vendor Code 4

FORMATIONS		TYPES	STANDARD (1)	"T" GAUGE (2)	GAUGE INSERT (3)	RLLR. SEAL BEARING (4)	FEATURES SEAL BRG. & GAUGE (5)	FRICTION SEAL BRG (6)	FRICTION BRG & GAUGE (7)	DIRECTIONAL (8)	OTHER (9)
SERIES											
MILLED TOOTH BITS	1. Soft Formations Having Low Compressive Strength and High Drillability	1.	DS			SDS					
		2.	DT	DTT		SDT			FDT	DJ	BHDJ
		3.	DG	DGT	DGH	SDG	SDGH		FDG		
		4.	K2		K2H	K2H					
	2. Medium to Medium-Hard Formations with High Compressive Strength	1.	V1		V1H						
		2.	V2		V2H	SV	SVH				
		3.	T2		T2H	ST2					
		4.									
	3. Hard Semi-Abrasive or Abrasive Formations	1.	L4		L4H	SL4	SL4H				
		2.									
		3.									
		4.	WC		WCH	SWC	SWCH	FWC			
	4. For Future Use	1.									
		2.									
		3.									
		4.									
INSERT BITS	5. Soft to Medium Formations with Low Compressive Strength	1.									
		2.					2JS		F2		
		3.					3JS		F3		
		4.									
	6. Medium Hard Formations of High Compressive Strength	1.					4JS		F4/F45		
		2.					5JS		F5		A1
		3.					57JS		F57		
		4.									
	7. Hard Semi-Abrasive and Abrasive Formations	1.					6JS		F6		
		2.					7JS		F7		
		3.					8JS		F8		
		4.									
	8. Extremely Hard & Abrasive Formations	1.									
		2.					9JS		F9		
		3.									
		4.									

Figure D-5. Tri-Cone Bit Code: Smith Tool, Vendor Code 5

265

| SERIES | FORMATIONS | TYPES | FEATURES |||||||||
			STANDARD (1)	"T" GAUGE (2)	GAUGE INSERT (3)	RLLR. SEAL BEARING (4)	SEAL BRG. & GAUGE (5)	FRICTION SEAL BRG (6)	FRICTION BRG & GAUGE (7)	DIRECTIONAL (8)	OTHER (9)
MILLED TOOTH BITS											
1.	Soft Formations Having Low Compressive Strength and High Drillability	1.	TS2			ES2	ES2K			BS2	
		2.	TS3		TS3K	ES3	ES3K			BS3	BS3K
		3.	TS5		TS5K	ES5	ES5K			BS5	
		4.	TS8			ES8					
2.	Medium to Medium-Hard Formations with High Compressive Strength	1.	TM2		TM2K	EM2	EM2K				
		2.	TM5		TM5K	EM5	EM5K				
		3.	TM8		TM8K	EM8	EM8K				
		4.									
3.	Hard Semi-Abrasive or Abrasive Formations	1.	TH3		TH3K	EH3	EH3K				
		2.	TH5		TH5K		EH5K				
		3.									
		4.									
4.	For Future Use	1.									
		2.									
		3.									
		4.									
INSERT BITS											
5.	Soft to Medium Formations with Low Compressive Strength	1.					ES6		LS6		
		2.					EM6		LM6		
		3.									
		4.									
6.	Medium Hard Formations of High Compressive Strength	1.					EM9		LM9		
		2.					EH6		LH6		
		3.									
		4.									
7.	Hard Semi-Abrasive and Abrasive Formations	1.					EH7		LH7		
		2.					EH8		LH8		
		3.									
		4.									
8.	Extremely Hard & Abrasive Formations	1.					EH9		LH9		
		2.									
		3.									
		4.									

Figure D-6. Tri-Cone Bit Code: S.M.F. (Creusot-Loire), Vendor Code 6

SERIES	FORMATIONS	TYPES	STANDARD (1)	"T" GAUGE (2)	GAUGE INSERT (3)	RLLR. SEAL BEARING (4)	SEAL BRG. & GAUGE (5)	FRICTION SEAL BRG (6)	FRICTION BRG & GAUGE (7)	DIREC-TIONAL (8)	OTHER (9)
MILLED TOOTH BITS											
1.	Soft Formations Having Low Compressive Strength and High Drillability	1.	3SS			3SS-Z					
		2.	3S			3S-Z					
		3.	3MSS		3MSST	3MSS-Z	3MSST-Z				
		4.									
2.	Medium to Medium-Hard Formations with High Compressive Strength	1.	3MS		3MST	3MS-Z	3MST-Z				
		2.	3MH		3MHT	3MH-Z	3MHT-Z				
		3.									
		4.									
3.	Hard Semi-Abrasive or Abrasive Formations	1.	3HS		3HSI	3HS-Z	3HST-Z				
		2.	3H		3HT	3H-Z	3HT-Z				
		3.									
		4.	3HR		3HRT	3HR-Z	3HRT-Z				
4.	For Future Use	1.									
		2.									
		3.									
		4.									
INSERT BITS											
5.	Soft to Medium Formations with Low Compressive Strength	1.					Z20		K20		
		2.					Z30		K30		
		3.									
		4.									
6.	Medium Hard Formations of High Compressive Strength	1.					Z40		K40		
		2.					Z50		K50		
		3.					Z60		K60		
		4.									
7.	Hard Semi-Abrasive and Abrasive Formations	1.					Z70		K70		
		2.					Z80		K80		
		3.									
		4.									
8.	Extremely Hard & Abrasive Formations	1.					Z90		K90		
		2.									
		3.									
		4.									

Figure D-7. Tri-Cone Bit Code: T.S.K. (Tsukamoto Seiki)

						FEATURES				
FORMATIONS		TYPES	STANDARD (1)	"T" GAUGE (2)	GAUGE INSERT (3)	RLLR. SEAL. BEARING (4)	SEAL. BRG. & GAUGE (5)	FRICTION SEAL BRG (6)	FRICTION BRG & GAUGE (7)	DIREC- TIONAL (8)	OTHER (9)
SERIES											
MILLED TOOTH BITS	1.	Soft Formations Hav- ing Low Compressive Strength and High Drillability	1.	ES1CA							
			2.	ES1C							
			3.								
			4.	ES2G							
	2.	Medium to Medium - Hard Formations with High Compressive Strength	1.	EM1V							
			2.	EM1C							
			3.								
			4.								
	3.	Hard Semi-Abrasive or Abrasive Form- ations	1.	EH1							
			2.	EH3							
			3.								
			4.								
	4.	For Future Use	1.								
			2.								
			3.								
			4.								
INSERT BITS	5.	Soft to Medium Formations with Low Compressive Strength	1.								
			2.								
			3.								
			4.			EH4					
	6.	Medium Hard Form- ations of High Com- pressive Strength	1.			EH6					
			2.			EH66					
			3.			EH7					
			4.								
	7.	Hard Semi-Abrasive and Abrasive Formations	1.			EH8					
			2.								
			3.								
			4.								
	8.	Extremely Hard & Abrasive Formations	1.			EH9					
			2.								
			3.								
			4.								

Figure D-8. Tri-Cone Bit Code: C.P. (Chicago Pneumatic)

APPENDIX E
EXLOG'S ROLE IN DRILLSTEM TESTING

INTRODUCTION

This appendix supplements the main text of the <u>Field Geologist's Training Guide</u>, not so much with technical descriptions, but suggests procedures to follow for obtaining and communicating information during formation testing.

THE SIGNIFICANCE OF TESTING

Logging geologists must fully understand everything that happens during testing, for there is a great deal of simple but critical information to be collated and communicated to the client. This information is equally important as any other on the log, the log being primarily one of hydrocarbon analysis and secondarily of geological description.

EXPLANATION OF PROCEDURES

Find out from the "company man" what the client wants, and be sure that you understand clearly. These requirements will probably include the points listed below. Even if these are not specifically requested, do them. Make tidy and orderly notes so that anyone can decipher them and obtain a clear picture of what happened. Note down the following:

- Zones being tested; depth of packers; how much water cushion.

- Time when tool is opened and closed. Duration of flow periods and shut-in periods.

- Blow description, similar to the following example:

Time	Blow
14:00	weak
14:15	weak
14:30	v weak
15:00	dead

On the log these descriptions can be condensed as in the examples under "Presentation of Test Results," but this must be done accurately.

- Exact times when fluids (water cushion, gas, oil) reach the surface. Rate of flow if fluid surfaces (Example 1).

- Amount of fluid recovered, if none comes to surface.

Conduct the following:

- Take gas samples at the bubble hose and analyze them in the chromatograph as explained later in "Formation Test Gas Analysis".

- Take oil samples at the separator if the well flows. Usually the "company man" will make specific request regarding frequency and numbers of samples to be taken. Record information as in Example 2, or as follows:

Time	Sample	Bbl	API	Temp°F
14:00	1	100	27	87
14:30	2	210	27	87

- Take rathole (pilot hole) mud samples at the conclusion of the test. If it is necessary to reverse out, sample the transition (say three samples) from water cushion to mud. Analyze them for chlorides and resistivity (of Example 1).

As the trip out of the hole is completed, always take rathole mud samples from:

— Two stands above the tool.
— One stand above the tool.
— Directly above the tool.

- When the pressure bomb has been recovered, record all pressures and the bottomhole temperature (BHT).

Once the test has been completed, type the information for the log as soon as possible. Type the originals on ordinary tracing paper.

FORMATION TEST GAS ANALYSIS

The following analysis is performed on DST gas samples from the bubble hose or separator. It can also be performed on gas from Wireline Formation Test tools and gas released from sidewall cores. However, as it is not possible to accurately determine the gas/air ratio of samples taken from sidewall cores, that particular analysis is only qualitative.

TEST PROCEDURE

1. Collect the gas sample in a balloon from the separator during a DST or from the sample chamber drain valve during a Wireline Formation Test. When there are no balloons available, collect the gas in a glass jar by inverting it in a bucket of water and allowing the gas from the bubble hose or separator to displace the water. The gas is trapped in the jar when the lid, which contains a small hole covered with tape, is securely screwed on.

2. With a syringe, take 10 cc of the gas sample from the balloon or glass jar and inject it into a squeeze bottle, after first applying positive pressure to the bottle to reduce the amount of air within it.

3. While injecting the sample, release the pressure on the bottle gradually. This method ensures that the total volume of air and sample in the bottle is equivalent to that of the bottle. It also ensures that no gas sample escapes from the bottle during this procedure.

4. Follow the chromatograph calibration procedure and keep in mind that the chromatograph is too sensitive to accept a pure sample of gas without diluting it with air.

CALCULATION

Procedure 1

A. Volume of squeeze bottle = X (check the squeeze bottle's volume before proceeding with the analysis).

B. Calculate the dilution factor:

$$\text{Dilution Factor} = \frac{\text{Total Volume}}{\text{Sample Injected}} = \frac{X}{10 \text{ cc}}$$

C. Obtain total chart division for each gas by multiplying the recorded divisions by the attentuation factor.

D. Convert to ppm by multiplying the total chart division for each gas by its calibration factor.

E. The original concentration of each gas in the 10 cc sample expressed in ppm can be obtained by multiplying ppm (D) by the dilution factor (B). Divide by 10,000 to convert to %.

    Example: volume of squeeze bottle = 385 cc
              Sample injected = 10 cc
              C1 chart division = 12 div
              C1 attenuation = 7
              C1 calibration factor = 21

Solution:

$$\text{dilution factor} = \frac{385}{10} = 38.5 \quad (B, \text{ above})$$

$$\text{total chart divisions} = 84.0 \quad (C, \text{ above})$$

$$C1 \text{ ppm} = 84 \times 21 = 1764.0 \quad (D, \text{ above})$$

Original concentration of C1 in the sample = 38.5 x 1764 = 67914 ppm = 6.79 percent.

Procedure 2

Another approach to the calculation procedure is as follows:

a. volume of squeeze bottle $= X$

b. sample taken and injected $= 10$ cc

c. total ppm of sample injected $= \frac{10}{X} \times 1{,}000{,}000$

d. concentration of C1 $= \dfrac{\text{ppm of C1}}{\text{total ppm of sample}}$

$= \dfrac{\text{C1 chart divisions} \times \text{calibration factor}}{\text{total ppm of sample}}$

e. multiply (d) by 1,000,000 to convert to ppm, or by 100 to convert to percent; for example,

$$\text{total ppm} = \frac{10}{385} \times 1{,}000{,}000 = 25{,}974.03 \text{ ppm} \quad (c, \text{ above})$$

$$\text{original concentration of C1} = \frac{12 \times 7 \times 21}{25974.03}$$

ppm = 0.67914 x 1,000,000 = 67914   (d, above)
percent = 0.67914 x 100 = 6.7914   (e, above)

Using Procedure 1 or 2, repeat for C2, C3, iC4, nC4.

PRESENTATION OF TEST RESULTS

The following are examples of actual log reports. They should be presented as a supplemental log, attached to the bottom of the mud log as in the example in Appendix A (Supplemental Log for Core Analysis and Test Results).

DST #1:   15684 to 15723 ft

Packer set at 15630 ft with 10000-ft water cushion. Tool opened initially 15 min, v weak blow. Shut in 1 hour. Tool reopened for 4 hr, initially weak blow. Shut in 1 hour. Tool reopened for 4 hr, initially weak blow, decreasing to v weak. Shut in 3 hr; reverse out.

IHP	9234	IFP	2349
IFP	3234	FFP	3681
ISIP	3116	FSIP	3543
		FHP	9895

(pressures in psi)

where:

IHP : Initial hydrostatic pressure.
IFP : Initial flow pressure.
ISIP : Initial shut-in pressure.
FFP : Final flow pressure.
FSIP : Final shut-in pressure.
FHP : Final hydrostatic pressure.

Bomb at 15630 ft, BHT 290°F.

Rathole mud samples

Sample	Saline/ppm Cl-	Rw/ohms	Temp/°F	Comments
1	260	x.xx	79	-
2	2500	x.xx	79	-
3	4100	x.xx	80	tr wh -bl flor
4-6 etc.				

Note: Samples 1, 2 & 3 taken during reverse circulation, spanning transition from fresh water to rathole mud. Samples 4 & 5 from 2 & 1 stands above the tool respectively, and 6 immediately above.

Recovered 67 bbl of oil-cut mud.

Example 1: A test in which fluid was recovered, but none flowed to the surface.

DST #2:   10003 to 10020 ft

Packer set at 9975 ft, with 2000-ft water cushion. Tool opened initially for 15 min with immediate moderate blow increasing to full bucket in 2 min. Strong full bucket blow in 2-1/2 min. Initial shut-in of 1 hr. Final tool open for 20-1/4 hr. Full bucket blow in 1-1/2 min, decreasing after 2-1/4 hr to 5" below water surface. Gas to surface after 2-1/2 hrs. final flow; oil to surface after 2 hr. 55 min. of final flow. Flowed oil for 17-1/2 hrs, final shut-in of 1 hr, reverse-out.

	Samples	Gravity API	Base Solid*	Water %
1.	First oil to surface	13.7	1/4	—
2.	3 hrs after oil to surface	16.6	1/2	1/2
3.	6 hrs after oil to surface	17.2	1/2	1/2
4.	9 hrs after oil to surface	17.6	1/2	TR
5.	12 hrs after oil to surface	17.5	1/4	1/4
6.	15 hrs after oil to surface	17.6	1/4	1/2
7.	17 hrs after oil to surface	17.8	1/2	2

*Note: Base solid % = percentage content of solids from shake-out. Water from shake-out of sample #7 measured for salinity & Rw. Salinity: 30000 ppm Cl, Rw 0.16 at 78°F.

8.	Sample taken during reverse-out	17.6	1/2	—
9.	Sample above tool-mud/oil			

Salinity: 12500 ppm Cl, Rw 0.40 at 82°F.

IHP 5298 (psi)

First Flow		Second Flow	
IFP	1874	IFP	2349
FFP	2069	FFP	3861
ISIP	4205	FSIP	out of time

Note: Insufficient pressure to operate separator, no accurate flow rates measured. Estimate flow rate of 4 bbl/hr.

Bomb Depth   9988 ft
Temperature   250° F

Example 2: A test in which oil flowed to the surface.

> DST #3:   8409 to 8432 ft
>
> Packers set at 8404 and 8450 ft in open hole. Tool opened initially 3 min w/med blow and shut-in for 45 min. Tool reopened for 1 hr 30 min w/strong blow. Gas to surface in 4 min. Oil to surface in 15 min. Gas flowed at 2.6 MMCFD and oil at 860 BOPD thru 5/8-inch choke. Recovered 2000 ft clean oil (est. 35 API gravity) and 3000 ft oil-cut mud with 1500-ft water cushion, salinity - 3200 ppm Cl.
>
> | | | | | |
> |---|---|---|---|---|
> | IHP | 3400 | IFP | 4700 | |
> | IFP | 4000 | FFP | 5300 | |
> | ISIP | 3850 | FSIP | 5600 | |
> | | | FHP | 3400 | (pressures in psi) |
>
> Chromatograph analysis of gases recovered from separator:
>
> | | | | | |
> |---|---|---|---|---|
> | C1 | 76000 ppm | iC4 | 8300 | ppm |
> | C2 | 18500 ppm | nC4 | 9100 | ppm |
> | C3 | 9700 ppm | iC5 | 4600 | ppm |

Example 3: A test with results similar to those of Example 1.

SPECIFIC GRAVITY (SG) CALCULATION

$$SG = \frac{(\text{mol wt C1})}{(\text{mol wt air})} \times \frac{\%C1}{100} + \frac{(\text{mol wt C2})}{(\text{mol wt air})} \times \frac{\%C2}{100} + \ldots C3, C4, C5, \text{etc.}$$

Using the following known data, the calculation is simple:

Gas	Molecular Weight	Specific Gravity gm/cc
air	28.91	1.0000
methane	16.04	0.5536
ethane	30.07	1.0379
propane	44.09	1.5219
i-butane	58.12	2.0062
n-butane	58.12	2.0062
i-pentane	72.15	2.4905
n-pentane	72.15	2.4905

The calculation using Example 3 results is as follows:

			then	gm/cc
C1	76,000 ppm	60.2%	0.554 x 0.602 =	0.334
C2	18,500	14.6%	1.038 x 0.146 =	0.152
C3	9,700	7.6%	1.522 x 0.076 =	0.116
iC4	8,300	6.6%	2.006 x 0.066 =	0.132
nC4	9,100	7.2%	2.006 x 0.072 =	0.144
iC5	4,600	3.6%	2.491 x 0.036 =	0.089
	126,200 ppm	99.8%		0.967

The specific gravity of the gas analyzed in the chromatograph is 0.967 gm/cc.

Note: This is only a rough approximation of the specific gravity, because the Exlog gas detectors cannot give an accurate quantitative value to gases such as carbon dioxide, hydrogen sulfide, nitrogen, helium, pentane or the higher hydrocarbons. However, it is a useful calculation for purposes of comparison, and a client occasionally requires this information.

HEATING VALUE CALCULATION (Btu)

The heat of combustion (Btu, British thermal units) of each component of the above sample can be calculated using the following known calorific values.

Calorific values at 60°F:

methane	1012	Btu/cu ft at 14.696 psia
ethane	1783	
propane	2557	
i-butane	3354	
n-butane	3369	
i-pentane	4001	
n-pentane	4009	

Calculate the heats of combustion as follows, remembering the percentages from the specific gravity calculation.

Let $x1$ = the heat of combustion of C1
     $x2$ = the heat of combustion of C2 etc.,

then

$x1$	=	.602	x	1012	=	609.22	Btu/cu ft
$x2$	=	.146	x	1783	=	260.32	
$x3$	=	.076	x	2557	=	194.33	
$ix4$	=	.066	x	3354	=	221.36	
$nx4$	=	.072	x	3369	=	242.57	
$ix5$	=	.036	x	4001	=	144.04	
						1671.84	Btu/cu ft

The dry basis heating value at 14.696 psia and 60°F is 1672 Btu/cu ft.

In passing this information to an operator, emphasize that it is a <u>dry basis</u> calculation with no correction for water vapor presence or saturation. Also mention that the Exlog chromatograph does not detect N2 or other gases, only hydrocarbons; therefore, the value is only an approximation. Always specify <u>all</u> conditions, and state units clearly.

CONCLUSION

Keep tidy notes, and keep all information together so that anyone who wants information can easily get it. Surprising as it may seem, the Logging Geologist may be the only person onsite with a complete record of what happened, when it happened, and results of all analyses.

GLOSSARY

Many terms and phrases used in the oilfield are unique to the oil industry, and to a geologist new to this environment they may be quite misleading. Furthermore, field geologists must "know the language" to gain and maintain credibility with experienced oilfield personnel. This glossary has been prepared so that, from the first day on the job, inexperienced personnel can quickly grasp and learn to communicate what is happening at the wellsite.

In addition to standard oilfield terms, some commonly used Exlog terminology is included.

AAPG: American Association of Petroleum Geologists.

acid job: Injecting HCl or HF into the formation to remove cement and improve permeability.

ADES:® Advanced Data Evaluation Service. An Exploration Logging service mark.

ALFA:℠ Advanced Logging and Formation Analysis. An Exploration Logging service mark for which registration is pending.

API: American Petroleum Institute.

API gravity: Unit of density (degrees API) of crude oils. Defined as:

$$^\circ API = \frac{141.5}{S.G.} - 131.5 \quad (S.G. = \text{specific gravity})$$

Water at $4^\circ C$ has an API gravity of 10; crude oil gravities range from 10° to 50°.

asphalt: Black to dark-brown solid or semisolid bitumens which gradually liquify when heated. Composed principally of the elements carbon and hydrogen, but contain appreciable quantities of nitrogen, sulfur and oxygen; largely soluble in carbon disulfide.

B: Billion (one thousand million).

back off: (1) The action of securing a section of drillpipe while another is unscrewed from it. The term applies to a fishing operation as well as to routine backing off. In fishing, an explosive charge is detonated against a pipe joint while it is in tension, and the resulting explosion may free the connection and allow the pipe to be unscrewed. (2) To pay out line from the drawworks and so lower the traveling block.

Bakerlock: An epoxy compound used to seal threads of guide shoe, float collar and lowermost joints of casing. Replaces traditional welding. Commonly used for many other repair jobs on the rig.

barrel(s) (bbl): Forty-two U.S. gallons.

BCPD: Barrels of condensate per day (from a gas well).

BDF: Below drillfloor. Depth reference.

belly buster: Otherwise known as a belly band. Both are slang terms for an adjustable safety belt worn by a drilling rig crewman working at height or in an otherwise dangerous position. The belt must be at least four inches wide and is attached to a safety line on the structure near where the crewman is working.

BHC: Bottomhole choke.

BHP: Bottomhole pressure.

BHT: Bottomhole temperature.

bitumen: Native substances of variable color, hardness and volatility; composed principally of the elements carbon and hydrogen. Sometimes associated with mineral matter, the nonmineral constituents being largely soluble in carbon disulfied.

blowout: An uncontrolled kick.

bomb: A small device placed in the DST tool, containing pressure temperature measuring equipment. The pressure record is made on a small chart which is read when the bomb is recovered from the tool after the test.

B.O.P.: Blowout preventer.

BOPD (bopd): Barrels of oil per day.

borehole: The wellbore, with or without casing. See "open hole."

bottomhole money: Contribution to a joint-venture well, payable only if a specified horizon is reached.

bottoms up: One complete circulation. When a drilling break is seen it may be decided to circulate "bottoms up" (i.e., all sample out of the hole) prior to proceeding.

bridge: Soft material swelling from the borehole wall to close off the annulus and trap the drillstring or prevent the bit from passing it on a trip back to bottom.

bridge plug: Packer run into the hole on drillpipe or wireline and left in the hole to seal it. Can be used while the B.O.P. is being changed or as a seat for cement when abandoning the well. Some types can be retrieved and reused, others are drilled up after use or abandoned in the hole.

British thermal unit (Btu): A unit of energy. One Btu (mean) is 1/180th of the energy required to raise the temperature of one pound of water 180 degrees, from $32^\circ F$ to $212^\circ F$. The value is equal to 1055.79 joules. The value of a natural gas is governed by its "Btu value."

BS&W: Borehole solids and water. The nonpetroleum fraction of a crude oil sample.

bubble hose: A small hose that is attached to the valve system on the rig floor at the top of the drillstem during a DST. The end of this hose is placed in a bucket of water; if bubbles emerge only from 1 inch below the surface it is described as a 1-inch blow, or, more generally, a weak blow. If bubbles emerge when the hose end is pushed to the bottom, it is described as a full bucket blow.

button bit: The hardest type of insert bit, and has small hemispherical inserts. It is a general term for insert bits.

BWPD: Barrels of water per day.

cable tool drilling: A method of drilling a hole whereby a heavy bit suspended on a wire or cable is driven into the ground under its own weight. This form of drilling has been replaced almost entirely by rotary drilling, especially offshore, though it is sometimes used onshore.

carbohydrates: Organic compounds with the approximate general formula $(CH_2O)_n$ where n is equal to or greater than 4. Sucrose (table sugar), glucose, starch and cellulose are typical carbohydrates.

casing: The steel pipe that supports the wellbore.

CD: Contract Depth. Maximum depth specified in a footage rate contract.

Centrifuge Test: Test to determine BS&W in oil sample.

cfd, cf/d, cu ft/d: Cubic feet per day.

chitin: A polysaccharide (carbohydrate polymer) with a structure similar to cellulose except two OH groups are replaced by CH_3ONH groups in each $(CH_2O)_6$ unit. Chitin forms the horny, hard outer cover of insects, crustaceans and parts of some other invertebrates.

choke: (1) a heavy steel nipple inserted into the production tubing that closes off the flow of oil except through an orifice in the nipple. Chokes are of various sizes. It is customary to refer to the production of a well as so many barrels through (or on) a (for example) 22/64th-inch choke. (2) A variable orifice used to control flow of mud out of the hole during well kill procedures.

Christmas tree: An assembly of valves mounted on the casinghead through which a well is produced. The Christmas tree also contains valves for testing the well and for shutting it in if necessary. Also referred to as the production tree.

closed in: Refers to a shut-in well that is capable of producing.

cloud point: Temperature at which, upon cooling, the crude oil will become cloudy to the appearance of solid hydrocarbons.

CMC: Carboxymethyl Cellulose. Used for viscosity and fluid loss control of drilling mud and cement slurries.

Compensator: See motion compensator.

condensate: A natural gas liquid with a low vapor pressure, compared with natural gasoline and liquified petroleum gas. It is produced from a deep well where the temperature and pressure are high. Gas condenses as it rises up the wellbore and reaches the surface as condensate. Similarly, condensate separates out naturally in pipelines or in a separation plant, by the normal process of condensation.

connate water: "Fossil" water which has always occupied pore spaces in permeable rock since it was first deposited. Unlike edge and bottomwater, which can help production, connate water present in an oil reservoir may adversely affect production, since oil is denied access to the pores and the total volume of oil recoverable per unit volume of rock may be low. Connate water is also called interstitial water.

COST well: Continental Offshore Stratigraphic Test well (USA).

Coupon: Metal strip exposed to mud stream to test effectiveness of corrosion inhibitors.

CPI: Computer-processed interpretation.

crude oil: A petroleum which is removed from the earth in liquid state or is capable of being so removed.

deadline: The end of the wire from the drawworks on a rotary drilling rig. It is fixed to the derrick structure so that, when the hoisting drum takes in wire, the traveling block is caused to rise toward the crown block. Conversely, the block is lowered by paying out wire, or backing off.

delineation well: An exploration well drilled as part of a carefully planned program with the objective of appraising the value of an oil or gas discovery. Delineation wells, or step-out wells, are drilled so that the probable outline of the oil or gas field may be delineated.

development drilling: Extension of a drilling program once a strike has been established and proved to be capable of economic production. Development wells, which may be vertical wells or deviated wells, are usually drilled from permanent platforms. Some development wells may be production wells, others injection wells; hence, the number and locations depend on the extent of the field and the drive mechanism employed.

deviated well or directional well: Borehole that is intentionally deviated from vertical in order to reach some remote or inaccessible subsurface target.

D-exponent: Drilling exponent. Unitless number derived from rate of penetration normalized for the effect of weight on bit, hole diameter, rotary speed and mudweight. Can be indicative of changes in formation pore pressure gradient or lithology.

DF: Drillfloor (see BDF).

distillate: The condensate collected when a liquid is boiled and the vapors are allowed to condense at a desired temperature. An original charge stock may be separated into distillates which are more violatile and those which are less volatile, by the process of distillation.

diverter: System to direct well flow away from the rig in shallow blowouts when the well cannot be shut in.

dog: To latch or fasten.

doghouse: This name was originally given to a shelter alongside a land-based rig, but offshore installations also have their doghouses — usually a small room on the edge of the rig floor used as office, tool store, and for coffee breaks. "Doghousing" refers to idle conversation, time-wasting at work.

dogleg: A sharp deviation in direction of the wellbore.

dope: Thread sealing compound used in connecting pipe fittings. On the rig it refers to the metalized grease used on drillpipe, collar and casing connections.

drift off: Floating rig moves off location due to thruster or anchor failure.

drive off: Floating rig is pushed off location due to thruster malfunction.

dry gas: Natural gas consisting principally of methane and devoid of readily condensable constituents such as gasoline. Dry gas contains less than 0.1 gal natural gas liquid vapors per 1000 cu ft.

DST: Drillstem test.

ECD: Effective circulating density. Apparently increased mud density in borehole due to circulating backpressures.

E-log: Logs run on a wireline into the borehole. Originally "Electric" Log, but now refers to all such logs (e.g., neutron, sonic, etc.).

ELOS:® Exploration Logging Operating System (software). An Exploration Logging service mark.

EXLOG:® Exploration Logging service mark.

farm in, farm out: An agreement between two exploration companies, whereby the prospecting rights of one are assigned to the other, in whole or in part, subject to the approval of the licensing authority. The original lessee agrees to farm out his holding and the new lessee farms in while usually taking on an obligation to do certain drilling work in return for a consideration, usually a share of any profits from a strike.

fish: Pieces of pipe stuck in the wellbore (see "junk"). A "fishing operation" is performed to retrieve pipe.

GEMDAS:® An Exploration Logging service mark: Geological and Engineering Monitoring and Data Acquisition System.

Geronimo line: An escape line from the monkey board to some safe distance away from the derrick.

GL: Ground level. Depth reference.

gpm: Gallons per minute.

guideline tensioner: Motion compensated line connecting rig to temporary guidebase which is used to guide the B.O.P. onto the wellhead.

gumbo: A slang expression meaning a sticky material. The name of a very sticky kind of clay found in some parts of the southern USA, gumbo is also applied to any viscous clay encountered while drilling.

hang off: Leave the drillstring hanging from the seabed B.O.P. while a floating rig drifts off location in bad weather.

hang-up: Some irregularity, either in the surface equipment or downhole, which is preventing drillpipe, casing or a logging tool from going to the bottom.

hydrocarbon: A compound composed only of the elements hydrogen and carbon. Bitumens such as petroleum are composed principally (but not only) of hydrocarbons.

IADC: International Association of Drilling Contractors.

id.: Identity, identify (ier) (e.g., of a well name).

i.d.: Inside diameter.

IFP: French Institute of Petroleum.

inert gas: A gas that is chemically unreactive. Also known as rare gas. Two common inert gases, helium (He) and argon (Ar), are used as breathing gases by divers, usually in a mixture of other substances. Both are stable compounds present in air and are colorless, odorless and tasteless. They are liquid at low temperatures and slightly soluble in water. Helium is extracted from nitrogen, while argon is obtained by distilling liquid air. (Nitrogen itself is a relatively unreactive gas much used in offshore work though it is not strictly an inert gas.) Other inert gases, neon (Ne), krypton (Kr), xenon (Xe) and radon (Rn), have not as yet any special application in offshore engineering and technology.

injection wells: Well in an oilfield via which water, gas or steam is injected into the reservoir to maintain reservoir pressure, improve relative permeability or decrease oil mobility.

interstitial water: Another name for connate water, water permanently held in the pore spaces of a rock.

IP: British Institute of Petroleum.

isomers: Molecules that have the same number and kinds of atoms, but are different substances. Structural isomers differ in the way atoms are linked together, viz., n-butane and isobutane.

junk: Fragments of tools, etc., that have broken off into the wellbore (see "fish").

kerogen: The disseminated organic matter in sedimentary rocks that is insoluble in acids and organic solvents. The organic matter initially deposited with unconsolidated sediments is not kerogen, but a precursor that is converted to kerogen during diagenesis. Sapropelic kerogens yield oil when heated while humic kerogens yield mainly gas. Kerogen includes both marine- and land-derived organic matter, the latter being the same as the components of coal.

kick: An uncontrolled flow of formation fluid(s).

kick-off: Initiation of planned well deviation.

kill: To control or stop a kick.

LCM: Lost-circulation material.

LCZ: Lost-circulation zone.

lignin: The substance that gives tensile strength to plant structures such as tree trunks. It is a polymer of propylbenzéne nuclei joined by oxygen and carbon linkages plus -OH and $-OCH_3$ side chains.

lipids: A broad term that includes all oil-soluble and water-soluble substances such as fats, waxes, fatty acids, sterols, pigments, and terpenoids.

LNG: Liquid natural gas.

LPG: Liquid petroleum gas.

manifold: An accessory to a piping system or other conductors which serves to divide a flow to any one of several possible destinations, e.g., pipe manifold, choke manifold.

MCF: Thousand cubic feet; the standard unit for

measuring volumes of natural gas. MMCF is one million cubic feet.

MIR: Moving in rig.

MO: Moving out.

moonpool: An exposed area on a drilling rig through which conductor pipe and other equipment are lowered into the sea, and drilling operations are conducted. On a floating rig the moonpool is located at the center of gravity where vessel motion is least felt, but its site on a jack-up rig is not critical. Another name for the area is "moonwell."

motion compensator: Device for maintaining constant weight on the bit while a floating drilling vessel moves with wave and tide motion.

mousehole: Like the rathole, a narrow recess in the floor of a drilling rig deck. The mousehole is used to stow the next joint of pipe until it is required for threading into the drillstring.

natural gas: A petroleum consisting of varying proportions of gaseous hydrocarbons such as methane, ethane, propane, butane, isobutane, and occasionally containing liquid hydrocarbons such as pentanes and hexanes, and nonhydrocarbon gases such as carbon dioxide, hydrogen sulfide, nitrogen, and helium.

nippling up: (1) Connecting or hooking up a blowout preventer stack. The operation is usually done from the nippling-up platform on a jack-up drilling unit, and the name probably stems from the cone-shaped mating flanges. (2) Any act of connecting two components together.

o.d.: Outside diameter. Tubular goods are always specified by the nominal o.d.

oil shale: A compact rock of sedimentary origin with an ash content of more than 33 percent. Contains organic matter that yields oil when destructively distilled, but not appreciably when extracted with petroleum solvents.

open hole: The lower section of the borehole that does not have casing.

OTC: Offshore Technology Conference.

packer: Sub, run in on the drillstring with an outer rubber sleeve. When weight is applied to the sub, it shortens, bulging the rubber sleeve outward and sealing the annulus. Used in DSTs and squeeze cementing.

pack-off: Pressure being applied to a sealed-off section of the annulus. This may be done using a packer in stimulation or squeeze cementing, or accidentally by continuing to pump after a bridge has formed in the hole.

payzone: A productive horizon in a formation which is sufficiently thick and contains a high enough concentration of petroleum to make production from it commercially viable.

P.E.G.: Pressure Evaluation Geologist (an Exlog term).

P.E.P.: Pressure Evaluation Package (an Exlog term).

petroleum: A species of bitumen composed principally of hydrocarbons and existing in the gaseous or liquid state in its natural reservoir.

pig (or rabbit): Full gauge plug pulled through casing to check inside diameter (i.d.) and straightness.

plugged and abandoned: The status of a borehole which has been abandoned. It signifies that the hole has been sealed with a plug, probably because it was dry, and is unlikely to be reopened. This status is usually given in the form of "p and a" or simply "abandoned."

POH: Pull out of the hole, not necessarily all the way (e.g., POH to casing).

possum belly: (1) A metal box built underneath a truck bed to hold pipeline repair tools (shovels, bars, tongs, chains and wrenches). (2) A narrow tank into which mud from the flowline flows before passing over the shakers. (3) Any tall narrow tank.

pour point: The temperature at which a liquid hydrocarbon ceases to flow, or at which it congeals. It is a function of the viscosities and melting points of the various hydrocarbons.

ppm: Parts per million.

prime mover: The driving element of the rig. Engines may drive the rig equipment directly or via an electricity-generating system, comprising prime mover and generator or alternator. In an offshore environment the choice of prime mover depends on (1) the availability of fuel and water, and (2) the nature of the duty for which it is required. The user may choose from an electric motor or converter, a steam turbine or a gas turbine, or a gas or diesel engine. The gas turbine is particularly suitable for an offshore platform where supplies of natural gas may be available for fuel, whereas some gas combustion engines can burn gas, crude oil or distillates. Large units can be coupled together; for example, four 12,000-hp units may drive a common generator — producing over 100 MW.

proteins: High-molecular-weight polymers of amino acids that constitute more than 50 percent of the dry weight of animals. The organic nitrogen and sulfur of living organisms is concentrated in the protein fraction. Gelatin, albumen, collagen (connective tissue), keratin (hair, hoofs, nails), serum globulins are typical proteins.

pup joint: An intermediate length of pipe that is used to make up a string to a given length.

PV: Petroleum vapors.

PV: Plastic viscosity. Used together with YP in the context of mud.

PVT: Pit Volume Totalizer.

rare gas: See inert gas.

rathole: (1) a narrow hole in the derrick floor, communicating with the cellar deck below. When not in use, the kelly is placed in the rathole. (2) An extension of a borehole drilled at a reduced diameter, a preliminary step sometimes taken before a drillstem test.

rathole mud: Any mud from the borehole during a DST, which enters the drillstem via the test tool or the reverse circulating valve.

relief well: A deviated well drilled into a structure for the purpose of relieving pressure in an adjacent well which has suffered a blowout.

reverse circulation: Pump heavy fluid into the annulus in order to displace lighter fluid from the drillstring. Used in testing and squeeze cementing.

rheology: (1) The science that deals with the flow of matter. Rheology in drilling refers to the makeup and handling of a drilling mud circulation system. (2) Drilling mud control and characteristics.

rigging up: The preparation of rig and drilling equipment by a rig's crew before beginning a drilling program.

RIH: Run into hole.

RKB: Rotary kelly bushing. Depth reference.

roughneck: A grade of laborer working on a drilling rig. Roughnecks display higher skills than roustabouts and they are generally employed to assist in the many facets of drilling operations, handling casing, pipe, tongs and other tools used on the rig floor. All roughnecks are shift-working members of a rig's crew. Also known as floorhands.

roustabout: A general-purpose laborer working on a drilling site. The most junior of all personnel, his duties include handling bulk supplies and assisting in most jobs not directly connected with drilling. The term strictly applies to men working on land rigs, but sometimes refers to the laborer or deckhand on an offshore rig.

sand line: Small hose with calibrated wireline used to run survey tools, grapples or small fishing tools into the hole.

satellite well: A secondary well drilled close to a discovery well; a step-out well. There may be as many as 60 closely spaced satellite wells terminating on a single production platform, most of which will be directional wells entering a field some distance from the original, vertical well.

scfh: Standard cubic feet per hour.

separator: A pressure vessel (either horizontal or vertical) used for the purpose of separating well fluids into gaseous and liquid components. Separators segregate oil, gas, and water, at times with the aid of chemical treatment and the application of heat.

S.G.: Specific gravity.

shakeout: To force the sediment in a sample of oil to the bottom of a test tube by whirling the sample at high speed in a centrifuge. After the sample has been whirled for 3 to 5 minues, the percent of BS&W (sediment and water) is read on the graduated test tube.

short trip: Pulling the drillstring partway out of the hole and running back to bottom. Serves to clean and open up the hole.

show: Significant occurrence of oil or gas while drilling a well.

shut-in: The status of a well which has been closed temporarily to become a sealed pressure vessel.

sidetrack: To by-pass an obstruction in a borehole. In order to sidetrack the obstruction, which may be an irretrievable "fish," the bottom of the hole is first cemented and a turbo-drill or other deviation tool is used to make a dogleg diversion.

slug the pipe: An expression meaning to fill the top pipe in a string with heavy mud. This serves to evacuate thinner fluid from the pipe, which prevents it from spilling onto the rig floor when the pipe is uncoupled.

snub: To tie back with a rope or line.

SOH: Strap out of hole. Measure drillstring with steel tape.

SOS: "Same old stuff." No lithology change.

sour gas: An acid gas containing a significant amount of hydrogen sulfide gas, as opposed to sweet gas.

sour oil: Crude oil containing a significant amount of hydrogen sulfide gas, as opposed to sweet oil.

sow belly: Shale shaker header tank or possum belly.

SPE: Society of Petroleum Engineers.

spider deck: Movable catwalk suspended below the drillfloor of an offshore rig, providing access and work areas when assembling B.O.P. stacks, risers and seabed assemblies (see "nippling up").

spotting: In the context of offshore drilling technology, spotting is the action of placing a small quantity of some material (usually cement or drilling fluid) in the desired position inside a borehole, casing or pipe, using a calculated pump displacement.

spud: The term means to dig with a spade, and was adapted to cable tool drilling, when the driller used a special kind of bit to begin a new hole. Now, "spud" simply means to start a new hole. It can also mean to push or drop a bit onto a soft formation and plug the bit.

SPWLA: Society of Professional Well Logging Analysts.

squeeze a well: A technique to seal off with cement a section of the wellbore where a leak or incursion of water or gas occurs, forcing cement to the bottom of the casing and up the annular space between the casing and the wall of the borehole to seal off a formation or plug a leak in the casing; a squeeze job.

SS: Sub-Sealevel. Depth reference.

stack: An assembly of equipment; for example, a B.O.P. stack or a flare stack. Also, to lay up or store, as a drilling rig may be "stacked" when it is not required for drilling.

stringer: Thin, usually discontinuous bed within a massive lithology, e.g., "shale with sandstone stringers."

suicide line: Wire or chain attached to the traveling block used to pull joints of pipe or collars onto the rig floor while lifting the drillstring out of the hole.

sweet gas: Natural gas containing only very small amounts of hydrogen sulfide gas and carbon dioxide. Sweet gas is desirable in the interests of reducing the concentration of sulfur dioxide in the atmosphere, so any sour gas is first treated (for example, with triethanolamine) to remove the undesirable elements before being passed to the consumer. Most North Sea gas is considered to be naturally sweet.

sweet oil: Crude oil containing only very small amounts of hydrogen sulfide gas and carbon dioxide. North Sea oil is generally sweet oil.

tar: A thick black or dark-brown viscous liquid obtained by the destructive distillation of coal, wood or peat. Tar is not a natural product, and it is a misnomer to refer to asphalt deposits or seeps as tars.

taking a picture: Running a deviation survey of the borehole.

TBHP: True bottomhole pressure.

TD: Total depth, or planned final depth.

telltale (or tattletale): Any form of warning or alarm device. Commonly a small hole in a mud pump cylinder from which mud will flow if the pump liner is damaged or worn.

TG: Total gas.

tight hole: A well in which the information obtained is restricted and passed only to those who may be authorized to receive information.

toolpusher: An experienced drilling engineer having direct responsibility for the work done by drillers and their crews on a drilling rig. In addition to supervising crews, the toolpusher is responsible to the client company's drilling superintendent for the efficient performance of all drilling operations on his rig.

tour: A turn or shift of duty lasting a specified number of hours. The word "tour" is preferred in drilling circles to "shift." It is normally pronounced "tower" which discriminates it from a "tour of duty" or long-term assignment. Onshore, rig crews normally work eight-hour tours, morning (8 a.m. to 4 p.m.), afternoon (4 p.m. to midnight) and graveyard (midnight to 8 a.m.). Offshore, twelve-hour tours are worked between midday and midnight.

tugger: Air hoist.

twist-off: Catastrophic failure at the joint of drillpipe and collar, usually caused by the application of excessive torque from the rotary table.

UTM: Universal Transverse Mercator system of geographical location. Position is given in feet or metres East and North of some fixed reference point.

vitrinite: A coal maceral group that is the dominant organic constituent of humic coals. Vitrinite forms the familiar brilliant black bands of coal. Macerals in the vitrinite group include telinite derived from plant cell walls and collinite from the cell filling. Vitrinite particles are found in about 80 percent of the clays and sands of sedimentary basins.

washout: Leakage of drilling fluid through a worn or damaged tool joint (on drillpipe or drill collar) or collar (on casing). The abrasive drilling fluid will further damage the tool joint, causing the drillstring to twist off or the casing string to leak formation fluids after setting. Joints can be pressure-tested by externally fitted equipment.

water cushion: During drillstem testing, the drillstem is partially filled with water (the water cushion) to prevent collapse of the drillpipe, but principally to avoid collapsing the formation. This could occur when the packers are set and tool opened. The formation adjacent to the tool is relieved of hydrostatic pressure, which would normally balance the formation pressure. If formation pressure is high, the sudden change can result in breakdown of the formation.

wellbore: The hole that has been drilled.

wellsite geologist: Oil company's supervising geologist for the drilling operation. Responsible for overall geological surveillance, supervising Exlog, Schlumberger and other geological services.

wet box: A cylindrical, hinged box which can be latched around tubing or drillpipe when pulling a wet string, to collect drilling fluid as pipe is uncoupled. A drain plug allows dumping through a hose to the shale shaker for reuse. Use of the wet box, which may be suspended over the rig floor, prevents spillage of fluid onto the floor during trips.

wet gas: Gas dissolved in heavier hydrocarbons. Natural gas is said to be wet when more than 300 gallons of propane, butane and other liquid hydrocarbons can be separated from every thousand cubic feet of gas.

wet oil: Oil that has not been processed to remove settlings (BS&W) as opposed to dry or clean oil.

wet string: A string of drillpipe or tubing that is full of drilling fluid.

wildcat well: Strictly an exploratory borehole drilled in an area which geological and geophysical surveys have shown to be promising for hydrocarbon extraction, but which is not known to be productive. Wildcat wells are drilled primarily to discover oil or gas, but often have secondary objectives. One is to confirm the geology of the structure, and to this end cores may be taken from the hole at intervals. Another is to measure temperature and pressure at various depths and provide information which will be of great value in drilling development wells later. Before starting a wildcat, an operator must ensure ample supplies of casing, mud, pipe and other requisites since borehole conditions will be unpredictable. The AAPG classifies wildcat wells as either new pools or new field wildcats.

WOC, WOO, WOW: Waiting on cement, waiting on orders, waiting on weather.

workover: A program of work performed on an existing well which has either ceased to produce or is not producing economically. Workover operations may involve reevaluation of the producing formation, clearing sand from the producing zones, jet lifting, replacing downhole equipment, deepening the well, acidizing or fracturing, or improving the drive mechanism.

X-line: Early brand name for sealed bearing bits. Commonly used as generic for sealed bearing bits.

XO: Cross-over.

YP: Yield point.

INDEX

Abbreviations
  standard, 234–236 (fig. A-3)
  for wireline logs, 237–238 (fig. A-4)
Abnormal formation pressure, 97, 98
Accelerators, 119
Active pit, 107
Agitator motor, 164
Angular unconformity, 9
Annular preventer, 82, 83
Anticlinal traps, 28
Anticlines, 8
API gravity, 36, 38, 42, 190
  factors affecting, 43
Aromatics, 38

Barge rig, 49
Barites, 101
Basin evaluation, 14
Bearings
  friction, 70
  grading of, 73
  sealed, 70
Bent sub, 134
Benzene, 38
BGT, 219
BHA, 67
BHCP, 96
Biogenic debris, 4
Bit
  cooling, 100
  diamond, 73
  drag, 67
  gauge insert, 69
  grading, 72
  jet, 135
  stabilization of, 65
  tri-cone, 67–69, 261–269
  types of, 67–75
Bit classification, 72
  IADC, 261–269
Bit code
  C.P., 269 (fig. D-8)
  Hughes Tool Company, 262 (fig. D-1)
  Reed Tool Company, 264 (fig. D-3)
  Security, 265 (fig. D-4)
  S.M.F., 267 (fig. D-6)
  Smith Tool, 266 (fig. D-5)
  tri-cone, 261–269
  T.S.K., 268 (fig. D-7)
  Varel, 263 (fig. D-2)
Bit data
  typing, 204–205
Bit data record, 242 (fig. A-7)
Bit sub, 67
Blender, 26
Blender cample chromatographic analysis, 257
Blind rams, 82, 83

Blowout, 92
Blowout prevention (B.O.P.), 48, 81–85
Boot basket, 123
Borehole geometry tool (BGT), 219
Bottom plug, 114 (fig. 4-15)
Bottomhole assembly (BHA), 67
Bottomhole circulating pressure (BHCP), 96
Breccia, 5
Bumper subs, 66
Buoyancy effect, 64
Butane, 36
Bypass valve, 222

Cable tool rig, 1
Cal seal, 119
Calcimetry, 187, 200, 250–251
Calcium carbide, 150–151
Calcium chloride, 119
Calcium lignosulfonate, 119
Caliper, 219
Carbide evaluation, 259
Carbon dioxide, detection of, 200
Carbonate staining, 244–246
Carbonates, 4, 187–190
  characteristics of, 24, 188–190
  determination of, 244–253
Casing, 108–113
  accessories, 110–112
  capacity of, 154 (fig. 5-2)
  displacement of, 154 (fig. 5-2)
  functions of, 108
  intermediate, 109
  production, 110
  running, 110
  surface, 109
Casing hanger, 109 (fig. 4-13)
Casing slip, 110
Catalytic chromatographs, 258
Catalytic gas detector, 166–168
Catalytic standard chromatograph, 172–173
Cavings, 182
Cement additives, 119–120
Cement Bond Log, 122
Cement classifications, 116–117
Cement contamination, 184
Cement plugs, 115
Cement slurry, 114 (fig. 4-15)
Cement unit, 121
Cementation, 21, 23, 187
Cementing, 113–122
  considerations, 122
  multi-stage, 116
  primary, 113–116
  purpose of, 113
  secondary, 116
  single-stage, 113–115
  techniques of, 113

Cementing heads, 112
Centralizer, 112, 114 (fig. 4-15)
Centrifuge, 104-105
Charging pump, 56 (fig. 3-7)
Chemical precipitates, 4
Chert, 4
Chicksan line, 121
Chloride test, 247
Choke line, 83
Chromatogram, 174
Chromatography, 162-164, 172-175
  blender sample, 257
  catalytic standard, 172-173, 258
  drafting, 202
  flame ionization detector (FID), 172, 173-174, 258
  typing, 206
Chronological sample taker (CST), 127
Circulating system, 60-75
Clastic material, 4, 5
Clay, 4
Coastal interdeltaic clastics, 5
Color, of samples, 185, 188
Compaction, 23
  abnormal reservoir pressure and, 32
Compressed gases, 102
Computational system, 199
Conceptual models, 15
Conductor pipe, 83, 108
Conglomerates, 5
Connate interstitial water, 33
Connection gas, 170
Connections, 87-88
Continental aeolian clastics, 5
Core
  cutting, 123-125
  drafting, 201
  handling of, 125-127
Core analysis, 200-201, 253, 259
Core barrels, types of, 123
Core log, 212
Coring, 123-128
  conventional, 123-127
  sidewall, 123, 127-128
Crossover subs, 66
Crown block, 48, 58-59
Crude oil
  degradation of, 43
  gravity of, 43
  products from, 45
CST, 127
Cut
  drafting, 202
  of oil-based muds, 256
Cuttings. *See also* Samples
  examination of, 182-184, 192-193
  gas analysis of, 12
  gas in, 13
  gasoline in, 13
  preparation of, 254-255
  recycled, 182-184
  removing, 99
  unwashed, 192-193
  washed, 192
Cuttings gas
  drafting, 201-202
  typing, 206
Cuttings lithology
  drafting, 201
  typing, 205-206
Cycloparaffins, 38

Darcy, 22
d'Arcy, Henry, 22
Deflection tools, 132-137
Degasser, 106
Deltaic clastics, 5
Depositional environments, 5 (fig. 2-3)
Depth, typing, 205
Depth recorder, 156-162
Derrick, 48, 49, 57
Desander, 104
Desilter, 104
Developed areas, 15
Deviation, preventing and correcting, 130
Diacel A, 119
Diamond bit, 73
Diamond bit core barrel, 123
Directional drilling, 128-138
Disconformity, 9
Ditch gas, recording, 168-169
Ditch gas analyser, 256-257
Dogleg, 128
Dolomite, 24, 188
Dolomitization, 24
Dome, 8
Downhole hydraulic motors, 133-134
Drag bit, 67
Drawworks, 48 (fig. 3-2), 49, 60
Drill bit. *See* Bit
Drill collars, 64
  capacity of, 154 (fig. 5-2)
  displacement of, 154 (fig. 5-2)
Drill monitor, 197
Drill rate recorder, 156-162
Drill returns logging, 1, 147-211. *See also* Mud logging
  applications of, 148-149
  gel strength and, 99
  mudweight and, 93
  theory of, 147-148
  viscosity and, 99
Drilling, 87-143
  directional, 128-138
  routine, 87-89
Drilling breaks, 159
Drilling fluid chemistry, 100-102
Drilling fluid conditioning equipment, 102-107
Drilling fluid technology, 92-100
Drilling fluids, 1
  inert solids, 101-102
  oil-based, 254-260
  purposes of, 92
  reactive solids, 101
Drilling fluids engineering, 92-107
Drilling hook, 58-60

Drilling line, 58–60
Drilling mud, 92. *See also* Drilling fluids
   examination of, 191
Drillpipe, 48, 63–64
   capacity of, 154 (fig. 5-2)
   displacement of, 154 (fig. 5-2)
Drillship, 54–55
Drillstem, 62
Drillstem testing (DST), 1, 146, 221–226
   calculation of, 272–273
   logging geologist's responsibilities for, 270–278
   presentation of test results, 273–276
   procedures, 270–272
Drillstring, 63–67
   compensator, 78–79
   cooling, 100
Dry gas, 36
DST. *See* Drillstem testing
Duplex pumps, 61
Dynamic positioning, 53

Earth movements, 8–9
Earth Resources Satellite, 11
ECD, 96
Effective circulating density (ECD), 96
Electromagnetic spectrum, 191 (fig. 5-21)
Elevators, 48, 88
Emulsion, 102
Equivalent mudweight, 94
ERSAT scans, 11
Escape line, 48
Ethane, 36
Evaporites, 184, 190

Fatigue failure, 138
Fault, 8–9
Fault traps, 28–29
FID chromatograph, 172, 173–174, 258
Filtercake, 100
Filtrate
   chloride content of, 175
   nitrate ion content of, 175–176
   resistivity of, 175
   testing of, 175–176
Fishing, 138–143
Fishing tools, 140
Fixed platform rig, 50–52
Flame ionization detector (FID) chromatograph, 172, 173–174
Flame ionization gas detector, 168
Float collar, 110, 114 (fig. 4-15)
Flowline temperature, 259
Fluid-loss additives, 120
Fluorescence, 190–191, 256
Flushing, 26
Folds, 8
Formation density, 218
Formation evaluation, 1, 145–229
Formation interval tester, 228
Formation pressure, 31–32, 97–98
Formation test gas analysis, 271–273
Formation tests, 221–229
Fractures, seepage along, 10

Free gas, 35
Free-point indicator, 143
Friction bearing, 70
Friction reducers, 120

Galena, 101
Gamma ray, natural, 219
Gas
   compressed, 102
   connection, 170
   determination of, 162–164
   dry, 36
   free, 35
   origin of, 41
   solution, 35
   trip, 170–171
   wet, 36
Gas analysis, 12, 256–259
Gas cap, 35
Gas detector, 162–170
   flame ionization (FID), 168
   total, 162–164
Gas detector, catalytic, 166–168
Gas readings, factors affecting, 170
Gas trap, 162–165
Gauge, grading of, 73
Gauge insert bit, 69
Gel strength, 99
Geochemical prospecting, 11
Geochemical temperature facies, 13
Geologic area, conceptual models of, 15
Geologic structures, 8
Geology. *See also* Petroleum geology
   oil exploration and, 10–15
   subsurface, 12–13
Gouge, 28
Grain character, 188
Grain shape, 185
Grain size, 185, 188
Gravimeter, 10
Gravity, AP. *See* API gravity
Gravity structures, 52
Ground water, 3
Guide shoe, 110, 114 (fig. 4-15)
Guideline tensioners, 80–81

HA-5, 119
HDT, 219
Heating value calculation, 277–278
Heavy hydrocarbons, 13
Heavyweight additives, 120
Helicoid hydraulic motor, 134
Hematite, 120
High-resolution dipmeter tool (HDT), 219
Hoist. *See* Drawworks
Hoisting system, 57–60
Hole
   capacity of, 154 (fig. 5-2)
   foreign objects in, 140
Hole opener, 75
Hole patterns, 131–132
Hook, 48
Hook wall packer test, 225

Hopper, mixing, 106
Hydrates, 14
Hydrocarbons
  analysis of, 11
  evaluation, 190–191
  generation of, 18
  heavy, 13
  hydrates, 14
  impurities associated with, 44–45
  petroleum composition and, 35–38
  properties of, 44 (fig. 2-42)
  saturated, 38
  size distribution, 36
  structure of, 37
  type distribution of, 37–38
  unsaturated, 38
Hydroclone, 104–105
Hydrodynamic conditions, traps for, 27
Hydrogen sulfide, 200
Hydrostatic conditions, traps for, 27
Hydrostatic pressure, 93

IADC bit classification, 72, 261–269
Inclination, measuring, 128
Inclusions, of sample, 187, 189–190
Induction logs, 215–216
Induration, of sample, 185
Inert solids, 101–102
Insert bit, 69
  grading of, 73
Intermediate casing, 109
International Association of Drilling Contractors. *See* IADC
Interpreted lithology, drafting, 203
Invert emulsion, 102
Isobutane, 36
Isoparaffins, 38

Jack-up rig, 49–51
Jar, 143
Jet bit, 135
Joints, 8
Journal, 70
Junk, fishing for, 140

Kelly, 63, 87
Kelly down, 87
Kelly drive bushings, 78
Kerogen, 13, 41
Kerogen analysis, 13
Kick, 92
Kill line, 83

Lag, determination of, 150–155
Land rigs, 47–49
Lateral faults, 8–9
LCM. *See* Lost circulation material
Lenticular traps, 30
Lightweight additives, 119
Limestone, 4, 5, 188
Liner string, 110
Links, 48

Lithology, 13
  measurement of, 217–219
  typing of, 207
Log heading, typing of, 203–204
Logging geologist, responsibilities of, 208–211, 270–278
Lost circulation material (LCM), 92, 101–102, 120
Lubraglide, 102
Luster, of sample, 186

Magnetometer, 10
Marine clastics, 5
Marine riser, 80–81
Mast. *See* Derrick
Master bushings, 77
Metal, in cuttings, 184
Methane, 36, 41
Microgas analyzer, 170, 257
Migration, 18–21
  primary, 18–20
  secondary, 20–21
  traps and, 26
Mill, 143
Milled tooth bit, 68
  grading of, 72
Millidarcy, 22
Mixing equipment, 121
Mixing hopper, 106
Monkey board, 48
Moon pool, 54
Motion compensation system, 66, 78–81
Motor
  agitator, 164
  positive displacement, 134
Mousehole, 87
Mud, 1
  gas analysis of, 12
  water/clay, 100–102
  water/oil/clay, 102
Mud chemicals, 184
Mud log
  drafting, 201–203
  examples of, 231–242
  presentation of, 201–207
  typical program, 232 (fig. A-1)
Mud logging, 1, 147–211. *See also* Drill returns logging
  applications of, 148–149
  defined, 146
  equipment, 156–181
  with oil-based drilling fluids, 254–260
  procedures, 179–181
  secondary equipment, 193–201
  techniques, 149–155, 179–181
  theory of, 147–148
Mud pits, 56
Mud press, 175–176
Mud pumps, 61–62
Mud resistivity, recording, 197
Mud temperature, recording, 196–197
Mudflow monitor, 197
Mudweight
  apparent, 94

drill returns logging and, 93
effective, 94–96
equivalent, 94
pressure and, 92
recording, 196

Naphthenes, 38
Neutron, 218–219
Nitrate ion test, of filtrate, 175–176, 225–226, 248–249
Noncarbonate clastics, description of, 185–187
Normal faults, 8–9
Normal formation pressure, 97

Offset, 71
Offshore rigs, 49–85
Oil evaluation, drafting, 201
Oil exploration
  conceptual models of, 15
  early methods of, 10
  geological concepts and, 10–15
  seismic method for, 10
  subsurface methods for, 12–13
  surface methods for, 11–12
Oil seeps, 10, 12
Oil-based mud samples, 254–260
  carbide evaluation of, 259
  chromatographic analysis of, 257–258
  core analysis of, 259
  ditch gas analysis of, 256–257
  examination of, 256
  external sensors, 260
  flowline temperature, 259
  gas analysis of, 256–259
  labeling, 255
  microgas analysis of, 257
  precautions, 260
  preparation of, 254–255
  shale density of, 259
  shale factor, 259
  wireline logs of, 259–260
Oil-water contact, 46
Oilfield fluids, 32–46
Oilwell, wildcat, 1, 14
Olefins, 38
Open-hole straddle packer test, 225
Organic carbon, 12
Organic debris, 4
Overburden pressure, 96
Overpressure, 97
Overshot, 143

Packed hole, 65
Packer, 222, 225
Paraffins, 38, 40
Pendulum effect, 64
Penetration rate, 233 (fig. A–2)
  drafting of, 201
  typing of, 204
Permeability, 3
  in carbonates, 24
  degree of, 22
  estimations of, 24–26
  in sandstones, 22

Petroleum
  accumulation of, 17–30
  composition of, 35–38
  maturation of, 42–43
  occurrence of, 41
  origin of, 17–18
  source of, 3, 17, 38
Petroleum geology, 3–32
Petroleum reservoir. *See* Reservoir
Piezometric surface, 97
Piled steel platform rig, 50–52
Pipe, fishing for, 140–143
Pipe dope, 191
Pipe rams, 82, 83
Pipe spinners, 78
Pit level indicators, 176–177
Pore pressure, 96
Porosity, 3
  in carbonates, 24
  degree of, 21–22
  estimations of, 24
  interparticle, 24
  intraparticle, 24
  measurement of, 217–219
  reservoir fluids and, 31
  in sandstones, 21–22
  vuggy, 24
Positive displacement motor, 134
Power tongs, 78
Pozzolans, 119
Pressure
  abnormal foundation, 97–98
  normal foundation, 97
  overburden, 96
  petroleum maturation and, 42
  pore, 96
  subnormal foundation, 98
  subsurface, 92–98
Pressure gradient, 93–94
Pressure Log, 211–212
Pressure-sealed core barrel, 123
Primary cementing, 113–116
Primary migration, 18–20
Production casing, 110
Propane, 36
Pump stroke counter, 156

Quartz, 4

Rate of penetration. *See* Penetration rate
Rathole, 88
Reactive solids, 100–101
Reamer, 75
Recycled cuttings, 182–184
Repeat formation tester, 227
Reservoir, 3, 17, 26–32
Reservoir fluid distribution, 46
Reservoir gas, 35–45
Reservoir oil, 35–45
Reservoir pressure, 31–32
Reservoir rock, 3, 17
  permeability of, 22, 24–25
  porosity of, 21, 24
  types of, 21–24

Reservoir trap. *See* Trap
Reservoir water, 32–35
Resistivity, 33
  from filtrate, 175
  measurement of, 214–217
  recording, 197
Retarders, 119
Reverse drilling breaks, 159
Reverse faults, 8–9
Rig
  barge, 49
  cable tool, 1
  components, 56–85
  fixed platform, 50–52
  gravity structure, 52
  jack-up, 50
  land, 47–49
  offshore, 49–85
  piled steel platform, 50–52
  rotary, 1
  semi-submersible, 52–53
  types of, 47
Riser tensioner, 56 (fig. 3–7), 80–81
Rock seal, 3
Rotary hose, 56 (fig. 3–7), 62
Rotary rig, 1
Rotary table, 48, 77
Rotating system, 75–81
Rotational faults, 9

Safety joint, 143
Salt cements, 120
Samples, 181–193
  collection of, 181
  description of, 184
  examination of, 182–184, 192–193
  labeling, 255
  oil-based muds, 254–260
  preparation of, 181–182, 254–255
  test procedures, 243–253
Sand line, 128
Sand trap, 104
Sandstone, 4, 5
  abnormally pressured, 32
  porosity of, 21–22
Saturated hydrocarbons, 38
Sealed bearings, 70
  grading of, 73
Secondary cementing, 116
Secondary equipment, 193–201
Secondary migration, 20–21
Sedimentary rock
  cemented, 3
  classification of, 4–5
  deposit of, 8
  geology of, 3–9
Seepage, 12
  along fractures, 10
  up-dip, 10
Seismograph, 10
Semi-submersible rigs, 52–53
Setting pit, 104
Shaker, 56 (fig. 3–7)

Shale, 4, 5
Shale density, 199, 259
Shale factor, 199, 252, 259
Shale shaker, 104
Shear rams, 83
Shock subs, 66
Show, description of, 193
Sidewall coring, 123, 127–128
Siltstone, 5
Single-stage cementing, 113–115
Slips, 78
Sodium chloride, 119
Solution gas, 35
Sonic, 217
Sorting, degree of, 185–186
Source material, 3, 15, 17–18
SP, 216–217
Specific gravity (SG) calculation, 276–277
Spill point, 26, 27 (fig. 2–24)
Spinning wrench, 78
Spontaneous potential (SP), 216–217
Spotting, 143
Stabilizers, 67
Stain kit, 200, 244
Standard abbreviations, 234–236 (fig. A–3)
  for wireline logs, 237–238 (fig. A–4)
Standpipe, 62
Stands, 88
Stratigraphic traps, 30
Structure, of sample, 187, 189
Stuck pipes, 139–140
Subnormal foundation pressure, 98
Subs, 66–67
Subsurface exploration, 12–13
Subsurface pressures, 92–98
Suction pit, 107
Surface casing, 109
Surface exploration, 11–12
Swivel, 48, 63
Synclines, 8

T-gauge bit, 69
Temperature, effect on petroleum maturation, 42–43
Test
  drafting, 201
  procedures, 243–253
Tester valve, 222
Texture, of samples, 185–186, 189
Three-cone roller bit core barrel, 123
Three-dimensional closure, 3
Thrust faults, 8–9
Tongs, 48, 78
Tool joints, 48
Top plug, 114 (fig. 4–15)
Total gas
  drafting, 201–202
  typing, 206
Total gas detector, 162–164
Transit time, 217
Trap, 3, 17, 26–30
  anticlinal, 28
  fault, 28–29

lenticular, 30
piercement, 29
stratigraphic, 30
Traveling block, 58-60
Tri-cone bit, 67-69, 261-269
Trip gas, 170-172
Triplex pumps, 61-62
Tripping in, 89
Tripping out, 89
Trips, 88-89
Turbine motor, 133
Turbodrill, 133

Ultraviolet-light box, 175
Unconformity, 9
Unsaturated hydrocarbons, 38
Upthrusts, 9

Vacuum system, 166
Vertical control, 128
Viscosity, 99
Vitrinite reflectance, 13
Vuggy porosity, 24

Wall-hook guide, 143
Wallcake scratchers, 112
Washover, 143
Water. *See also* Reservoir water
  connate interstitial, 33
  contact with oil, 46
  resistivity of, 33
Water/clay muds, 100-102
Water injection, 32
Water/oil/clay muds, 102
Well
  casing, 108-113
  cementing, 113-122
  drilling of, 87-143
  wildcat, 14
Well completion, final report for, 210
Well log, 146
Well logging, 145-221
Wellhead, 112-113
Wellsite services, 89-91
Well stimulation, 228-229
Wet gas, 36
Whipstock, 135-136
  circulating, 136
  permanent casing, 136
Wildcat oilwell, 1, 14
Wireline core barrel, 123
Wireline coring, 123
Wireline formation testing, 226-228
Wireline log, 1, 13, 212-221, 259-260
  interpretation of, 220-221
  resistivity, 214-216
  standard abbreviations for, 237-238 (fig. A-4)